AF388964

Roles of MET in Cancer Development and Treatment

Roles of MET in Cancer Development and Treatment

Editors

Jan Trøst Jørgensen
Jens Mollerup

Basel • Beijing • Wuhan • Barcelona • Belgrade • Novi Sad • Cluj • Manchester

Editors
Jan Trøst Jørgensen
Dx-Rx Institute
Fredensborg, Denmark

Jens Mollerup
Agilent Technologies
Denmark ApS
Glostrup, Denmark

Editorial Office
MDPI
St. Alban-Anlage 66
4052 Basel, Switzerland

This is a reprint of articles from the Special Issue published online in the open access journal *Cancers* (ISSN 2072-6694) (available at: https://www.mdpi.com/journal/cancers/special_issues/MET_DT).

For citation purposes, cite each article independently as indicated on the article page online and as indicated below:

Lastname, A.A.; Lastname, B.B. Article Title. *Journal Name* **Year**, *Volume Number*, Page Range.

ISBN 978-3-0365-9530-6 (Hbk)
ISBN 978-3-0365-9531-3 (PDF)
doi.org/10.3390/books978-3-0365-9531-3

Cover image courtesy of Jens Mollerup

Contents

About the Editors

Jan Trøst Jørgensen

Jan Trøst Jørgensen holds a Master's in Pharmaceutical Science and a Ph.D. in Clinical Pharmacy from the University of Copenhagen. He has more than four decades of experience in clinical research from various pharmaceutical, biotech, and diagnostic companies, including Novartis, Novo Nordisk, and Dako/Agilent. Currently, he serves as the Director of the Dx-Rx Institute in Fredensborg, Denmark. Over the past two decades, his primary focus has been drug diagnostic co-development and companion diagnostics in oncology. Dr. Jørgensen strongly advocates individualized pharmacotherapy and has authored and edited numerous scientific papers on companion diagnostics, drug diagnostic co-development, and precision medicine. He also serves as an Editorial board member for several medical journals and a board member of the Danish Society of Cyto and Histochemistry.

Jens Mollerup

Jens Mollerup holds a Master's and PhD in Biology from The August Krogh Institute, University of Copenhagen. He has 10 years of experience as an Assistant Professor and Supervisor at the Institute of Molecular Biology, University of Copenhagen, with a research focus on membrane transport, calcium-binding proteins, leukotriene receptors, and HSP90 p23-chaperone functions. He has 15 years of industry experience with late and early product development in the pathology area of cancer diagnostics from the Clinical and R&D Departments at Dako Denmark and Agilent Technologies.

 cancers

 MDPI

Editorial

The Different Roles of MET in the Development and Treatment of Cancer

Jens Mollerup [1,*] and Jan Trøst Jørgensen [2]

[1] Pathology Division, Agilent Technologies Denmark ApS, Produktionsvej 42, 2600 Glostrup, Denmark
[2] Department: Medical Sciences, Dx-Rx Institute, Baunevaenget 76, 3480 Fredensborg, Denmark; jan.trost@dx-rx.dk
* Correspondence: jens.mollerup@agilent.com

This Special Issue features contributions from leading international researchers in the field of MET (hepatocyte growth factor (HGF) receptor) biology and therapeutics. Recent discoveries regarding non-small cell lung cancer (NSCLC) and gastric cancer, as well as advancements in the detection of *MET* aberrations and new targeted therapies for MET-driven cancers and resistance mechanisms are explored. Aberrations in the *MET* gene leading to impaired MET-dependent signaling have been identified as primary and secondary drivers of cancer development. To optimally detect and potentially counteract these effects with mono- or combination therapy, it is crucial to understand the underlying cellular mechanisms involved in MET-dependent cancer cell development and growth.

Dysregulated MET signaling, which predisposes cells to cancer development, can occur due to MET overexpression, *MET* gene amplification, MET kinase mutations, mutations resulting in *MET* exon 14 skipping, *MET* rearrangements, and *MET* fusions [1]. A variety of technologies are used to detect aberrations linked to MET in clinical samples, including immunohistochemistry (IHC), next-generation sequencing (NGS) of DNA or RNA, Sanger sequencing of RNA, reverse transcription–polymerase chain reaction (RT-PCR), digital droplet PCR (ddPCR), nanostring nCounter, in situ hybridization (ISH), and mass spectrometry [1–5].

Das et al. [3] revealed two novel noncanonical *MET* splice variants leading to *MET* exon 14 skipping in NSCLC. Their study highlights the importance of recognizing non-canonical splice events by integrating next-generation sequencing data with in silico predictions in order to assess the potential impact of mutations. Additionally, they demonstrated the potential of routinely using cytology slides for RNA-based NGS testing.

Feldt et al. [6] report that the progression of NSCLC following treatment with epidermal growth factor receptor (EGFR) tyrosine kinase inhibitor (TKI) frequently involves changes in the *MET* gene, including *MET* amplification. In NSCLC patients who had progressed on osimertinib, early clinical trials have shown promising antitumor activity following a combination therapy with the third-generation EGFR TKI lazertinib and the MET-EGFR bispecific antibody amivantamab.

Gamerith et al. [7] utilized ISH and NGS to examine genetic alterations in lung cancer patients exposed to radon. Their study unexpectedly revealed a higher frequency of *MET* alterations in radon-exposed patients compared with the control group.

In a systematic review following PRISMA guidelines, Bodén et al. [8] examined 22 published papers relating to clinical trials on MET, lung cancer, and targeted MET therapies from the Embase and PubMed databases between 2013 and February 2023. Six clinical trials indicated favorable outcomes of MET inhibitor treatment in terms of progression-free survival (PFS) and the overall response rate (ORR), while two clinical trials failed to show a beneficial effect of adjunctive MET-targeted therapy.

MET amplification is known as a pivotal biomarker, but establishing the optimal thresholds for recognizing *MET* amplification in patient samples is challenging [4,9]. The

Citation: Mollerup, J.; Jørgensen, J.T. The Different Roles of MET in the Development and Treatment of Cancer. *Cancers* **2023**, *15*, 5087. https://doi.org/10.3390/cancers15205087

Received: 17 October 2023
Accepted: 19 October 2023
Published: 21 October 2023

determination of the *MET* copy number can be achieved through ISH and NGS, and according to Kumaki et al. [4], a significant challenge is the distinction between focal amplification and polysomy. As ISH exhibits higher sensitivity compared with NGS, ISH is considered the gold standard for copy number determinations and detection of *MET* amplification [5]. With regard to metastatic NSCLC, Qin et al. [10] elaborated on how *MET* amplification plays a key role in resistance to tyrosine kinase inhibitors (TKI) and discuss strategies to overcome this.

By employing a proteomics approach, Jie et al. [2] presented evidence that the four plasma biomarkers MYH9, GNB1, ALOX12B, and HSD17B4 could substitute or complement response prediction by using fluorescence ISH (FISH) or IHC in patients receiving MET inhibitors.

To gain further insights into the mechanisms underlying drug resistance, Cecchi et al. [11] investigated the path to rilotumumab resistance in a glioblastoma cell line that was dependent on autocrine signaling via HGF and MET. Rilotumumab is an investigational, fully human monoclonal antibody that binds HGF and prevents HGF-mediated activation of MET. Resistance towards rilotumumab was found to depend on *MET* and *HGF* amplification, excessive production and misfolding of HGF, induction of endoplasmic reticulum stress-response signaling, and an increased uptake and degradation of rilotumumab. Collectively, these mechanisms enable resistant glioblastoma cells to sustain adequate HGF-dependent MET signaling, thereby promoting survival and cell growth.

In a narrative review by Hsu et al. [12], the development of capmatinib from preclinical studies to its approval for treatment of MET-driven NSCLC was presented. Capmatinib received FDA approval in 2022 for advanced non-small cell lung cancer (NSCLC) with *MET* exon 14 skipping mutations. Hsu et al. specify that ongoing clinical research aims to improve the treatment efficacy and explore new indications for capmatinib, including addressing *MET* amplification that has developed following EGFR TKI resistance. Combination therapies with capmatinib and other agents are also under investigation. Based on data from various clinical trials, Hsu et al. compared the efficacy outcomes of the three approved MET TKIs—capmatinib, tepotinib, and savolitinib—in the treatment of patients with metastatic NSCLC with *MET* exon 14 skipping mutations [12]. These clinical trials demonstrated an ORR range of 41% to 68%, dependent on the patient type and previous treatment history, and a corresponding PFS range of 6.8 to 12.4 months.

Similarly, the review by Zhu et al. [13] examined the development of the highly selective MET-TKI, savolitinib. Savolitinib obtained conditional approval in China in 2021 for the treatment of NSCLC with *MET* exon 14 skipping mutations, and this review outlines preclinical models, phase I studies in Chinese patients, and the TATTON study combining savolitinib with osimertinib. The authors conclude that both preclinical and clinical evidence support the efficacy and tolerability of savolitinib in treating advanced NSCLC patients with *MET* exon 14 skipping mutations. Furthermore, when using savolitinib in conjunction with EGFR-TKIs, the authors indicate that it demonstrates potential in terms of overcoming treatment resistance stemming from both *MET* amplification and MET overexpression.

The review by Van Herpe and Van Cutsem [9] focuses on the role of MET in gastric cancer and discusses the clinical significance of MET-targeted therapies. The review also explores various diagnostic assays, such as immunohistochemistry, FISH, H-score, and NGS. The authors highlight the challenges of identifying patients who will benefit from treatment with MET inhibitors due to the variability in diagnostic assays. They note that the success of MET-targeted therapy in gastric cancer appears to be limited, with consistent limitations such as the number of patients, differences in inclusion criteria, and diagnostic assays for patient selection in clinical trials with TKIs. However, the VIKTORY umbrella trial stands out as an exception, where a cohort of gastric cancer patients with *MET* amplification received treatment with savolitinib and achieved an ORR of 50%. The authors emphasize that a major challenge remains in establishing clinically significant cut-off values for *MET* amplification and MET overexpression to guide treatment-related decisions.

The development of companion diagnostic assays for targeted cancer therapies requires a profound understanding of the pathophysiology and the drug's mechanism of action [5]. In the case of MET inhibitors, unexplored avenues requiring further investigation remain. Over the past decade, intensive research has been carried out to develop MET-targeting drugs, including both small-molecule inhibitors and antibody-based drugs. As described in this Special Issue of *Cancers*, only a few MET inhibitors have obtained regulatory approval, and this is so far limited to the treatment of metastatic NSCLC patients. A key challenge in the development of MET inhibitors appears to be related to identifying the appropriate predictive biomarker to guide drug use. In NSCLC, a small number of MET TKIs have demonstrated efficacy when patients are selected based on *MET* exon 14 skipping mutations. Another potential predictive biomarker for MET-targeted therapy is *MET* amplification, identified as a resistance mechanism in patients with EGFR-mutated NSCLC. However, their full potential in both NSCLC and gastric cancer remains to be fully explored, which we hope to witness in the coming years.

Conflicts of Interest: Jens Mollerup is an employee of Agilent Technologies Denmark ApS and a shareholder of Agilent Technologies Inc. Jan Trøst Jørgensen is an employee of the Dx-Rx Institute and has worked as a consultant for Agilent Technologies, Alligator Biosciences, Argenx, Azanta, Biovica, Euro Diagnostica, Leo Pharma, and Oncology Venture, and has given lectures at meetings sponsored by AstraZeneca, Merck Sharp & Dohme, and Roche.

References

1. Heydt, C.; Ihle, M.A.; Merkelbach-Bruse, S. Overview of Molecular Detection Technologies for MET in Lung Cancer. *Cancers* **2023**, *15*, 2932. [CrossRef] [PubMed]
2. Jie, G.-L.; Peng, L.-X.; Zheng, M.-M.; Sun, H.; Wang, S.-R.; Liu, S.-Y.M.; Yin, K.; Chen, Z.-H.; Tian, H.-X.; Yang, J.-J.; et al. Longitudinal Plasma Proteomics-Derived Biomarkers Predict Response to MET Inhibitors for MET-Dysregulated NSCLC. *Cancers* **2023**, *15*, 302. [CrossRef] [PubMed]
3. Das, R.; Jakubowski, M.A.; Spildener, J.; Cheng, Y.-W. Identification of Novel MET Exon 14 Skipping Variants in Non-Small Cell Lung Cancer Patients: A Prototype Workflow Involving in Silico Prediction and RT-PCR. *Cancers* **2022**, *14*, 4814. [CrossRef] [PubMed]
4. Kumaki, Y.; Oda, G.; Ikeda, S. Targeting MET Amplification: Opportunities and Obstacles in Therapeutic Approaches. *Cancers* **2023**, *15*, 4552. [CrossRef] [PubMed]
5. Jørgensen, J.T.; Mollerup, J. Companion Diagnostics and Predictive Biomarkers for MET-Targeted Therapy in NSCLC. *Cancers* **2022**, *14*, 2150. [CrossRef] [PubMed]
6. Feldt, S.L.; Bestvina, C.M. The Role of MET in Resistance to EGFR Inhibition in NSCLC: A Review of Mechanisms and Treatment Implications. *Cancers* **2023**, *15*, 2998. [CrossRef] [PubMed]
7. Gamerith, G.; Kloppenburg, M.; Mildner, F.; Amann, A.; Merkelbach-Bruse, S.; Heydt, C.; Siemanowski, J.; Buettner, R.; Fiegl, M.; Manzl, C.; et al. Molecular Characteristics of Radon Associated Lung Cancer Highlights MET Alterations. *Cancers* **2022**, *14*, 5113. [CrossRef] [PubMed]
8. Bodén, E.; Sveréus, F.; Olm, F.; Lindstedt, S. A Systematic Review of Mesenchymal Epithelial Transition Factor (MET) and Its Impact in the Development and Treatment of Non-Small-Cell Lung Cancer. *Cancers* **2023**, *15*, 3827. [CrossRef] [PubMed]
9. Van Herpe, F.; Van Cutsem, E. The Role of cMET in Gastric Cancer—A Review of the Literature. *Cancers* **2023**, *15*, 1976. [CrossRef] [PubMed]
10. Qin, K.; Hong, L.; Zhang, J.; Le, X. MET Amplification as a Resistance Driver to TKI Therapies in Lung Cancer: Clinical Challenges and Opportunities. *Cancers* **2023**, *15*, 612. [CrossRef] [PubMed]
11. Cecchi, F.; Rex, K.; Schmidt, J.; Vocke, C.D.; Lee, Y.H.; Burkett, S.; Baker, D.; Damore, M.A.; Coxon, A.; Burgess, T.L.; et al. Rilotumumab Resistance Acquired by Intracrine Hepatocyte Growth Factor Signaling. *Cancers* **2023**, *15*, 460. [CrossRef] [PubMed]
12. Hsu, R.; Benjamin, D.J.; Nagasaka, M. The Development and Role of Capmatinib in the Treatment of MET-Dysregulated Non-Small Cell Lung Cancer - A Narrative Review. *Cancers* **2023**, *15*, 3561. [PubMed]
13. Zhu, X.; Lu, Y.; Lu, S. Landscape of Savolitinib Development for the Treatment of Non-Small Cell Lung Cancer with MET Alteration - A Narrative Review. *Cancers* **2022**, *14*, 6122. [PubMed]

Review

Overview of Molecular Detection Technologies for *MET* in Lung Cancer

Carina Heydt *,†, Michaela Angelika Ihle † and Sabine Merkelbach-Bruse

Faculty of Medicine, Institute of Pathology, University Hospital Cologne, University of Cologne, Kerpener Str. 62, 50937 Cologne, Germany
* Correspondence: carina.heydt@uk-koeln.de
† These authors contributed equally to this work.

Simple Summary: A variety of MET aberrations that lead to the dysregulation of the *MET* oncogene and thus the activation of various signaling pathways have been described. These include MET overexpression, the activation of *MET* mutations comprising exon 14 skipping mutations, *MET* gene amplifications, and *MET* fusions. Patients with such aberrations can be treated using a targeted inhibitor such as crizotinib, cabozantinib, tepotinib, and capmatinib. Therefore, the implementation of high-quality and sensitive methods for the detection of the various *MET* aberrations is essential.

Abstract: MET tyrosine kinase receptor pathway activation has become an important actionable target in solid tumors. Aberrations in the *MET* proto-oncogene, including MET overexpression, the activation of *MET* mutations, *MET* mutations that lead to *MET* exon 14 skipping, *MET* gene amplifications, and *MET* fusions, are known to be primary and secondary oncogenic drivers in cancer; these aberrations have evolved as predictive biomarkers in clinical diagnostics. Thus, the detection of all known *MET* aberrations in daily clinical care is essential. In this review, current molecular technologies for the detection of the different *MET* aberrations are highlighted, including the benefits and drawbacks. In the future, another focus will be on the standardization of detection technologies for the delivery of reliable, quick, and affordable tests in clinical molecular diagnostics.

Keywords: MET; NSCLC; *MET* exon 14 skipping mutation; *MET* gene amplification; *MET* fusion

Citation: Heydt, C.; Ihle, M.A.; Merkelbach-Bruse, S. Overview of Molecular Detection Technologies for *MET* in Lung Cancer. *Cancers* **2023**, *15*, 2932. https://doi.org/10.3390/cancers15112932

Academic Editors: Jan Trøst Jørgensen and Jens Mollerup

Received: 1 April 2023
Revised: 22 May 2023
Accepted: 23 May 2023
Published: 26 May 2023

1. Introduction

The *MET* gene (MET proto-oncogene, receptor tyrosine kinase), which consists of 21 exons separated by 20 introns, is located on chromosome 7q21-31 and encodes the MET receptor tyrosine kinase (190 kDa). Together with its ligand, hepatocyte growth factor (HGF), MET plays an important role in tumor proliferation, angiogenesis, and migration [1,2]. A variety of *MET* aberrations that lead to the dysregulation of the *MET* oncogene and thus the activation of various signaling pathways such as MAPK, PI3K-AKT, and JAK-STAT have already been described. These include MET overexpression, the activation of *MET* mutations comprising *MET* exon 14 skipping mutations, *MET* gene amplifications, and *MET* fusions [3,4]. Patients with these types of aberrations can be treated by using an inhibitor targeting either *MET*, such as capmatinib and tepotinib, or by using multikinase inhibitors, such as crizotinib and cabozantinib [5]. Capmatinib (Tabrecta®, Novartis Pharma GmbH, Basel, Switzerland) and tepotinib (Tepmetko®, Merck KGaA, Darmstadt, Germany) have been approved for the treatment of patients with advanced non-small cell lung carcinoma (NSCLC) and *MET* exon 14 skipping mutation who are undergoing systemic therapy after platinum-based chemotherapy or require treatment using immunotherapy. Thus, the implementation of high-quality and sensitive detection methods is essential for the identification of the various *MET* aberrations.

2. MET Receptor

The MET receptor functions as a disulfide-linked heterodimer tyrosine kinase receptor and is composed of an extracellular, transmembrane, and intracellular domain. The extracellular domain is the binding site of its ligand, HGF, and consists of the semaphorin (SEMA), plexin semaphorin integrin (PSI), and immunoglobulin plexin transcription factor (IPT) domains. The intracellular domain consists of the juxtamembrane (JM) domain with the E3 ubiquitin ligase casitas B-lineage lymphoma (c-CBL) binding site, tyrosine kinase (TK) domain, and C-terminal multifunctional docking site (Figure 1) [2,6–8]. Binding of the ligand HGF to the SEMA domain on the extracellular portion of the MET receptor induces MET homodimerization and the subsequent autophosphorylation of the tyrosine residues at codon Y1234 and Y1235 (NM_000245 (Y1252 and Y1253 NM_0001127500)) in the intracellular TK domain, thus leading to the activation of the kinase domain. This is followed by the phosphorylation of Y1349 and Y1356 (NM_000245 (Y1367 and Y1374 NM_0001127500)) in the multifunctional docking site, which opens and forms a docking site for intracellular adaptors that recruit SRC (SRC proto-oncogene, non-receptor tyrosine kinase) adapter protein; additionally, the subsequent activation of several downstream pathways occurs, such as the PI3K/AKT pathway, RAS mitogen activated protein kinase (MAPK) cascade, signal transducer and activator of transcription (STAT), and NF-κB pathway [4]. These signaling pathways play an important role in proliferation, organogenesis, liver regeneration, embryogenesis, wound healing, and cell motility [6,7,9].

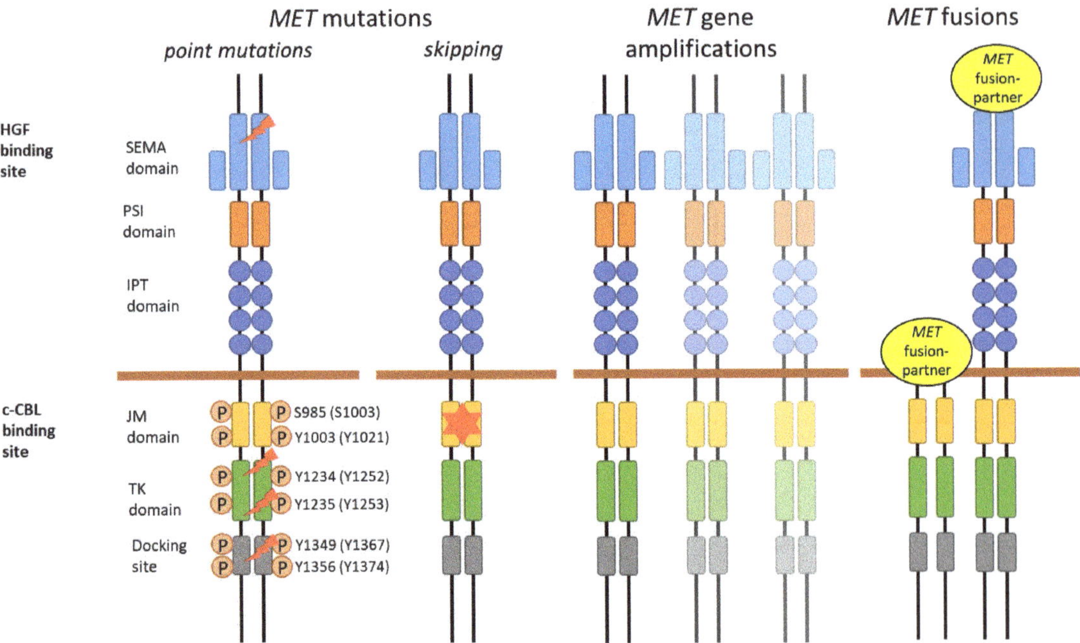

Figure 1. Schematic representation of *MET* aberrations in the MET receptor.

A variety of *MET* aberrations have been described, including *MET* mutations, *MET* gene amplifications, and *MET* fusions. *MET* point mutations (lightning) can occur in the tyrosine kinase (TK) domain, in the Sema domain, and at the multifunctional docking site.

The main amino acid residues (transcript NM_000245 and NM_0001127500) involved in MET regulation through phosphorylation (P) are depicted. *MET* exon 14 skipping mutations (star) are located in the juxtamembrane (JM) domain, which contains the E3 ubiquitin ligase casitas B-lineage lymphoma (c-CBL) binding site. *MET* fusions are very heterogeneous and can result in different fusion proteins.

3. MET Aberrations in Lung Cancer

3.1. MET Overexpression

MET overexpression was discovered to be one of the first mechanisms of dysregulation of MET; since its discovery, it has been detected in a variety of cancers [10–12]. MET overexpression increases ligand-independent phosphorylation and activation of signaling pathways, and it has been linked to metastases, enhanced tumor invasion, and poor survival [13]. In non-small cell lung cancers (NSCLC), MET overexpression has been found with a varying frequency of 35–72% through immunohistochemistry (IHC) [14–16].

Several MET IHC antibodies have been developed, including monoclonal and polyclonal antibodies and antibodies against phosphorylated MET [17–20]. Most commonly, the anti-c-MET (SP44) rabbit monoclonal primary antibody (Ventana Medical Systems Inc.) is used; however, comparative studies of the different antibodies are still missing. IHC slides are evaluated by pathologists, and MET protein expression is semi-quantitatively measured (Table 1). At present, a variety of MET IHC scoring systems and cutoff values have been published. Mostly, staining intensities are classified as negative (0), weak (1+), moderate (2+), and strong (3+), and a staining intensity of 2+ in at least 50% of tumor cells is classified as MET overexpression [15,17].

Studies that use MET protein expression as a biomarker for MET-targeted therapies with monoclonal antibodies and MET tyrosine kinase inhibitors have not been successful up to this point. This poses the question of whether the cutoff values for patient enrollment have been chosen sufficiently [17,21,22] or whether MET overexpression determined by IHC may have a low predictive value for MET activation and ET tumor dependency [23] in contrast to, for example, ALK overexpression in *ALK* fusion-positive NSCLCs [20]. This is supported by growing evidence that MET IHC cannot be used as a screening tool for MET activation by *MET* exon 14 skipping mutations or *MET* gene amplifications in NSCLC [15]. One study revealed that only 3% of MET IHC positive samples had *MET* exon 14 skipping mutations, and only 1% showed a *MET* gene amplification [15]. Another study showed an only 16.1% concordance between IHC and *MET* DNA alterations (*MET* exon 14 skipping mutations and *MET* gene amplifications) [24]. Thus, MET overexpression as a primary biomarker and oncogenic driver remains unclear and thus has not reached clinical use.

Table 1. Detection techniques for *MET* aberrations.

MET Aberration	Detection Technique	Tested Material	Evaluation Criteria	Advantages	Disadvantages
MET overexpression [14–23]	IHC antibodies	FFPE slide	Semi-quantitative score 0–3+	Technique widely used and available, fast and cheap	Observer-dependent, tissue sectioning artefacts, new FFPE slide for every analysis, no consensus on scoring system and cutoff
MET exon 14 skipping [25–39]	RNA NGS (amplicon-, AMP-, or hybridization-based)	RNA from FFPE or fresh frozen material	Mutation, coverage, MAF, fusion product of exon 13 and 15	Sensitive, reliable, direct detection of alternative splicing, multiplexing	RNA degradation, underlying mutation cannot be determined
	RT-PCR	RNA from FFPE or fresh frozen material	Fusion product of exon 13 and 15	Sensitive, reliable, direct detection of alternative splicing, fast turnaround time, widely used and available	RNA degradation, underlying mutation cannot be determined, targeted mutations only
MET exon 14 skipping mutations and point mutations [33,35,37,40–44]	DNA NGS (amplicon- or hybridization-based)	DNA from FFPE, fresh frozen material, or liquid biopsy	Mutation, coverage, VAF	Sensitive, reliable, detection of exact mutation, multiplexing	No assessment of splicing effect
	Sanger sequencing	DNA from FFPE, fresh frozen material, or liquid biopsy	Mutation, VAF	Detection of exact mutation, fast turnaround time, widely used and available	Sensitivity, single assay for each target, no assessment of splicing effect
MET amplifications [5,41,45–61]	FISH	FFPE slides	*MET* GCN, *MET*/CEN7 ratio	Technique widely used and available, detection of focal amplification, polysomy, and chromosome duplications	Observer-dependent, tissue sectioning artefacts, new FFPE slide for every analysis, no consensus on scoring system and cutoff
	DNA NGS (amplicon- or hybridization-based)	DNA from FFPE, fresh frozen material, or liquid biopsy	Mutation, coverage, VAF	Sensitive, DNA from FFPE easily accessible, multiplexing	High number of false negatives, no standardized cutoff or bioinformatics, no morphological correlation
	Other DNA-based technologies (ddPCR, NanoString nCounter technology)	DNA from FFPE, fresh frozen material, or liquid biopsy	Expression, GCN	DNA from FFPE easily accessible	High number of false negatives, no morphological correlation, no standardized cutoff, large amounts of DNA needed
MET Aberration	Detection Technique	Tested Material	Evaluation Criteria	Advantages	Disadvantages
MET fusions [35,39,62–71]	RNA NGS (AMP- or hybridization-based)	RNA from FFPE or fresh frozen material	Fusionreads, 3′-5′ imbalance	Sensitive, reliable detection of known and novel fusion partners, multiplexing	RNA degradation
	DNA NGS (Hybridization-based)	DNA from FFPE	Fusionreads, 3′-5′ imbalance, coverage,	DNA from FFPE easily accessible, detection of known and novel fusion partners if region is covered, multiplexing	False negative results, novel fusions are problematic due to location of fusion break point
	FISH	FFPE slides	n.a. break-apart events	Technique widely used and available	No standardized assay available, observer-dependent, tissue sectioning artefacts, new FFPE slide for every analysis
	RT-PCR	RNA from FFPE	Fusion product	Technique widely used and available	No standardized assay available, only for known fusion partners, RNA degradation

IHC: immunohistochemistry; FISH: fluorescence in situ hybridization; FFPE: formalin-fixed paraffin-embedded; RT-PCR: quantitative real-time polymerase chain reaction; ddPCR: digital droplet PCR; n.a.: not available; GCN: gene copy number; CEN7: centromere of chromosome 7; AMP: anchored multiplex polymerase chain reaction; NGS: next-generation sequencing; VAF: variant allele frequency.

3.2. MET Mutations

3.2.1. *MET* Exon 14 Skipping

MET exon 14 skipping mutations occur in 3–4% of NSCLC and are very heterogeneous [3]. *MET* exon 14 skipping mutations gained importance when Frampton et al. and Paik et al. first reported large studies that featured patients with stage IV lung adenocarcinomas harboring a variety of *MET* exon 14 splice variants, which resulted in *MET* exon 14 skipping and showed clinical sensitivity towards MET inhibitors [25,27]. To date, over 100 different mutations that can lead to *MET* exon 14 skipping have been described [26]. Mutations leading to *MET* exon 14 skipping interfere with the normal regulation of MET transcription. These include numerous point mutations, insertions, and deletions disrupting the splice acceptor at the branch point or the polypyrimidine tract site in exon 14, the 5′ end of exon 14, or the splice donor site at the 3′ end [28]. Furthermore, silent mutations in the splice sites can lead to *MET* exon 14 skipping. As a consequence, the spliceosome skips transcribing exon 14, leading to the loss of the entire exon and thus the JM domain encompassing 141 base pairs with the c-CBL binding site (Y1003 (NM_000245); Y1021 (NM_0001127500)). Without the c-CBL binding site, ubiquitination by CBL and the subsequent lysosomal degradation of MET is impaired; thus, its downstream signaling pathway is constitutively activated [72]. Additionally, the JM domain contains a second phosphorylation site at codon S985 NM_000245 (S1003 NM_0001127500). Phosphorylation of S985 negatively regulates kinase activity [7].

MET skipping mutations are mutually exclusive with mutations in *EGFR*, *KRAS*, and *ERBB2* and fusions in *ALK*, *RET*, and *ROS* [25,27,28,73,74]. Some of the previously reported *MET* mutations seem to represent SNPs; thus, their clinical importance is highly questionable [74]. This problem especially arises when two different mRNA transcripts are frequently used in the literature and when the mutations are reported without the corresponding transcript (transcript NM_000245 and transcript NM_0001127500). For example, in transcript NM_0001127500, a rare polymorphism at codon 1010 (T1010I) is reported; however, in transcript NM_000245, the splice site of exon 14 is located at codon 1010, which can lead to a misconception of the detected mutation in the clinic. The same has been reported with two different *MET* resistance mutations, which are identical variants reported on different transcripts [75]. Thus, mutations should always be reported with the utilized mRNA transcript number [76].

There are several methods available for the detection of *MET* exon 14 skipping mutations (Table 1). They can be either detected on the DNA or RNA level. Increasingly, for the detection of *MET* exon 14 skipping mutations, amplicon- or hybridization-based next-generation sequencing (NGS) methods are used. Although single gene analyzes such as Sanger sequencing or quantitative real-time polymerase chain reaction (RT-PCR) can detect these aberrations, they are used less often and are considered impractical due to the large number of biomarkers that have to be tested simultaneously, especially in lung cancer, and at the same time the low availability of material. The following paragraphs highlight the different methods available.

Parallel Sequencing (NGS) Multigene Assays

DNA-based NGS assays analyze the DNA variant that underlies the skipping of exon 14 on the RNA level. For this, *MET* exon 14 as well as the intronic regions up- and downstream of exon 14 have to be covered by NGS assays sufficiently, as *MET* exon 14 skipping alterations are very heterogenous and deletions can reach far into the intronic sites of exon 14 [25]. At the DNA level, the exact mutation can be specified according to the human genome variation society (HGVS) nomenclature, which is not possible on RNA level. However, the splicing effect cannot be directly assessed at the DNA level. To verify the splicing effect, a confirmatory RNA-based assay is necessary or the effect should be substantiated by the literature [77].

There are two main types of NGS assays, amplicon-based and hybridization-based NGS [29,33,35,37,43]. Either custom panels or commercially available panels can be used. However, commercially available targeted as well as whole exome sequencing panels sometimes lack the intronic regions around *MET* exon 14 and are unable to detect all relevant *MET* exon 14 skipping mutations. Amplicon-based assays use a defined set of primers for the enrichment of the target regions via multiplex PCR followed by library preparation and sequencing. The advantage of an amplicon-based approach is the faster turn-around time, demand of smaller DNA amounts, and ability to use even chemically modified and fragmented DNA derived from formalin-fixed and paraffin-embedded (FFPE) tissue [34]. The disadvantages are the limitation of the total target size, possible allele dropout and thus false-negative results if either the mutation is localized within the primer binding site, or the occurrence of primer mismatches, especially in repetitive sequences [77]. Studies have shown that *MET* exon 14 skipping alterations can be missed by amplicon-based sequencing if the assay is not optimized for this purpose [31,36]. When using hybridization-based panels, DNA is initially sheared. During library preparation, target-specific biotinylated capture probes are hybridized to target regions, and the probes are enriched by streptavidin beads before sequencing [29,33,37]. The advantages of this method are that it circumvents allelic dropout and that duplicate sequences can be removed. The disadvantages are that larger amounts of DNA are needed and that the data analysis is highly complex [29,31,33,37]. Furthermore, some assays have poor intronic coverage and off-target sequencing reads reduce the sequencing coverage. In general, hybrid-capture assays often fail to detect larger deletions if the bioinformatic analysis does not enable their detection [77].

In contrast to DNA-based assays, RNA-based assays for the detection of *MET* exon 14 skipping mutations permit the direct detection of alternative splicing of *MET* exon 14, resulting in a fusion of exons 13 and 15. This method's limitation is that the underlying mutation cannot be determined. However, only the splicing effect is clinically relevant and qualifies for targeted therapy [3,31]. For RNA-based assays, amplicon-based and hybridization-based panels can also be used. Additionally, an anchored multiplex polymerase chain reaction (AMP) approach can be utilized, which has shown promising results [39]. This technology uses a single-primer extension approach without predefined amplicon sizes. With this technology, fusions and splice variants can be detected without knowledge of the fusion partner, as only one target primer is included [4]. RNA-based assays, however, are highly dependent on the RNA integrity. RNA quality should be closely monitored, especially when using FFPE material. Ideally both DNA- and RNA-based approaches should be simultaneously used. However, as many laboratories perform the DNA extraction first followed by the DNA-based NGS analysis, often no material is left for RNA-based sequencing, especially when using lung cancer biopsies [31,77]. Thus, ideally a combined DNA and RNA extraction from the same tissue slides should be performed.

Single Gene Analyzes

Sanger sequencing can also be used for the detection of *MET* exon 14 skipping mutations. It is material-consuming, as a single DNA fragment is sequenced at the time and shows low sensitivity because of it only being able to detect mutations with an allele frequency above 20%. Thus, a higher tumor cell content is needed than for other sequencing technologies. On the other hand, it is easy to implement and allows for the detection of previously unknown alterations if the region is covered. At the RNA level, the RT-PCR technique is widely used in laboratories for the detection and confirmation of *MET* exon 14 skipping variants, as this method is a cost-effective and fast approach to test FFPE material with high sensitivity [30,38]. However, these assays fail to detect additional mutations and should only be used as a pre-screening or confirmatory tool. Newly developed multigene RT-PCR assays from Biocartis NV (Mechelen, Belgium), Diatech Pharmacogenetics S.R.L. (Jesi AN, Italy), or AmoyDx (Xiamen, China) overcome this issue and allow for the detection of a variety of gene fusions and splice variants at the same time in an easy-to-use and time-sensitive manner [32].

3.2.2. Other MET Mutations

In addition to *MET* exon 14 skipping mutations, activating point mutations in the TK, JM, and extracellular domains have been reported in cancer, leading to ligand-independent receptor phosphorylation and signaling (Figure 1 and Table 1) [78–80]. A variety of these mutations, including H1094Y/R/L (NM_000245; H1112, NM_0001127500) and D1228H/N (NM_000245; D1246, NM_0001127500), were first described in hereditary papillary renal cell carcinoma (HPRCC) [44] and were later also found in sporadic papillary renal cell carcinoma (PRCC) with up to a 15% frequency [40,42].

In NSCLC, *MET* mutations in the TK domain are rare and mainly emerge as an acquired resistance mechanism to MET tyrosine kinase inhibitors or as a resistance mechanism to combinational therapy with EGFR and MET TKI in *MET* exon 14 skipping positive patients or *EGFR*-mutant and *MET* gene amplification positive patients. Mutations such as Y1230C (NM_000245; Y1248C, NM_0001127500), Y1230H (NM_000245; Y1248H, NM_0001127500), D1228H (NM_000245; D1246H, NM_0001127500), and D1228N (NM_000245; D1246N, NM_0001127500) were found to mediate resistance by disrupting the drug binding site of crizotinib [3,81].

MET mutations in the SEMA domain and extracellular compartment have also been reported to possibly affect ligand binding. However, the functional significance and relevance of these mutations is still unknown [13]. Thus, it remains important to evaluate the functional consequences of other *MET* mutations and their clinical implications.

3.3. MET Gene Amplification and Gene Copy Number Alterations

MET gene amplifications and copy number alterations have been reported in 1–6% of NSCLC [61]. Therapy approaches for NSCLC with *MET* gene amplifications, which both occur as the primary driver aberration and as the resistance mechanism to other kinase inhibitors, have currently been evaluated in numerous studies. Studies have shown that MET inhibitors are particularly effective in highly amplified (high level) tumors with a gene copy number [GCN] $\geq$ 10, were *MET* gene amplification acts as an oncogenic driver [5,55,61]. Thus, patient selection and the exact analysis of the *MET* gene amplification status is crucial.

MET gene amplifications and copy number alterations arise from the focal or regional amplification of the *MET* genomic region or from polysomy (Figure 1). In cases with focal amplification, *MET* GCN gains occur without chromosome 7 duplication, whereas *MET* GCN gains due to polysomy arise from the duplication of parts or the entire chromosome 7; thus, multiple parts of chromosome 7 are present [3,47].

MET gene amplifications can be assessed using a variety of methodologies, which determine either the average *MET* GCN and/or the ratio to the centromeric region of chromosome 7 (Table 1). However, the cutoff point for setting MET positivity is still very variable. Different clinical studies have used a variety of thresholds for the definition of amplifications, from amplification positive only to defining an exact *MET* GCN [22,46,49,56,61]. Depending on the method used, thresholds are set at different levels, which causes problems in the interpretation of the potential of *MET* GCN as true biomarker [47].

Additionally, true gene amplifications without chromosome 7 duplication are more likely to lead to oncogene addiction [3,61].

3.3.1. Fluorescence In Situ Hybridization (FISH)

MET gene amplifications and *MET* GCN can be detected using various methods. The gold standard for the detection of *MET* gene amplifications and the most accurate detection method is still fluorescence in situ hybridization (FISH), which is currently still superior to DNA-based methods and mainly used in clinical trials. Bicolor FISH probes label both the *MET* gene and the centromere of chromosome 7 (CEN7). The number of signals identified in a nucleus represent the number of copies present. The signals in a predefined number of cell nuclei are counted and scored based on evaluation criteria such as the *MET* gene/CEN7 ratio and/or the average *MET* GCN. According to the resulting score, cases are divided into different groups [48,50,53,54,58]. Various evaluation scores have been published so far. Thus far, publications have defined *MET* gene amplification by GCN only, either as 5 or more copies per cell [48], or as *MET* GCN ≥ 6, ≥10, or 15 [54,55]. Other studies have also included the number of chromosomes present by calculating *MET*/CEN7 ratio; thus, true amplification can be distinguished from polysomy. A *MET*/CEN7 ratio of ≥2.0 is commonly defined as *MET* gene amplification [50,53,58]. Other studies have categorized the degree of amplification into low, intermediate, high-level, and top-level amplified cases. A top-level amplification was classified as an average *MET* GCN per cell of ≥10. a high-level amplification was defined in tumors with a *MET*/CEN7 ratio ≥2.0 or an average *MET* GCN per cell of ≥6. An intermediate level of GCN gain means that ≥50% of cells contain ≥5 *MET* signals. A low level of GCN gain was defined as ≥40% of tumor cells showing ≥4 *MET* signals [55,58].

The FISH technique is especially useful in cases with low tumor cell content, tumor heterogeneity, and focal amplifications, as FISH is performed on slides and evaluated under the microscope [52,58,60]. However, in situ-based approaches like FISH are also thereby hampered. The evaluation is observer-dependent, and tissue sectioning artefacts can impact the analysis. Furthermore, a new slide of material must be used for each additional parameter that is tested by FISH, which can be problematic when using small biopsies [52,60].

3.3.2. DNA-Based Methods

Another option for the detection of *MET* gene amplifications are a variety of DNA-based methods that work with extracted nucleic acids, such as digital droplet PCR (ddPCR), next-generation sequencing (NGS), or the NanoString nCounter technology. The detection of *MET* gene amplifications by GCN changes using DNA-based methods is still under evaluation. These methods allow for an easier quantification of GCN in comparison to FISH but do not allow for morphological correlation. At present, the performance of NGS-based assays for the detection of *MET* gene amplification have been characterized the best. Data for the other DNA-based methods mentioned above are very limited [52]. Studies that have compared NGS and FISH assays showed low consistency between both methods. Currently, only high-level and top-level amplified samples with GCN ≥ 10 and negative samples determined by FISH can be reliably detected by an NGS analysis [3,52,57,59,60]. Comparative studies have further shown that *MET* gene amplifications can be missed by NGS assays due to a variety of reasons. On the one hand, the tumor material itself can pose problems in the evaluation due to low tumor purity (inclusion of normal, necrotic, and inflammatory cells), low tumor cell content, the overall amount of material present, the FFPE DNA quality, tumor heterogeneity, or focal amplifications and polysomy [52,57,59,60]. On the other hand, analyzing large genomic alterations can be very challenging and can create computational challenges, such as call accuracy and noise reduction. Additionally, a defined set of normal samples or standardized set of controls and the tumor cell content of the samples have to be used for bioinformatic analyses [41,45,51].

In molecular diagnostics, amplicon-based as well as hybridization-based NGS assays are used for *MET* GCN detection. Hybridization-based NGS assays can assess *MET* GCN variations more accurately than amplicon-based NGS assays, as the sequencing bias is reduced, duplicate reads can be filtered out, and the true mean coverage used for GCN determination is less affected by DNA quality, tumor complexity such as tumor purity, and heterogeneity [3,60,78–80]. However, to date, there are no methodologically or clinically defined cutoffs for the definition of *MET* positivity when utilizing NGS assay, nor is there an accepted standard or general consensus regarding the protocols and bioinformatics used, which inevitably leads to discordant results across studies.

3.4. MET Fusions

MET gene fusions are rare oncogenic driver alterations in a variety of cancers, including hepatocellular carcinoma, gastric carcinoma, sarcoma, and NSCLC. Only in glioblastomas, *MET* fusions are described in 12% of cases [3]. The frequency of *MET* fusions in NSCLC is < 0.5%, and they are found to be mutually exclusive with other oncogenic drivers [66]. The first *MET* fusion identified in lung cancer was the *TPR-MET* fusion [82]. Since then, several fusion partners have been characterized, such as *KIF5B, CLIP2, TFG, STARD3NL, ATXN7L1, PTPRZ*, and *CD74* [63,64,66,67,71]. *MET* fusions occur through inter- or intra-chromosomal rearrangement and mostly include the kinase domain on exon 15 and downstream, resulting in ligand-independent constitutive MET activation (Figure 1) [65–68]. The *TPR-MET* fusion, however, does not include exon 14 of *MET* and can show the same oncogenic behavior as NSCLCs with *MET* exon 14 skipping [70]. Fusions such as *KIF5B-MET* and *PTPRZ-MET* that include exon 14 appear to be less oncogenic than the *TRP-MET* fusion [65]. In the *PTPRZ-MET* fusion protein, the *MET* gene is present in the full length, including in the dimerization domain in exon 2, resulting in MET over-expression and increased activation [65]. Therefore, the knowledge of the exact fusion break point seems to be important for the success of MET TKIs. Currently, clinical trials are evaluating the efficiency of MET TKIs in *MET* fusion-positive cancers.

For the detection of *MET* fusions, an RNA-based NGS approach that uses either AMP- or hybridization-based technologies is the first choice. In this way, both unknown fusion partners and the involved exons can be determined [62]. Alternatively, FISH, RT-PCR, and DNA-based NGS techniques can be used. However, an RNA-based NGS panel analysis is the most sensitive approach for such rare and novel events, as all relevant fusions and *MET* exon 14 skipping mutations can be detected in just one assay, thus making FISH and RT-PCR inadequate detection tools. Additionally, intrachromosomal rearrangements may lead to false negative FISH results, as the distance between the 5′ and 3′ probes are too short [35,39,62,71]. DNA-based hybrid-capture NGS approaches that can detect both mutations and fusions have proven to be unreliable in the past and often lead to false negative results for fusion detection, especially in cases of novel fusions. This is due to the localization of fusion breakpoints in large intronic regions with repetitive sequences, which are difficult to cover using capture probes [35,54,69].

4. Conclusions

In recent years, the large, growing number of detected *MET* alterations in NSCLC and other carcinomas as well as the better understanding of the diverse biology driving MET dysregulation in cancer has shown the important role of this kinase for targeted therapy approaches.

Particularly, since the FDA and EMA approval of MET inhibitors for NSCLCs with *MET* exon 14 skipping mutations, testing for all *MET* alterations, e.g., MET expression, *MET* mutations, *MET* gene amplifications, and *MET* fusions, should be routine standard of care for patients with NSCLC. However, there is still the need for the further development of quality assured and sensitive molecular detection methods, especially under the new In Vitro Diagnostics Regulation (IVDR) and when it comes to the detection of *MET* gene amplifications, as these are still widely analyzed reliably by only FISH while the cutoffs for other technologies are lacking. Additionally, inconsistent nomenclature of somatic variants and the different transcripts used in the literature are still of concern, as this can lead to a misconception of the detected mutations in the clinic and thus therapeutic failure. As a final note, quality assured and sensitive molecular detection methods are especially important in Europe, as laboratories are free to choose the diagnostic method used for the detection of the different *MET* alterations.

In the future, the number of targetable biomarkers will increase more and more, and the amount of tissue, effort, and time required to complete complex diagnostic tests will become even more limiting. As molecular targets and therapeutic approaches are continuously changing, the ongoing development and implementation of high-quality molecular testing, and the continuous adaption to the latest findings in cancer research will become increasingly important while limiting economic costs at the same time (Figure 2).

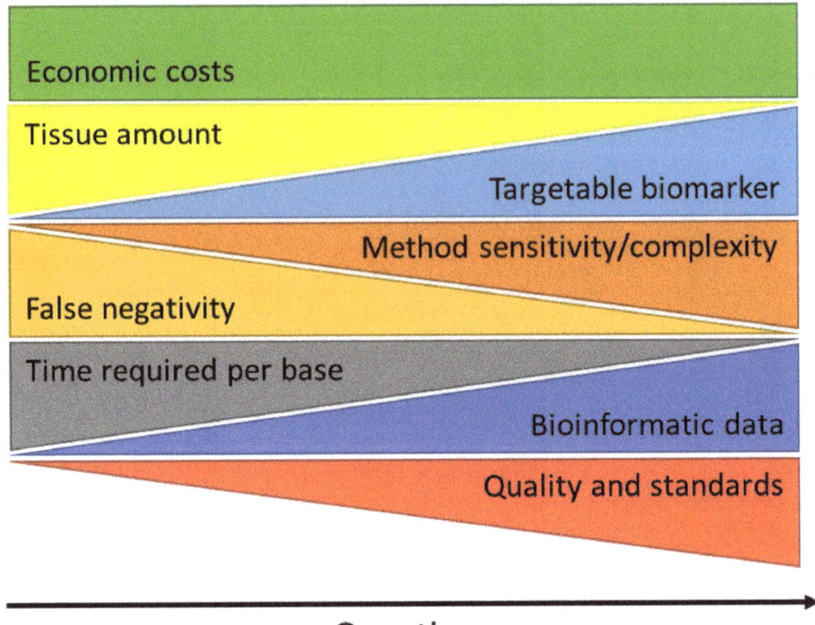

Figure 2. Clinical utility of the analysis of tumor material in molecular pathology diagnostics over time.

Author Contributions: Conceptualization, C.H.; investigation, C.H., M.A.I. and S.M.-B., data curation, C.H., M.A.I. and S.M.-B.; writing—original draft preparation, C.H., M.A.I. and S.M.-B.; writing—review and editing, C.H., M.A.I and S.M.-B.; supervision, C.H. All authors have read and agreed to the published version of the manuscript.

Funding: This research received no external funding.

Conflicts of Interest: Carina Heydt has received honoraria from AstraZeneca, BMS, Illumina, and Molecular Health; Michaela Angelika Ihle has received honoraria from AstraZeneca, BMS, Novartis, and Merck; Sabine Merkelbach-Bruse has received honoraria and travel support from Amgen, AstraZeneca, Bayer, BMS, GSK, Janssen, Merck, MSD, Molecular Health, Novartis, Pfizer, QuIP, Roche, and Targos.

References

1. Fujino, T.; Suda, K.; Mitsudomi, T. Lung Cancer with MET exon 14 Skipping Mutation: Genetic Feature, Current Treatments, and Future Challenges. *Lung Cancer* **2021**, *12*, 35–50. [CrossRef] [PubMed]
2. Gelsomino, F.; Rossi, G.; Tiseo, M. MET and Small-Cell Lung Cancer. *Cancers* **2014**, *6*, 2100–2115. [CrossRef] [PubMed]
3. Guo, R.; Luo, J.; Chang, J.; Rekhtman, N.; Arcila, M.; Drilon, A. MET-dependent solid tumours-molecular diagnosis and targeted therapy. *Nat. Rev. Clin. Oncol.* **2020**, *17*, 569–587. [CrossRef]
4. Lee, M.; Jain, P.; Wang, F.; Ma, P.C.; Borczuk, A.; Halmos, B. MET alterations and their impact on the future of non-small cell lung cancer (NSCLC) targeted therapies. *Expert Opin. Targets* **2021**, *25*, 249–268. [CrossRef] [PubMed]
5. Jørgensen, J.T.; Mollerup, J. Companion Diagnostics and Predictive Biomarkers for MET-Targeted Therapy in NSCLC. *Cancers* **2022**, *14*, 2150. [CrossRef]
6. Organ, S.L.; Tsao, M.S. An overview of the c-MET signaling pathway. *Ther. Adv. Med. Oncol.* **2011**, *3*, S7–S19. [CrossRef]
7. Cecchi, F.; Rabe, D.C.; Bottaro, D.P. Targeting the HGF/Met signaling pathway in cancer therapy. *Expert Opin. Targets* **2012**, *16*, 553–572. [CrossRef]

8. Kawaguchi, M.; Kataoka, H. Mechanisms of Hepatocyte Growth Factor Activation in Cancer Tissues. *Cancers* **2014**, *6*, 1890–1904. [CrossRef]
9. Eder, J.P.; Vande Woude, G.F.; Boerner, S.A.; LoRusso, P.M. Novel therapeutic inhibitors of the c-Met signaling pathway in cancer. *Clin. Cancer Res. Off. J. Am. Assoc. Cancer Res.* **2009**, *15*, 2207–2214. [CrossRef]
10. Catenacci, D.V.; Ang, A.; Liao, W.L.; Shen, J.; O'Day, E.; Loberg, R.D.; Cecchi, F.; Hembrough, T.; Ruzzo, A.; Graziano, F. MET tyrosine kinase receptor expression and amplification as prognostic biomarkers of survival in gastroesophageal adenocarcinoma. *Cancer* **2017**, *123*, 1061–1070. [CrossRef]
11. Gayyed, M.F.; Abd El-Maqsoud, N.M.; El-Hameed El-Heeny, A.A.; Mohammed, M.F. c-MET expression in colorectal adenomas and primary carcinomas with its corresponding metastases. *J. Gastrointest. Oncol.* **2015**, *6*, 618–627. [CrossRef] [PubMed]
12. Graveel, C.R.; Tolbert, D.; Vande Woude, G.F. MET: A critical player in tumorigenesis and therapeutic target. *Cold Spring Harb. Perspect. Biol.* **2013**, *5*, a009209. [CrossRef] [PubMed]
13. Recondo, G.; Che, J.; Janne, P.A.; Awad, M.M. Targeting MET Dysregulation in Cancer. *Cancer Discov.* **2020**, *10*, 922–934. [CrossRef] [PubMed]
14. Watermann, I.; Schmitt, B.; Stellmacher, F.; Muller, J.; Gaber, R.; Kugler, C.; Reinmuth, N.; Huber, R.M.; Thomas, M.; Zabel, P.; et al. Improved diagnostics targeting c-MET in non-small cell lung cancer: Expression, amplification and activation? *Diagn. Pathol.* **2015**, *10*, 130. [CrossRef]
15. Guo, R.; Berry, L.D.; Aisner, D.L.; Sheren, J.; Boyle, T.; Bunn, P.A., Jr.; Johnson, B.E.; Kwiatkowski, D.J.; Drilon, A.; Sholl, L.M.; et al. MET IHC Is a Poor Screen for MET Amplification or MET Exon 14 Mutations in Lung Adenocarcinomas: Data from a Tri-Institutional Cohort of the Lung Cancer Mutation Consortium. *J. Thorac. Oncol.* **2019**, *14*, 1666–1671. [CrossRef] [PubMed]
16. Salgia, R. MET in Lung Cancer: Biomarker Selection Based on Scientific Rationale. *Mol. Cancer Ther.* **2017**, *16*, 555–565. [CrossRef]
17. Spigel, D.R.; Edelman, M.J.; O'Byrne, K.; Paz-Ares, L.; Mocci, S.; Phan, S.; Shames, D.S.; Smith, D.; Yu, W.; Paton, V.E.; et al. Results From the Phase III Randomized Trial of Onartuzumab Plus Erlotinib Versus Erlotinib in Previously Treated Stage IIIB or IV Non-Small-Cell Lung Cancer: METLung. *J. Clin. Oncol. Off. J. Am. Soc. Clin. Oncol.* **2017**, *35*, 412–420. [CrossRef]
18. Ryan, C.J.; Rosenthal, M.; Ng, S.; Alumkal, J.; Picus, J.; Gravis, G.; Fizazi, K.; Forget, F.; Machiels, J.P.; Srinivas, S.; et al. Targeted MET inhibition in castration-resistant prostate cancer: A randomized phase II study and biomarker analysis with rilotumumab plus mitoxantrone and prednisone. *Clin. Cancer Res. Off. J. Am. Assoc. Cancer Res.* **2013**, *19*, 215–224. [CrossRef]
19. Iveson, T.; Donehower, R.C.; Davidenko, I.; Tjulandin, S.; Deptala, A.; Harrison, M.; Nirni, S.; Lakshmaiah, K.; Thomas, A.; Jiang, Y.; et al. Rilotumumab in combination with epirubicin, cisplatin, and capecitabine as first-line treatment for gastric or oesophagogastric junction adenocarcinoma: An open-label, dose de-escalation phase 1b study and a double-blind, randomised phase 2 study. *Lancet. Oncol.* **2014**, *15*, 1007–1018. [CrossRef]
20. Srivastava, A.K.; Hollingshead, M.G.; Weiner, J.; Navas, T.; Evrard, Y.A.; Khin, S.A.; Ji, J.J.; Zhang, Y.; Borgel, S.; Pfister, T.D.; et al. Pharmacodynamic Response of the MET/HGF Receptor to Small-Molecule Tyrosine Kinase Inhibitors Examined with Validated, Fit-for-Clinic Immunoassays. *Clin. Cancer Res. Off. J. Am. Assoc. Cancer Res.* **2016**, *22*, 3683–3694. [CrossRef]
21. Neal, J.W.; Dahlberg, S.E.; Wakelee, H.A.; Aisner, S.C.; Bowden, M.; Huang, Y.; Carbone, D.P.; Gerstner, G.J.; Lerner, R.E.; Rubin, J.L.; et al. Erlotinib, cabozantinib, or erlotinib plus cabozantinib as second-line or third-line treatment of patients with EGFR wild-type advanced non-small-cell lung cancer (ECOG-ACRIN 1512): A randomised, controlled, open-label, multicentre, phase 2 trial. *Lancet. Oncol.* **2016**, *17*, 1661–1671. [CrossRef] [PubMed]
22. Schuler, M.H.; Berardi, R.; Lim, W.-T.; Geel, R.V.; Jonge, M.J.D.; Bauer, T.M.; Azaro, A.; Gottfried, M.; Han, J.-Y.; Lee, D.H.; et al. Phase (Ph) I study of the safety and efficacy of the cMET inhibitor capmatinib (INC280) in patients (pts) with advanced cMET+ non-small cell lung cancer (NSCLC). *J. Clin. Oncol.* **2016**, *34*, 9067. [CrossRef]
23. Finocchiaro, G.; Toschi, L.; Gianoncelli, L.; Baretti, M.; Santoro, A. Prognostic and predictive value of MET deregulation in non-small cell lung cancer. *Ann. Transl. Med.* **2015**, *3*, 83. [CrossRef] [PubMed]
24. Tong, J.H.; Yeung, S.F.; Chan, A.W.; Chung, L.Y.; Chau, S.L.; Lung, R.W.; Tong, C.Y.; Chow, C.; Tin, E.K.; Yu, Y.H.; et al. MET Amplification and Exon 14 Splice Site Mutation Define Unique Molecular Subgroups of Non-Small Cell Lung Carcinoma with Poor Prognosis. *Clin. Cancer Res. Off. J. Am. Assoc. Cancer Res.* **2016**, *22*, 3048–3056. [CrossRef]
25. Frampton, G.M.; Ali, S.M.; Rosenzweig, M.; Chmielecki, J.; Lu, X.; Bauer, T.M.; Akimov, M.; Bufill, J.A.; Lee, C.; Jentz, D.; et al. Activation of MET via diverse exon 14 splicing alterations occurs in multiple tumor types and confers clinical sensitivity to MET inhibitors. *Cancer Discov.* **2015**, *5*, 850–859. [CrossRef]
26. Kim, S.Y.; Yin, J.; Bohlman, S.; Walker, P.; Dacic, S.; Kim, C.; Khan, H.; Liu, S.V.; Ma, P.C.; Nagasaka, M.; et al. Characterization of MET Exon 14 Skipping Alterations (in NSCLC) and Identification of Potential Therapeutic Targets Using Whole Transcriptome Sequencing. *JTO Clin. Res. Rep.* **2022**, *3*, 100381. [CrossRef]
27. Paik, P.K.; Drilon, A.; Fan, P.D.; Yu, H.; Rekhtman, N.; Ginsberg, M.S.; Borsu, L.; Schultz, N.; Berger, M.F.; Rudin, C.M.; et al. Response to MET inhibitors in patients with stage IV lung adenocarcinomas harboring MET mutations causing exon 14 skipping. *Cancer Discov.* **2015**, *5*, 842–849. [CrossRef]
28. Onozato, R.; Kosaka, T.; Kuwano, H.; Sekido, Y.; Yatabe, Y.; Mitsudomi, T. Activation of MET by gene amplification or by splice mutations deleting the juxtamembrane domain in primary resected lung cancers. *J. Thorac. Oncol.* **2009**, *4*, 5–11. [CrossRef]

29. Cheng, D.T.; Mitchell, T.N.; Zehir, A.; Shah, R.H.; Benayed, R.; Syed, A.; Chandramohan, R.; Liu, Z.Y.; Won, H.H.; Scott, S.N.; et al. Memorial Sloan Kettering-Integrated Mutation Profiling of Actionable Cancer Targets (MSK-IMPACT): A Hybridization Capture-Based Next-Generation Sequencing Clinical Assay for Solid Tumor Molecular Oncology. *J. Mol. Diagn. JMD* **2015**, *17*, 251–264. [CrossRef]

30. Das, R.; Jakubowski, M.A.; Spildener, J.; Cheng, Y.W. Identification of Novel MET Exon 14 Skipping Variants in Non-Small Cell Lung Cancer Patients: A Prototype Workflow Involving in Silico Prediction and RT-PCR. *Cancers* **2022**, *14*, 4814. [CrossRef]

31. Davies, K.D.; Lomboy, A.; Lawrence, C.A.; Yourshaw, M.; Bocsi, G.T.; Camidge, D.R.; Aisner, D.L. DNA-Based versus RNA-Based Detection of MET Exon 14 Skipping Events in Lung Cancer. *J. Thorac. Oncol.* **2019**, *14*, 737–741. [CrossRef] [PubMed]

32. Depoilly, T.; Garinet, S.; van Kempen, L.C.; Schuuring, E.; Clave, S.; Bellosillo, B.; Ercolani, C.; Buglioni, S.; Siemanowski, J.; Merkelbach-Bruse, S.; et al. Multicenter Evaluation of the Idylla GeneFusion in Non-Small-Cell Lung Cancer. *J. Mol. Diagn. JMD* **2022**, *24*, 1021–1030. [CrossRef] [PubMed]

33. Drilon, A.; Wang, L.; Arcila, M.E.; Balasubramanian, S.; Greenbowe, J.R.; Ross, J.S.; Stephens, P.; Lipson, D.; Miller, V.A.; Kris, M.G.; et al. Broad, Hybrid Capture-Based Next-Generation Sequencing Identifies Actionable Genomic Alterations in Lung Adenocarcinomas Otherwise Negative for Such Alterations by Other Genomic Testing Approaches. *Clin. Cancer Res. Off. J. Am. Assoc. Cancer Res.* **2015**, *21*, 3631–3639. [CrossRef] [PubMed]

34. Heydt, C.; Fassunke, J.; Kunstlinger, H.; Ihle, M.A.; Konig, K.; Heukamp, L.C.; Schildhaus, H.U.; Odenthal, M.; Buttner, R.; Merkelbach-Bruse, S. Comparison of pre-analytical FFPE sample preparation methods and their impact on massively parallel sequencing in routine diagnostics. *PLoS ONE* **2014**, *9*, e104566. [CrossRef]

35. Heydt, C.; Wölwer, C.B.; Velazquez Camacho, O.; Wagener-Ryczek, S.; Pappesch, R.; Siemanowski, J.; Rehker, J.; Haller, F.; Agaimy, A.; Worm, K.; et al. Detection of gene fusions using targeted next-generation sequencing: A comparative evaluation. *BMC Med. Genom.* **2021**, *14*, 62. [CrossRef]

36. Poirot, B.; Doucet, L.; Benhenda, S.; Champ, J.; Meignin, V.; Lehmann-Che, J. MET Exon 14 Alterations and New Resistance Mutations to Tyrosine Kinase Inhibitors: Risk of Inadequate Detection with Current Amplicon-Based NGS Panels. *J. Thorac. Oncol.* **2017**, *12*, 1582–1587. [CrossRef]

37. Suh, J.H.; Johnson, A.; Albacker, L.; Wang, K.; Chmielecki, J.; Frampton, G.; Gay, L.; Elvin, J.A.; Vergilio, J.A.; Ali, S.; et al. Comprehensive Genomic Profiling Facilitates Implementation of the National Comprehensive Cancer Network Guidelines for Lung Cancer Biomarker Testing and Identifies Patients Who May Benefit From Enrollment in Mechanism-Driven Clinical Trials. *Oncologist* **2016**, *21*, 684–691. [CrossRef]

38. Sui, J.S.Y.; Finn, S.P.; Gray, S.G. Detection of MET Exon 14 Skipping Alterations in Lung Cancer Clinical Samples Using a PCR-Based Approach. *Methods Mol. Biol.* **2021**, *2279*, 145–155. [CrossRef]

39. Zheng, Z.; Liebers, M.; Zhelyazkova, B.; Cao, Y.; Panditi, D.; Lynch, K.D.; Chen, J.; Robinson, H.E.; Shim, H.S.; Chmielecki, J.; et al. Anchored multiplex PCR for targeted next-generation sequencing. *Nat. Med.* **2014**, *20*, 1479–1484. [CrossRef]

40. Albiges, L.; Guegan, J.; Le Formal, A.; Verkarre, V.; Rioux-Leclercq, N.; Sibony, M.; Bernhard, J.C.; Camparo, P.; Merabet, Z.; Molinie, V.; et al. MET is a potential target across all papillary renal cell carcinomas: Result from a large molecular study of pRCC with CGH array and matching gene expression array. *Clin. Cancer Res. Off. J. Am. Assoc. Cancer Res.* **2014**, *20*, 3411–3421. [CrossRef]

41. Chen, Y.-C.; Seifuddin, F.; Nguyen, C.; Yang, Z.; Chen, W.; Yan, C.; Chen, Q.; Wang, C.; Xiao, W.; Pirooznia, M.; et al. Comprehensive Assessment of Somatic Copy Number Variation Calling Using Next-Generation Sequencing Data. *bioRxiv* **2021**. [CrossRef]

42. Pal, S.K.; Ali, S.M.; Yakirevich, E.; Geynisman, D.M.; Karam, J.A.; Elvin, J.A.; Frampton, G.M.; Huang, X.; Lin, D.I.; Rosenzweig, M.; et al. Characterization of Clinical Cases of Advanced Papillary Renal Cell Carcinoma via Comprehensive Genomic Profiling. *Eur. Urol.* **2018**, *73*, 71–78. [CrossRef] [PubMed]

43. Pfarr, N.; Stenzinger, A.; Penzel, R.; Warth, A.; Dienemann, H.; Schirmacher, P.; Weichert, W.; Endris, V. High-throughput diagnostic profiling of clinically actionable gene fusions in lung cancer. *Genes Chromosomes Cancer* **2016**, *55*, 30–44. [CrossRef] [PubMed]

44. Schmidt, L.; Duh, F.M.; Chen, F.; Kishida, T.; Glenn, G.; Choyke, P.; Scherer, S.W.; Zhuang, Z.; Lubensky, I.; Dean, M.; et al. Germline and somatic mutations in the tyrosine kinase domain of the MET proto-oncogene in papillary renal carcinomas. *Nat. Genet.* **1997**, *16*, 68–73. [CrossRef]

45. Budczies, J.; Pfarr, N.; Stenzinger, A.; Treue, D.; Endris, V.; Ismaeel, F.; Bangemann, N.; Blohmer, J.U.; Dietel, M.; Loibl, S.; et al. Ioncopy: A novel method for calling copy number alterations in amplicon sequencing data including significance assessment. *Oncotarget* **2016**, *7*, 13236–13247. [CrossRef]

46. Camidge, D.R.; Otterson, G.A.; Clark, J.W.; Ignatius Ou, S.H.; Weiss, J.; Ades, S.; Shapiro, G.I.; Socinski, M.A.; Murphy, D.A.; Conte, U.; et al. Crizotinib in Patients With MET-Amplified NSCLC. *J. Thorac. Oncol.* **2021**, *16*, 1017–1029. [CrossRef]

47. Caparica, R.; Yen, C.T.; Coudry, R.; Ou, S.I.; Varella-Garcia, M.; Camidge, D.R.; de Castro, G., Jr. Responses to Crizotinib Can Occur in High-Level MET-Amplified Non-Small Cell Lung Cancer Independent of MET Exon 14 Alterations. *J. Thorac. Oncol.* **2017**, *12*, 141–144. [CrossRef]

48. Cappuzzo, F.; Marchetti, A.; Skokan, M.; Rossi, E.; Gajapathy, S.; Felicioni, L.; Del Grammastro, M.; Sciarrotta, M.G.; Buttitta, F.; Incarbone, M.; et al. Increased MET gene copy number negatively affects survival of surgically resected non-small-cell lung cancer patients. *J. Clin. Oncol. Off. J. Am. Soc. Clin. Oncol.* **2009**, *27*, 1667–1674. [CrossRef]

49. Dagogo-Jack, I.; Moonsamy, P.; Gainor, J.F.; Lennerz, J.K.; Piotrowska, Z.; Lin, J.J.; Lennes, I.T.; Sequist, L.V.; Shaw, A.T.; Goodwin, K.; et al. A Phase 2 Study of Capmatinib in Patients With MET-Altered Lung Cancer Previously Treated With a MET Inhibitor. *J. Thorac. Oncol.* **2021**, *16*, 850–859. [CrossRef]

50. Go, H.; Jeon, Y.K.; Park, H.J.; Sung, S.W.; Seo, J.W.; Chung, D.H. High MET gene copy number leads to shorter survival in patients with non-small cell lung cancer. *J. Thorac. Oncol.* **2010**, *5*, 305–313. [CrossRef]

51. Gusnanto, A.; Wood, H.M.; Pawitan, Y.; Rabbitts, P.; Berri, S. Correcting for cancer genome size and tumour cell content enables better estimation of copy number alterations from next-generation sequence data. *Bioinformatics* **2011**, *28*, 40–47. [CrossRef]

52. Heydt, C.; Becher, A.K.; Wagener-Ryczek, S.; Ball, M.; Schultheis, A.M.; Schallenberg, S.; Rüsseler, V.; Büttner, R.; Merkelbach-Bruse, S. Comparison of in situ and extraction-based methods for the detection of MET amplifications in solid tumors. *Comput. Struct. Biotechnol. J.* **2019**, *17*, 1339–1347. [CrossRef] [PubMed]

53. Jardim, D.L.; Tang, C.; Gagliato Dde, M.; Falchook, G.S.; Hess, K.; Janku, F.; Fu, S.; Wheler, J.J.; Zinner, R.G.; Naing, A.; et al. Analysis of 1,115 patients tested for MET amplification and therapy response in the MD Anderson Phase I Clinic. *Clin. Cancer Res. Off. J. Am. Assoc. Cancer Res.* **2014**, *20*, 6336–6345. [CrossRef] [PubMed]

54. Lee, H.E.; Kim, M.A.; Lee, H.S.; Jung, E.J.; Yang, H.K.; Lee, B.L.; Bang, Y.J.; Kim, W.H. MET in gastric carcinomas: Comparison between protein expression and gene copy number and impact on clinical outcome. *Br. J. Cancer* **2012**, *107*, 325–333. [CrossRef] [PubMed]

55. Overbeck, T.R.; Cron, D.A.; Schmitz, K.; Rittmeyer, A.; Korber, W.; Hugo, S.; Schnalke, J.; Lukat, L.; Hugo, T.; Hinterthaner, M.; et al. Top-level MET gene copy number gain defines a subtype of poorly differentiated pulmonary adenocarcinomas with poor prognosis. *Transl. Lung Cancer Res.* **2020**, *9*, 603–616. [CrossRef]

56. Paik, P.K.; Felip, E.; Veillon, R.; Sakai, H.; Cortot, A.B.; Garassino, M.C.; Mazieres, J.; Viteri, S.; Senellart, H.; Van Meerbeeck, J.; et al. Tepotinib in Non-Small-Cell Lung Cancer with MET Exon 14 Skipping Mutations. *N. Engl. J. Med.* **2020**, *383*, 931–943. [CrossRef]

57. Peng, L.X.; Jie, G.L.; Li, A.N.; Liu, S.Y.; Sun, H.; Zheng, M.M.; Zhou, J.Y.; Zhang, J.T.; Zhang, X.C.; Zhou, Q.; et al. MET amplification identified by next-generation sequencing and its clinical relevance for MET inhibitors. *Exp. Hematol. Oncol.* **2021**, *10*, 52. [CrossRef]

58. Schildhaus, H.U.; Schultheis, A.M.; Ruschoff, J.; Binot, E.; Merkelbach-Bruse, S.; Fassunke, J.; Schulte, W.; Ko, Y.D.; Schlesinger, A.; Bos, M.; et al. MET amplification status in therapy-naive adeno- and squamous cell carcinomas of the lung. *Clin. Cancer Res. Off. J. Am. Assoc. Cancer Res.* **2015**, *21*, 907–915. [CrossRef]

59. Schmitt, C.; Schulz, A.A.; Winkelmann, R.; Smith, K.; Wild, P.J.; Demes, M. Comparison of MET gene amplification analysis by next-generation sequencing and fluorescence in situ hybridization. *Oncotarget* **2021**, *12*, 2273–2282. [CrossRef]

60. Schubart, C.; Stohr, R.; Togel, L.; Fuchs, F.; Sirbu, H.; Seitz, G.; Seggewiss-Bernhardt, R.; Leistner, R.; Sterlacci, W.; Vieth, M.; et al. MET Amplification in Non-Small Cell Lung Cancer (NSCLC)-A Consecutive Evaluation Using Next-Generation Sequencing (NGS) in a Real-World Setting. *Cancers* **2021**, *13*, 5023. [CrossRef]

61. Wolf, J.; Seto, T.; Han, J.Y.; Reguart, N.; Garon, E.B.; Groen, H.J.M.; Tan, D.S.W.; Hida, T.; de Jonge, M.; Orlov, S.V.; et al. Capmatinib in MET Exon 14-Mutated or MET-Amplified Non-Small-Cell Lung Cancer. *N. Engl. J. Med.* **2020**, *383*, 944–957. [CrossRef] [PubMed]

62. Benayed, R.; Offin, M.; Mullaney, K.; Sukhadia, P.; Rios, K.; Desmeules, P.; Ptashkin, R.; Won, H.; Chang, J.; Halpenny, D.; et al. High Yield of RNA Sequencing for Targetable Kinase Fusions in Lung Adenocarcinomas with No Mitogenic Driver Alteration Detected by DNA Sequencing and Low Tumor Mutation Burden. *Clin. Cancer Res. Off. J. Am. Assoc. Cancer Res.* **2019**, *25*, 4712–4722. [CrossRef] [PubMed]

63. Duplaquet, L.; Kherrouche, Z.; Baldacci, S.; Jamme, P.; Cortot, A.B.; Copin, M.C.; Tulasne, D. The multiple paths towards MET receptor addiction in cancer. *Oncogene* **2018**, *37*, 3200–3215. [CrossRef]

64. Gow, C.H.; Liu, Y.N.; Li, H.Y.; Hsieh, M.S.; Chang, S.H.; Luo, S.C.; Tsai, T.H.; Chen, P.L.; Tsai, M.F.; Shih, J.Y. Oncogenic Function of a KIF5B-MET Fusion Variant in Non-Small Cell Lung Cancer. *Neoplasia* **2018**, *20*, 838–847. [CrossRef] [PubMed]

65. International Cancer Genome Consortium PedBrain Tumor Project. Recurrent MET fusion genes represent a drug target in pediatric glioblastoma. *Nat. Med.* **2016**, *22*, 1314–1320. [CrossRef] [PubMed]

66. Kim, P.; Jia, P.; Zhao, Z. Kinase impact assessment in the landscape of fusion genes that retain kinase domains: A pan-cancer study. *Brief. Bioinform.* **2018**, *19*, 450–460. [CrossRef]

67. Pan, Y.; Zhang, Y.; Ye, T.; Zhao, Y.; Gao, Z.; Yuan, H.; Zheng, D.; Zheng, S.; Li, H.; Li, Y.; et al. Detection of Novel NRG1, EGFR, and MET Fusions in Lung Adenocarcinomas in the Chinese Population. *J. Thorac. Oncol.* **2019**, *14*, 2003–2008. [CrossRef]

68. Plenker, D.; Bertrand, M.; de Langen, A.J.; Riedel, R.; Lorenz, C.; Scheel, A.H.; Muller, J.; Bragelmann, J.; Dassler-Plenker, J.; Kobe, C.; et al. Structural Alterations of MET Trigger Response to MET Kinase Inhibition in Lung Adenocarcinoma Patients. *Clin. Cancer Res. Off. J. Am. Assoc. Cancer Res.* **2018**, *24*, 1337–1343. [CrossRef]

69. Siemanowski, J.; Heydt, C.; Merkelbach-Bruse, S. Predictive molecular pathology of lung cancer in Germany with focus on gene fusion testing: Methods and quality assurance. *Cancer Cytopathol.* **2020**, *128*, 611–621. [CrossRef]

70. Vigna, E.; Gramaglia, D.; Longati, P.; Bardelli, A.; Comoglio, P.M. Loss of the exon encoding the juxtamembrane domain is essential for the oncogenic activation of TPR-MET. *Oncogene* **1999**, *18*, 4275–4281. [CrossRef]

71. Zhu, Y.C.; Wang, W.X.; Xu, C.W.; Zhang, Q.X.; Du, K.Q.; Chen, G.; Lv, T.F.; Song, Y. Identification of a novel crizotinib-sensitive MET-ATXN7L1 gene fusion variant in lung adenocarcinoma by next generation sequencing. *Ann. Oncol. Off. J. Eur. Soc. Med. Oncol.* **2018**, *29*, 2392–2393. [CrossRef] [PubMed]
72. Peschard, P.; Fournier, T.M.; Lamorte, L.; Naujokas, M.A.; Band, H.; Langdon, W.Y.; Park, M. Mutation of the c-Cbl TKB domain binding site on the Met receptor tyrosine kinase converts it into a transforming protein. *Mol. Cell.* **2001**, *8*, 995–1004. [CrossRef] [PubMed]
73. Awad, M.M.; Oxnard, G.R.; Jackman, D.M.; Savukoski, D.O.; Hall, D.; Shivdasani, P.; Heng, J.C.; Dahlberg, S.E.; Janne, P.A.; Verma, S.; et al. MET Exon 14 Mutations in Non-Small-Cell Lung Cancer Are Associated With Advanced Age and Stage-Dependent MET Genomic Amplification and c-Met Overexpression. *J. Clin. Oncol. Off. J. Am. Soc. Clin. Oncol.* **2016**, *34*, 721–730. [CrossRef] [PubMed]
74. Krishnaswamy, S.; Kanteti, R.; Duke-Cohan, J.S.; Loganathan, S.; Liu, W.; Ma, P.C.; Sattler, M.; Singleton, P.A.; Ramnath, N.; Innocenti, F.; et al. Ethnic differences and functional analysis of MET mutations in lung cancer. *Clin. Cancer Res. Off. J. Am. Assoc. Cancer Res.* **2009**, *15*, 5714–5723. [CrossRef]
75. Tsai, J.M.; Hata, A.N.; Lennerz, J.K. MET D1228N and D1246N are the Same Resistance Mutation in MET Exon 14 Skipping. *Oncol.* **2021**, *26*, e2297–e2301. [CrossRef]
76. Li, M.M.; Datto, M.; Duncavage, E.J.; Kulkarni, S.; Lindeman, N.I.; Roy, S.; Tsimberidou, A.M.; Vnencak-Jones, C.L.; Wolff, D.J.; Younes, A.; et al. Standards and Guidelines for the Interpretation and Reporting of Sequence Variants in Cancer: A Joint Consensus Recommendation of the Association for Molecular Pathology, American Society of Clinical Oncology, and College of American Pathologists. *J. Mol. Diagn.* **2017**, *19*, 4–23. [CrossRef]
77. Davies, K.D.; Ritterhouse, L.L.; Snow, A.N.; Sidiropoulos, N. MET Exon 14 Skipping Mutations: Essential Considerations for Current Management of Non-Small-Cell Lung Cancer. *J. Mol. Diagn. JMD* **2022**, *24*, 841–843. [CrossRef]
78. Feng, Y.; Thiagarajan, P.S.; Ma, P.C. MET signaling: Novel targeted inhibition and its clinical development in lung cancer. *J. Thorac. Oncol.* **2012**, *7*, 459–467. [CrossRef]
79. Ma, P.C.; Tretiakova, M.S.; MacKinnon, A.C.; Ramnath, N.; Johnson, C.; Dietrich, S.; Seiwert, T.; Christensen, J.G.; Jagadeeswaran, R.; Krausz, T.; et al. Expression and mutational analysis of MET in human solid cancers. *Genes Chromosomes Cancer* **2008**, *47*, 1025–1037. [CrossRef]
80. Jeffers, M.; Schmidt, L.; Nakaigawa, N.; Webb, C.P.; Weirich, G.; Kishida, T.; Zbar, B.; Vande Woude, G.F. Activating mutations for the met tyrosine kinase receptor in human cancer. *Proc. Natl. Acad. Sci. USA.* **1997**, *94*, 11445–11450. [CrossRef]
81. Heist, R.S.; Sequist, L.V.; Borger, D.; Gainor, J.F.; Arellano, R.S.; Le, L.P.; Dias-Santagata, D.; Clark, J.W.; Engelman, J.A.; Shaw, A.T.; et al. Acquired Resistance to Crizotinib in NSCLC with MET Exon 14 Skipping. *J. Thorac. Oncol.* **2016**, *11*, 1242–1245. [CrossRef] [PubMed]
82. Park, M.; Dean, M.; Cooper, C.S.; Schmidt, M.; O'Brien, S.J.; Blair, D.G.; Vande Woude, G.F. Mechanism of met oncogene activation. *Cell* **1986**, *45*, 895–904. [CrossRef] [PubMed]

Article

Longitudinal Plasma Proteomics-Derived Biomarkers Predict Response to MET Inhibitors for MET-Dysregulated NSCLC

Guang-Ling Jie [1,2,†], Lun-Xi Peng [3,†], Mei-Mei Zheng [2], Hao Sun [2], Song-Rong Wang [2], Si-Yang Maggie Liu [4], Kai Yin [2], Zhi-Hong Chen [2,5], Hong-Xia Tian [2,5], Jin-Ji Yang [2], Xu-Chao Zhang [2,5], Hai-Yan Tu [2], Qing Zhou [2], Catherine C. L. Wong [6,*] and Yi-Long Wu [1,2,*]

[1] School of Medicine, South China University of Technology, Guangzhou 510006, China
[2] Guangdong Lung Cancer Institute, Guangdong Provincial People's Hospital (Guangdong Academy of Medical Sciences), Southern Medical University, Guangzhou 510080, China
[3] Department of Clinical Skills Training Center, Zhujiang Hospital, Southern Medical University, Guangzhou 510282, China
[4] Department of Hematology, First Affiliated Hospital, Jinan University, Guangzhou 510632, China
[5] Guangdong Provincial Key Laboratory of Translational Medicine in Lung Cancer, Guangdong Provincial People's Hospital (Guangdong Academy of Medical Sciences), Southern Medical University, Guangzhou 510080, China
[6] Clinical Research Institute, State Key Laboratory of Complex Severe and Rare Diseases, Peking Union Medical College Hospital, Chinese Academy of Medical Science & Peking Union Medical College, Beijing 100730, China
[*] Correspondence: catclw321@126.com (C.C.L.W.); syylwu@live.cn (Y.-L.W.)
[†] These authors contributed equally.

Citation: Jie, G.-L.; Peng, L.-X.; Zheng, M.-M.; Sun, H.; Wang, S.-R.; Liu, S.-Y.M.; Yin, K.; Chen, Z.-H.; Tian, H.-X.; Yang, J.-J.; et al. Longitudinal Plasma Proteomics-Derived Biomarkers Predict Response to MET Inhibitors for MET-Dysregulated NSCLC. *Cancers* **2023**, *15*, 302. https://doi.org/10.3390/cancers15010302

Academic Editor: Maxim V. Berezovski

Received: 24 November 2022
Revised: 24 December 2022
Accepted: 29 December 2022
Published: 1 January 2023

Simple Summary: Targeted therapy has revolutionized the treatment of non-small cell lung cancer (NSCLC) and MET inhibition is a promising therapy for MET-dysregulated NSCLC. However, due to the lack of effective biomarkers, the clinical efficacy is unsatisfactory. This study aims to investigate the clinical utility of plasma proteomics-derived biomarkers for MET-dysregulated NSCLC (including *MET* amplification and MET overexpression). We analyzed 89 longitudinal plasma samples from MET-dysregulated advanced-stage NSCLC patients treated with MET inhibitors by the method of mass spectrometry. The results showed that the peripheral plasma proteomic characteristics were associated with the outcomes of patients treated with MET inhibitors. Through biomarker selection, we found a four plasma protein signature (MYH9, GNB1, ALOX12B, and HSD17B4 proteins) could predict the response and progression-free survival of patients treated with MET inhibitors with high accuracy. This study highlighted the clinical utilization of plasma biomarkers to scream patients to receive MET inhibitors.

Abstract: MET inhibitors have shown promising efficacy for MET-dysregulated non-small cell lung cancer (NSCLC). However, quite a few patients cannot benefit from it due to the lack of powerful biomarkers. This study aims to explore the potential role of plasma proteomics-derived biomarkers for patients treated with MET inhibitors using mass spectrometry. We analyzed the plasma proteomics from patients with MET dysregulation (including *MET* amplification and MET overexpression) treated with MET inhibitors. Thirty-three MET-dysregulated NSCLC patients with longitudinal 89 plasma samples were included. We classified patients into the PD group and non-PD group based on clinical response. The baseline proteomic profiles of patients in the PD group were distinct from those in the non-PD group. Through protein screening, we found that a four-protein signature (MYH9, GNB1, ALOX12B, HSD17B4) could predict the efficacy of patients treated with MET inhibitors, with an area under the curve (AUC) of 0.93, better than conventional fluorescence in situ hybridization (FISH) or immunohistochemistry (IHC) tests. In addition, combining the four-protein signature with FISH or IHC test could also reach higher predictive performance. Further, the combined signature could predict progression-free survival for MET-dysregulated NSCLC ($p < 0.001$). We also validated the performance of the four-protein signature in another cohort of plasma using an enzyme-linked immunosorbent assay. In conclusion, the four plasma protein signature (MYH9, GNB1, ALOX12B, and HSD17B4 proteins) might play a substitutable or complementary role to conventional MET FISH or IHC tests. This exploration will help select patients who may benefit from MET inhibitors.

Keywords: Non-small cell lung cancer; MET dysregulation; proteomics; MET inhibitor; biomarker

1. Introduction

The *MET* proto-oncogene has been known to play an important role in promoting tumor cell proliferation, tumor invasion, and metastasis in non-small cell lung cancer (NSCLC) either as a primary oncogenic driver or as a co-driver in the context of acquired resistance to tyrosine kinase inhibitors (TKIs) [1–3]. Activation of the MET pathway can be caused by *MET* amplifications, protein overexpression, gene mutations, and fusions [4]. The prevalence of *MET* amplification of NSCLC is 1–5% and 5–20% for *MET de novo* and acquired amplification, respectively [1,5]. MET overexpression is more common in NSCLC, with approximately 20% to 25% of patients identified by immunohistochemistry (IHC) [6,7]. Previous studies have demonstrated that multiple MET inhibitors showed promising efficacy for NSCLC patients with *MET* amplification or MET protein overexpression with an objective response rate of approximately 67% and 68% (IHC3+), respectively [8–10].

Fluorescence in situ hybridization (FISH) is a standard method to detect *MET* amplification for NSCLC patients. It can distinguish MET focal amplification from *MET* polysomy by calculating both the copies of *MET* per cell and the ratio of *MET* to chromosome (*MET*/CEP7) [11,12]. However, it remains challenging to define an optimal *MET* copy number and *MET*/CEP7 threshold to select eligible patients to receive MET inhibitors. Many FISH-selected patients cannot benefit from MET inhibitors [9,13]. MET overexpression is another potential biomarker for screening patients to be treated with MET inhibitors. Several clinical trials have shown promising efficacy for patients with MET overexpression treated with MET inhibitors plus epidermal growth factor receptor-TKIs (EGFR-TKIs) in the setting of acquired resistance to EGFR-TKIs [8,9]. However, the correlation between MET overexpression and *MET* amplification is poor [14,15]. Thus far, MET overexpression by IHC served as a biomarker for predicting response to MET inhibitor remains controversial. Together, the clinical practice of MET inhibitors is limited by ambiguous diagnostic criteria. Quite a few patients cannot benefit from MET inhibitors owing to the lack of predictive biomarkers with sufficient accuracy to select potentially beneficial patients to receive MET inhibitors. There is an emergent need to find more powerful and easier predictive biomarkers to identify eligible patients who would benefit from MET inhibitors.

Mass spectrometry (MS)-based proteomics is a high-through and unbiased method for characterizing oncogenic mechanisms and identifying potential prognostic and predictive biomarker [16]. It can detect and quantify tens of thousands of proteins with high specificity, making it an ideal approach for the study of biomarkers identification [17]. A large-scale study investigating the proteogenomics of lung adenocarcinoma revealed the signatures of oncogenesis and successfully identified several novel prognostic and therapeutic biomarker candidates [18]. In addition, MS-based proteomics can also detect plasma proteome by dynamic monitoring, and therefore act as an excellent tool to screen biomarker candidates for multiple diseases. A study used plasma proteomics to identify panels of biomarkers for anti-PD-(L)1 response prediction in NSCLC with an area under the curve (AUC) value of 94.1% [19]. Another study integrating a plasma and paired tissue proteomics approach also identified several noninvasive proteomic biomarkers panels for alcohol-related liver disease with an AUC value of 0.92 [20]. Recent advances in MS-based proteomics technology have greatly extended its application in clinical and translational research [21].

Herein, we conducted a MS-based, data-independent acquisition (DIA) quantitative proteomic approach to explore the blood-based proteomic profiles to determine predictive biomarkers for MET-dysregulated NSCLC patients treated with MET inhibitors. The selected biomarker candidates were further validated by enzyme-linked immunosorbent assay (ELISA) tests in the validation cohort.

2. Materials and Methods

2.1. Patient Enrollment and Sample Collection

Advanced-stage NSCLC patients with MET dysregulation treated with MET inhibitors were enrolled from 1 October 2014, to 10 April 2019, at Guangdong lung cancer institute. MET dysregulation consisted of MET protein overexpression with MET IHC score $\geq$270 and *MET* amplification by FISH with mean gene copy number greater or equal to five, and a *MET* to centromere of chromosome 7 (*MET*/CEP7) ratio of 2 or more [12,22]. Tumor response and time to progression were evaluated according to RECIST 1.1. The cut-off date for the last follow-up was 23 June 2020. Samples were collected up to 3 days before MET inhibitors treatment, the best response (about 8–12 weeks) after the initial MET inhibitors treatment, and the disease progression time point. The best response is recorded when patients have the largest tumor shrinkage during treatment, with 30% as partial response and −20–30% as stable disease according to RECIST 1.1 criteria. The best responses often occurred 8–12 weeks after treatment initiation in most of the patients. Progression-free survival (PFS) was defined as the time between the patient receiving treatment in the study and the date of disease progression or censored at the date of the last follow-up according to RECIST 1.1.

Plasma samples were collected in pro-coagulation vacuum tubes using standard venipuncture protocols and were then extracted by centrifugation for 15 min at 2500 rpm. The Plasma samples were stored at −80 °C before use.

2.2. Plasma Sample Preparation for Spectral Library Generation

All plasma samples were processed by the Agilent 1290 Infinity II liquid chromatography system coupled with the Multi Affinity Removal Column, Human-14 to remove abundant proteins. About 10 μL each sample was taken out and mixed. The mixed sample and all the 89 samples were precipitated by trichloroacetic acid (TCA) solution for about 4 h at 4 °C. After centrifuging at 16,000× g for 30 min at 4 °C, the pellets were washed with 500 μL cold acetone three times and dried with a vacuum concentrator (Labconco, Kansas, MO, USA). The dried pellets were dissolved in 40 μL 8 M Urea in 500 mM Tris-HCl buffer (pH 8.5) and ultrasonically treated for 10 min. The samples were reduced with 20 mM (2-carboxyethyl) phosphine hydrochloride (TCEP) (500 mM in 100 mM Tris/HCl pH 8.5) at room temperature for 20 min and alkylated with 40 mM IAA at room temperature in the dark for 30 min. The mixtures were diluted with 200 μL 100 mM Tris-HCl buffer (pH 8.5) followed by adding trypsin at a 1:20 ratio for 16 h. The peptides were desalted and re-dissolved with 50 μL Mil-li-Q water with 0.1 vol% formic acids (FA). The indexed Retention Time (iRT) calibration peptides were spiked into the 89 peptide samples for DIA analysis later. The mixed sample without iRT peptides was separated into two samples, one of which was used for High-PH reversed-phase fraction and quality control (QC) of the DIA analysis later, respectively. The QC sample was added with iRT before analysis.

2.3. High-pH Reversed-Phase Fractionation

The mixed peptide sample fractioning was performed on a Chromatographic column (BEH C18, 300A, 1.7 μm, 1 mm × 150 mm) coupled to a Waters XevoTM AC-QUITY UPLC (Waters, Milford, MA, USA) with an 80 min liquid phase gradient. We collected the first 4 min of liquid as the first fraction, the liquid of the 64–68 min as the last fraction, and discarded the liquid of the last 12 min. We collected the liquid sample every minute during the gradient of 4–64 min. The first fraction was mixed with the last one and the rest were mixed in pairs every 30 fractions. Finally, 31 fractions were obtained and vacuum-centrifuge dried. All 31 fractions were reconstituted in 10 μL Milli-Q water with 0.1 vol% formic acids (FA). IRT peptides were spiked before the data-dependent analysis (DDA).

2.4. Liquid Chromatography

All the peptide samples were separated on an EASY-nLC1200 liquid chromatography system (ThermoFisher, San Jose, CA, USA) coupled with a 25 cm × 75 μm home-packed analytical column (1.5 μm ReproSil-Pur 120 C18-AQ particles (Dr. Maisch)). Mobile phases A and B were water and 80% ACN with 0.1 vol% formic acids. Samples were analyzed with a 120 min gradient at a flow rate of 300 nL/min and the concentration of B% was increased from 4 to 10% within 4 min, followed by an increase to 30% at 4–103 min and a further increase to 100% at 103–113 min and kept 100% B for the last 7 min.

2.5. Mass Spectrometry

All the samples were analyzed on Thermo QExctive HF-X (ThermoFisher, San Jose, CA, USA). The 31 fraction samples obtained through high-Ph reversed-fraction processing were operated in data-dependent mode which was used for the spectral library generation. All 89 plasma samples were analyzed in data-independent mode and the data was used for bioinformatic analysis later. We add a technical QC every 12 samples.

For the DDA runs, the full MS scan was performed with a scan range (m/z) between 300 and 1500 m/z. The MS/MS had a resolution of 60,000. The automatic gain control (AGC) target was 3e6 with a maximum injection time of 50 ms. The HCD dd-MS2 scan selected top 30 intensity peptides and was performed with the following parameters: resolution = 15,000; AGC target = 5e5; maximum injection time = 40 ms, NCE = 30, isolation window = 1.7 m/z.

For the DIA analysis, the full MS-SIM had a resolution of 60,000 and a scan range between 350–1200 m/z. The AGC target was 3e6 and the maximum injection time was set to 50 ms. Each full MS was followed by 64 narrow isolation widths which were named DIA windows. The resolution was set to 30,000 and the AGC target was 1e6.

2.6. Generation of Spectral Libraries and DIA Data Analysis

Spectral libraries were generated from the acquired data of the 31 fractions using Spectronaut version 14.0 (Biognosys) with the default parameters. MS/MS spectra were matched against the database which was downloaded from human UNIPROT (only reviewed entries, human 20,421 entries).

2.7. Enzyme-Linked Immunosorbent Assay (ELISA)

Human protein ELISA kits were used to detect and quantify plasma levels of specific proteins according to manufacturers' instructions (SAB signalway ELISA Kit for MYH9 and HSD17B4, and Abebio ELISA Kit for GNB1 and ALOX12B). A total of 100 μL of plasma sample and standard dilutions were added to the precoated plates, and the plates were incubated at 37 °C for 2 h. After three times washing, 100 μL diluted Biotin-Conjugate was added, and the plates were then incubated at 37 oC for 1 h. After washing, 100 μL Streptavidin conjugated Horseradish Peroxidase (HRP) was added and incubated at 37 °C for 1 h. 100 μL of Substrate Solution was added and incubated at 37 °C for 10 min. Finally, we added 50 μL of Stop Solution and detected the OD values at 450 nm using microplate spectrophotometer (BIO-RAD, xMark). The determination of OD values from serial dilutions of the standard samples was used to generate a standard curve of each protein and the relative concentrations of samples were calculated.

2.8. Statistical Analysis

Data analysis including data imputation, normalization, and principal component analysis (PCA) was performed in R software (version 4.1.2). The missing value was replaced with a median value. Fold-change of 1.5 and *p*-value of 0.05 were used to filter differentially expressed proteins using the limma package in R. Dot plots of Kyoto Encyclopedia of Genes and Genomes (KEGG) and Gene Ontology (GO) enriched functional pathways were plotted by ClusterProfiler package in R. Significant proteins were used for protein-protein interaction network analysis and network visualization was performed using Cytoscape (version 3.9.1). The Student's t-test was used to compare the protein levels in the plasma between the two groups. Fisher's exact test was used to compare two categorical variables. Receiver operator characteristic curves (ROC) analyses were used to assess the overall performance of a test and to compare the performance of two or more other tests. ROC analyses in this study were conducted in pROC package in R using response outcomes and protein intensity values. The AUC value was calculated by the area under the ROC curve and was used to assess the performance of the predictive models. An AUC value of more than 0.8 was considered good. The Youden index, which integrates sensitivity and specificity information, was used to identify the optimal thresholds. The predictive model was constructed using logistic regression in R software. The probability of response was calculated using four protein intensities as the following formula listed.

$$\text{Logit}(p) = \log(p/(1 - p)) = -0.087 \times \text{MYH9} + 0.497 \times \text{GNB1} + 2.015 \times \text{ALOX12B} - 0.936 \times \text{HSD17B4} - 21.520$$

The predictive p value was used to conduct the ROC analysis for the four-protein signature and the corresponding AUC value was calculated. The cut-off value ($p = 0.68$) of the predictive model was calculated by the Youden index. The p value of more than 0.68 was considered as the low-risk group in the progression-free survival analyses.

Survival analysis was performed using Kaplan–Meier survival plot and log-rank test p-value were calculated. The hazard ratio was calculated by Cox proportional hazards regression and was used to estimate the ratio of the hazard rate in the two groups. A hazard ratio of 1 indicated that no difference was detected in survival between the two groups. A hazard ratio of greater than one or less than one indicated that survival was worse or better in one of the groups. In the present study, all tests of significance were two-sided, and p-value < 0.05 was considered statistically significant.

3. Results

3.1. Patient Characteristics and Samples Collection

A total of 33 advanced NSCLC patients diagnosed with MET dysregulation were enrolled in our study including *MET* amplification by FISH test (n = 16) and MET overexpression by IHC test (n = 23). Six patients were positive in both *MET* amplification and MET overexpression. All the patients were treated with MET inhibitors. The clinicopathological characteristics and treatment strategies of the enrolled patients were summarized in Table 1. Of the patients with co-occurrence *EGFR* mutations and MET dysregulation, 39.4% were treated with EGFR-TKIs plus MET inhibitors. No confounders were found between the PD and non-PD groups (Table S1). The disease control rates (DCRs) of patients with *MET* amplification or overexpression were 93.8% and 86.4%, respectively (Table S2). We collected a total of 89 longitudinal peripheral plasma samples at baseline before MET inhibitors treatment (n = 33), best response after treatment (n = 23), and disease progression time point (n = 33, Figure 1). We classified 10 patients who had primary drug resistance to MET inhibitors into the PD group and 23 patients who obtained partial response (PR) or stable disease (SD) into the non-PD group.

3.2. Global Proteomic Analysis of Peripheral Plasma and Predictive Biomarkers Selection for Patients Received MET Inhibitors

We performed high-resolution mass spectrometry using a DIA method for the peripheral plasma sample. A total of 1619 proteins were identified from all plasma samples and approximately 1106 unique proteins (range: 914–1296 proteins) were identified in each sample (Figure S1A). The patients in the PD group and non-PD group were clustered independently in unsupervised hierarchical clustering and principal component analysis (PCA), indicating the distinct peripheral proteomic profiles between the two groups at baseline (Figures 2 and S1B). Furthermore, we found a total of 463 differentially expressed proteins and the number of up-regulated proteins was comparable with down-regulated proteins (220 up-regulated vs. 243 down-regulated proteins) (Figure 2B,C). GO and KEGG enrichment analyses revealed that the differentially expressed proteins were enriched in the glycolysis, angiogenesis, Rap1 signaling pathway, cell adhesion, and gap junction, which may contribute to cancer metabolism, migration, and growth (Figure S2). To elucidate the correlation between the differentially expressed proteins and the MET dysregulation pathway, we conducted a protein-to-protein interaction network analysis through the STRING database (Figure 2D). We found a large number of proteins interacted with or regulated by the MET pathway. After manual screening, we found four proteins had greatly higher or lower fold change with significant p value in the PD group than the non-PD group (MYH9 = 4.00, p = 0.003; GNB1 = 2.53, $p \leq$ 0.003; ALOX12B = 2.40, $p \leq$ 0.001 and HSD17B4 = 0.46, $p \leq$ 0.001).

Table 1. The clinicopathological characteristics and treatment strategies of the enrolled patients.

Clinical characteristics		Overall (n = 33)
Age		
	Median [Range]	58.4 [29.3–73.5]
Gender (%)		
	Female	8 (24.2%)
	Male	25 (75.8%)
Smoking history (%)		
	No	14 (42.4%)
	Yes	19 (57.6%)
Pathology (%)		
	Adenocarcinoma	32 (97.0%)
	Pulmonary sarcomatoid carcinoma	1 (3.0%)
Stage (%)		
	III	1 (3.0%)
	IV	32 (97.0%)
Performance status score (%)		
	1	32 (97.0%)
	2	1 (3.0%)
Brain metastasis (%)		
	No	23 (69.7%)
	Yes	10 (30.3%)
EGFR mutation (%)		
	19DEL	5 (15.2%)
	L858R	8 (24.2%)
	Negative	20 (60.6%)
MET FISH (%)		
	Negative	7 (21.2%)
	Positive	16 (48.5%)
	NA	10 (30.3%)
MET IHC (%)		
	Negative	11 (33.3%)
	Positive	22 (66.7%)
Treatment (%)		
	MET inhibitor + EGFR-TKI	12 (36.4%)
	MET inhibitor	21 (63.6%)
Treatment line (%)		
	1	7 (21.2%)
	$\geq$2	26 (78.2%)

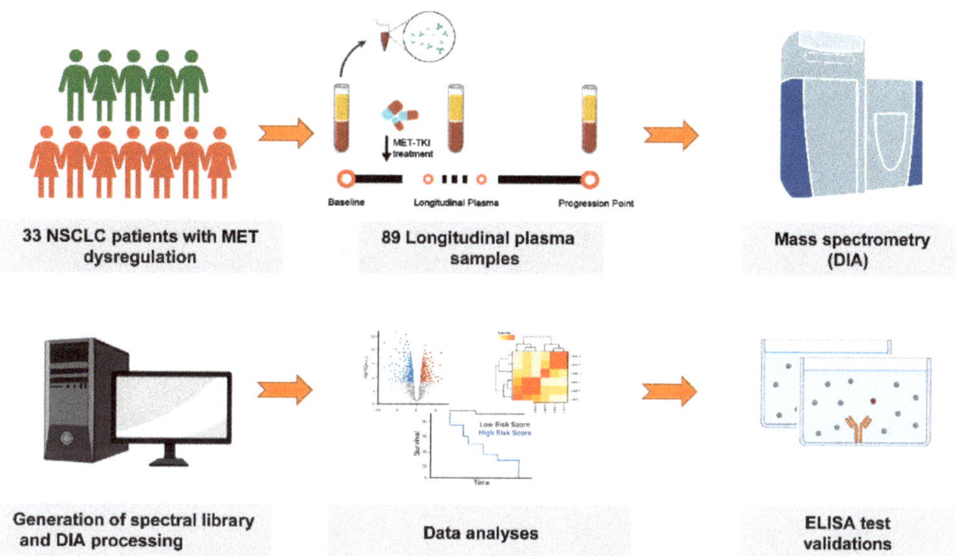

Figure 1. Summary of MET dysregulated NSCLC patients and study workflow.

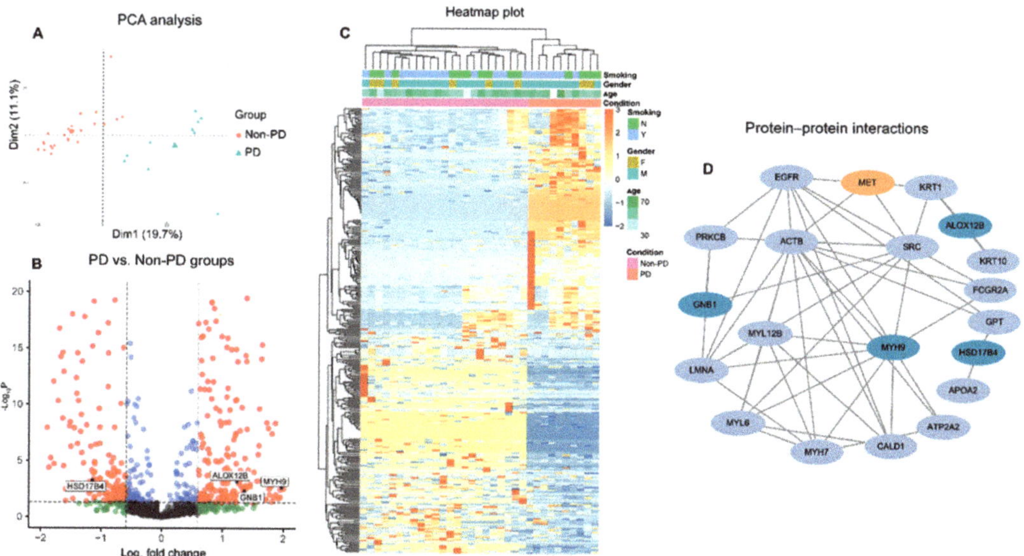

Figure 2. Plasma-based proteomics landscape and biomarkers selection at baseline. (**A**) Principal component analysis of 33 patients; the bigger points in the PD and non-PD groups represent the median value of each group. A volcano plot (**B**) and heatmap plot (**C**) of differentially expressed proteins between PD and non-PD groups; the red points in the volcano plot indicate the proteins that foldchange > 1.5 and *p* value < 0.5; the green points indicate the proteins that foldchange >1.5 and *p* value ≥ 0.5; the blue points indicate the proteins that foldchange ≤1.5 and *p* value < 0.5; the grey points indicate the proteins that foldchange ≤1.5 and *p* value ≥ 0.5. (**D**) Protein-protein interaction network of differentially expressed proteins.

3.3. The Predictive Performance of Biomarkers for Response to MET Inhibitors in MET-Dysregulated NSCLC Patients

We compared the relative protein intensities between the PD and non-PD groups at baseline plasma (Figure 3A). The results showed that MYH9, GNB1, and ALOX12B proteins had significantly higher intensities in the PD group versus those in the non-PD group, representing their potential relation with poor response to MET inhibitors. Another protein, called HSD17B4, was significantly downregulated in the PD group. The predictive performances of the four proteins at baseline were measured by the ROC analysis with AUC values of 0.809, 0.874, 0.878, and 0.796 for the MYH9, GNB1, ALOX12B, and HSD17b4 individual proteins, respectively (Figure 3B). After combining four proteins, the AUC value reached 0.930, which was higher than that of individual proteins and conventional FISH and IHC methods (AUC values: 0.763 and 0.858, respectively; Figure 3C,D). Besides, the addition of four-protein signature to FISH or IHC outperformed the individual FISH or IHC methods, with AUC values of 0.971 and 0.965, respectively (Figure 3C–E).

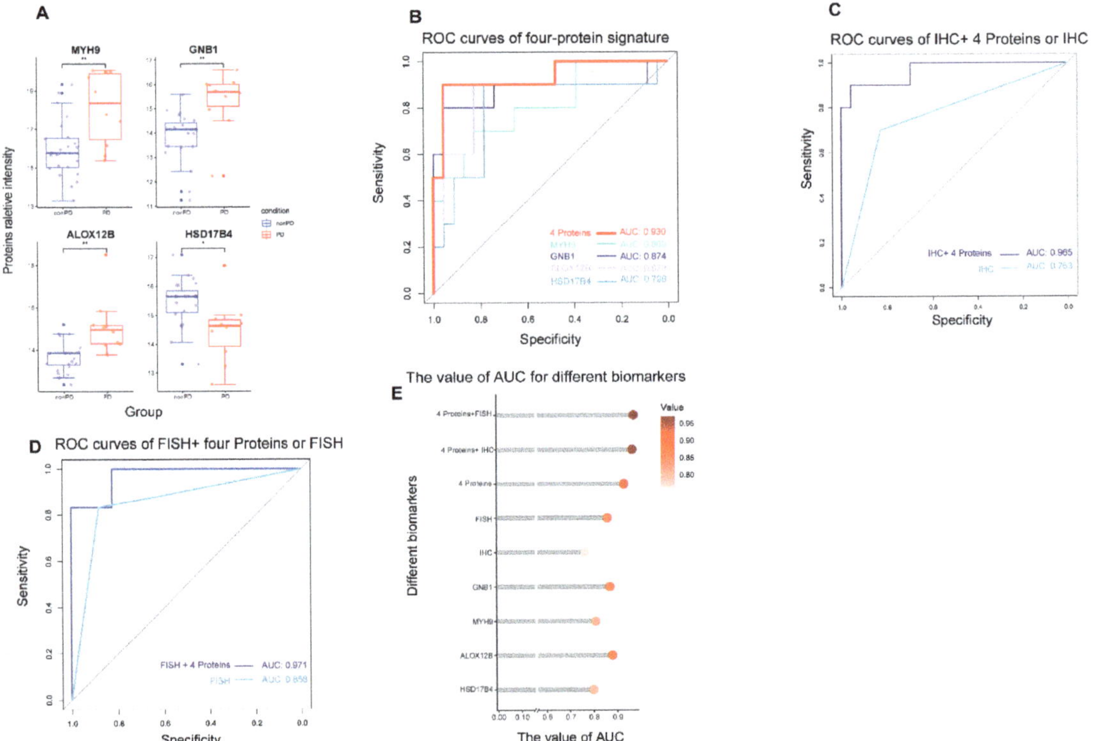

Figure 3. The predictive performance of different models for response to MET inhibitor. (**A**) Boxplots of the relative intensity of four selective protein biomarkers between PD and non-PD groups. * *p* value < 0.05, ** *p* value < 0.01. (**B**) The ROC curves for the performance of four proteins, (**C**) IHC, IHC plus four proteins, (**D**) FISH, and FISH plus four proteins in the prediction of response to MET inhibitors for MET-dysregulated lung cancer patients. (**E**) The AUC values of different models.

We further explored the performance of nine proteomic-based models (MYH9, GNB1, ALOX12B, HSD17B4, FISH, IHC, four-proteins signature, four proteins + IHC, four proteins + FISH) in the prediction of PFS in patients who received MET inhibitors (Figure 4A). Based on the ROC analysis and Youden index calculations, the patients were divided into the low-risk group and high-risk group in each of the models. The patients in the low-risk group meant they were more likely to benefit from MET inhibitors and survived longer. The four individual proteins can significantly stratify the PFS of patients treated with MET inhibitors with the hazard ratios (HRs) of MYH9 (HR = 2.35, *p* = 0.024), GNB1 (HR = 2.63, *p* = 0.009), ALOX12B (HR = 2.55, *p* = 0.012), and HSD17B4 (HR = 0.45, *p* = 0.031), respectively. The four-protein signature showed improved predictive performance with an HR of 12.66, 95%CI (4.34, 36.95), P <0.001, better than FISH (HR = 1.99, *p* = 0.13) and IHC (HR = 6.42, *p* = <0.001) methods (Figure 4B–D). The median PFS was 1.2 months for the high-risk group and 7.4 for the low-risk group in the four-protein signature model (Figure 4B). When four proteins were combined with the FISH or IHC test, the models reached higher predictive performances, with HR of 15.39, *p* = <0.001, and HR of 9.1, *p* = <0.001, respectively (Figure 4E,F).

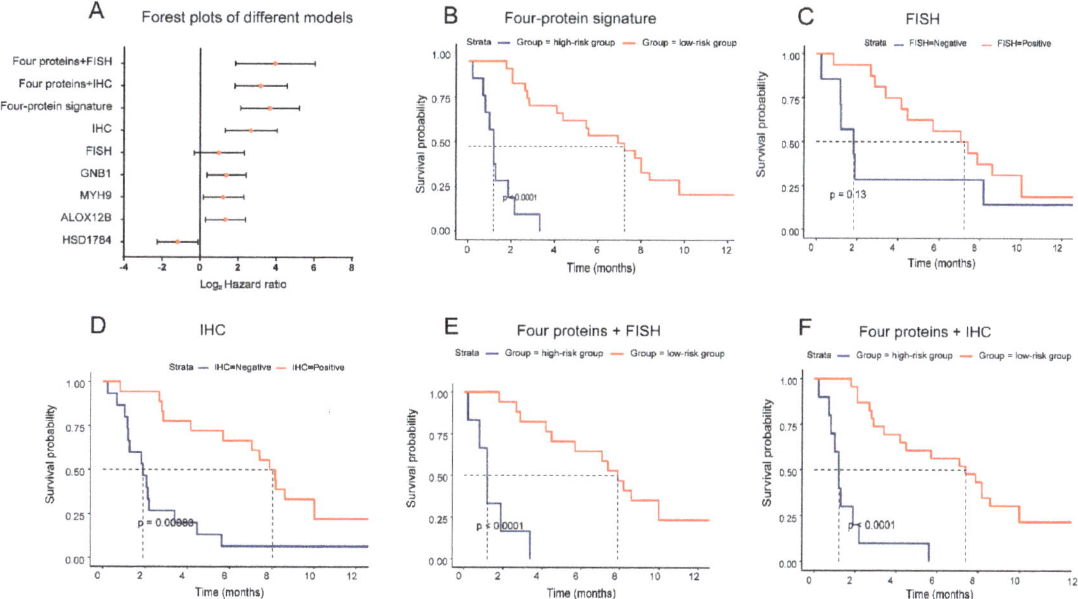

Figure 4. The predictive performance of different models for progression-free survival of patients treated with MET inhibitors. (**A**) Forest plots of hazard ratios and 95% confidence interval in different predictive models. Kaplan–Meier plots of progression-free survival based on the combination of four proteins (**B**), *MET* FISH (**C**), MET IHC (**D**), Four proteins plus FISH (**E**), and four proteins + IHC (**F**).

3.4. Dynamic Change and Validation of the Four Biomarker Candidates in Plasma following MET Inhibitors Treatment

In an attempt to investigate the correlation of four biomarkers with clinical efficacy to MET inhibitors, we also monitored the dynamic change of these four proteins in peripheral plasma (Figure 5A). In non-PD group, three biomarkers (MYH9, GNB1, and ALOX12B) have higher expression levels at baseline; then the intensities dropped at the best response and elevated at the progression. These phenomena indicated that the dynamic changes in the three proteins were negatively associated with the efficacy of MET inhibitors. Regarding the HSD17B4 protein, its intensity was low at baseline, then increased at the best response. In the PD group, the concentrations of the four proteins at baseline and disease progression did not change significantly, indicating the primary resistance to MET inhibitors for these patients. Although we could not exclude the effect of MET inhibitors on the change in protein levels, these phenomena indicated that the dynamic changes of proteins may be largely dependent on the efficacy of MET inhibitors. In addition, the addition of EGFR TKI did not affect the proteomics results (Figure S3).

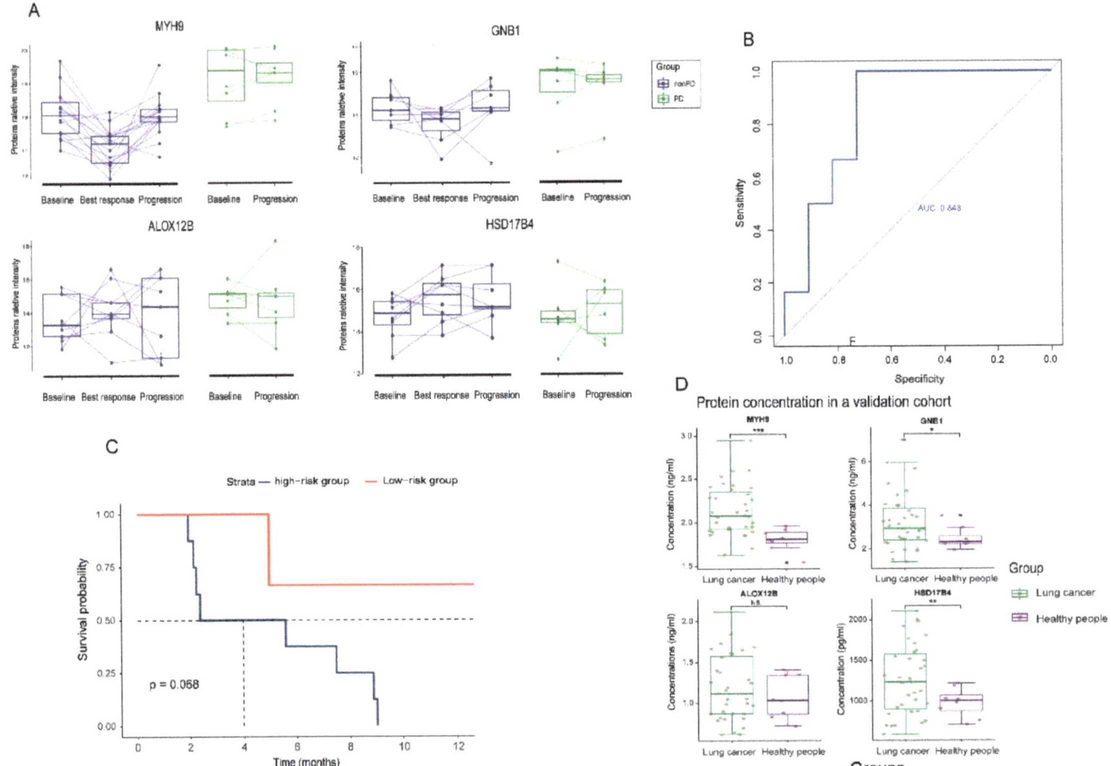

Figure 5. Dynamic change and validation of four biomarker candidates in the prediction of response to MET inhibitors. (**A**) Longitudinal relative proteins intensity at baseline, best response, and progression between non-PD and PD groups. (**B**) Validation of four biomarker candidates in another cohort of MET dysregulated NSCLC patients using plasma ELISA method. (**C**) Kaplan–Meier plots of progression-free survival based on the four-protein signature in the validation group. (**D**) Boxplots of concentrations of the four proteins in lung cancer patients and healthy people. * *p* value < 0.05, ** *p* value < 0.01, *** *p* value < 0.001.

We detected the concentration of the four proteins (MYH9, GNB1, ALOX12B, and HSD17B4) in a validation cohort of 17 patients using the ELISA kit. The clinical characteristics of the patients in a validation cohort was described in Table S3. All four proteins can be successfully detected in plasma. Consistent with the results above, the concentrations of MYH9, GNB1 and ALOX12B proteins were higher in the PD group (Figure S4A). However, no statistical significance was observed due to the small sample size. Further, the four-protein signature could predict the response and PFS for patients who received MET inhibitors with an AUC value of 0.848 and HR of 5.82 (*p* = 0.06, Figure 5B,C). The concentrations of the four proteins in lung cancer are significantly higher than those in healthy people, suggesting the change in the four proteins may result in tumor progression (Figure 5D). In 230 patients with adenocarcinoma from the TCGA cohort, higher expressions of MYH9, GNB1, and ALOX12B were associated with poor overall survival outcomes, while higher expression of HSD17B4 was associated with better survival outcomes (Figure S4B) [23].

4. Discussion

To our knowledge, this is the first study to explore the novel non-invasive predictive biomarkers of the efficacy of MET inhibitors for MET-dysregulated NSCLC patients using MS-based proteomics. We found that the plasma proteomic profiles at baseline were associated with the outcomes of patients treated with MET inhibitors. The combined four-protein signature (MYH9, GNB1, ALOX12B, HSD17B4) in plasma might effectively predict the responses and PFS outcomes of patients who received MET inhibitors, with a high AUC value of 0.930 and an HR of 12.66, p <0.001. This study highlights that the four-protein signature might play an alternative or complementary role to MET FISH or IHC method.

Several methods have been examined to select eligible patients for MET inhibitors, including FISH, next-generation sequence (NGS), droplet digital PCR (ddPCR), and IHC methods [11,22,24–26]. FISH was currently used to detect *MET* amplification in the clinic, but no consensus regarding the threshold of *MET* signals and *MET*/CEP7 value was defined to date [27]. Besides, the MET signal was distributed variably and the signal clustered or overlapped, making the counting signal difficult [25]. In terms of MET overexpression, the concordance of MET expression and *MET* amplification was low, and its correlation with treatment outcomes remained controversial [14,15]. Both FISH and IHC methods required tissue biopsies, which were not always feasible and put patients at risk. In addition, the heterogeneity of tumor and semi-quantitative FISH and IHC methods are prone to bias and depend on the experience of the pathologist [28]. Other diagnostic methods like NGS and ddPCR cannot distinguish true *MET* amplification from *MET* polysomy and the purity of tumor DNA also affected the results [26,29].

To overcome the flaws of traditional diagnostic methods discussed above, we proposed a novel MS-based proteomic method to select the predictive biomarker candidates for patients treated with MET inhibitors. Although DNA biomarkers have been used to guide personalized oncology, most of the small-molecule inhibitors target proteins instead of DNA, such as EGFR-TKIs [30,31]. In this study, we did not screen the biomarkers for patients with *MET* amplification or MET overexpression separately. Instead, we screened the differentially expressed protein biomarkers in plasma that participated in the MET signal pathway which were also associated with the treatment outcome. Previous studies have demonstrated that a subset of proteins in plasma can be secreted from or interact with the primary tumor [32–34]. Consistent with previous results, the proteomics profiles in our study were distinct between the patients in the PD group and those in the non-PD group. Further, through thousands of protein screenings, we identified four proteins (MYH9, GNB1, ALOX12B, HSD17B4) that can predict the response to MET inhibitors for MET-dysregulated lung cancer patients. The four-protein signature showed a higher predictive performance than the FISH or IHC methods, with AUC values of 0.930 vs. 0.858 or 0.763. The FISH-positive and IHC-positive patients showed the DCRs of 93.8% and 86.4%, which were consistent with results in the INSIGHT trial [9]. The positive group in our four-protein signature demonstrated a higher DCR of 95.7%, indicating the higher predictive performance of our model. The model also outperformed FISH and IHC methods in the prediction of PFS for patients treated with MET inhibitors, with a hazard ratio of 12.66 vs. 1.99 or 6.42. The addition of the four-protein signature to FISH or IHC methods could reach higher predictive performance, from the AUC values of 0.763 and 0.858 to 0.971 and 0.965, respectively. Therefore, the four-protein signature in our study not only represented an independent biomarker, but also a complementary biomarker to the FISH and IHC methods.

We integrated the downstream proteins as a predictive signature, as the single protein dysregulation may not fully represent the abnormality of a pathway. The biological functions of the four proteins have been reported to be associated with cancer development and progression. Previous studies showed that the MYH9 protein could act as a promoter of tumor stemness that facilitates tumor pathogenesis through the regulation of Wnt-β-catenin-STAT3 signaling, which can further interact with the MET pathway [35].

High expression of MYH9 conferred a poor prognosis for hepatocellular carcinoma, which was consistent with our results [36]. In our study, MYH9 enriched in angiogenesis and cell-cell junction, indicating its role in tumor progression. GNB1 protein played an important role in the PI3K/mTOR-related anti-apoptosis pathway, conferring transformed and resistance phenotypes across a range of human tumors. Acquiring mutations in the *GNB1* gene could cause resistance to tyrosine kinase inhibitors for leukemia [37]. ALOX12B was involved in lipid deoxygenation and the breakdown of amino acids. It promoted cell proliferation via regulation of the PI3K/ERK1 signaling pathway and was associated with survival outcomes in cancers [38,39]. HSD17B4 is a molecule involved in the peroxisome pathway and epithelial cell development [40]. Previous studies showed that HSD17B4 was highly expressed in most human cancers and was significantly associated with treatment efficacy [41,42]. Together, the four proteins are downstream molecules of the MET pathway, indicating the biological connection to the MET signaling.

To confirm the predictive performance of the four-protein signature, we monitored the dynamic change of circulated plasma-based proteomics, especially focusing on the four biomarker candidates. The results showed that protein intensities were associated with the efficacy of the patient treated with MET inhibitors. MYH9, GNB1, and ALOX12B were negatively associated with the tumor response, which represented biomarkers of poor outcomes, while HSD17B4 was positively associated with the tumor response. We also used the ELISA test to validate the concentration of four proteins in a validation cohort. The results showed a similar tendency with an AUC value of 0.848. The model also showed an encouraging performance in the prediction of PFS despite the small sample size. In addition, the results from ELISA also demonstrated the clinical utilities of the four-protein signature as a convenient, non-invasive tool to screen eligible patients for MET inhibitors as ELISA was readily available in most molecular laboratories. Overall, this study can select those patients that did not benefit from MET inhibitors and give them other treatments (such as chemotherapy, angiogenesis inhibitors, or immunotherapy), which can improve their survival outcomes.

Among the limitation of this study, firstly the sample sizes are relatively small with only 33 patients enrolled in our study, which may cause an overfit in our predictive models. Some results in the validation cohort were not significant may also attribute to the small sample size. It is hard to enroll large-scale patients with thorough clinical characteristics, serial plasma sample collection, and longtime follow-up. However, the dynamic change of the four proteins and the ELISA results can consolidate our findings. Secondly, the sample collections had heterogeneity, with a long period of collection time from 2014 to 2019, which may cause discrepant results in our study. Thirdly, the relationship of the four proteins with primary lung cancer tissue remained unknown. Due to the limited tumor tissue, we cannot perform the IHC test for primary cancer tissue to verify the origins of the four proteins.

5. Conclusions

The peripheral plasma proteomic characteristics were associated with the outcomes of MET-dysregulated patients treated with MET inhibitors. A combination of plasma MYH9, GNB1, ALOX12B, and HSD17B4 proteins could effectively and robustly predict the responses and PFS of patients receiving MET inhibitors, with a substitutable or complementary role to conventional MET FISH or IHC tests. This exploration will help select patients who may benefit from MET inhibitors.

Supplementary Materials: The following supporting information can be downloaded at: https://www.mdpi.com/article/10.3390/cancers15010302/s1, Figure S1: (A) Bar plot for the detected numbers of each sample at baseline. (B) Hierarchical cluster of the samples at baseline. Figure S2: GO and KEGG enrichment analyses of differentially expressed proteins between the PD and non-PD groups. Figure S3: The relative intensities of four selective protein biomarkers between patients treated with MET inhibitor with or without EGFR-TKI. Figure S4: (A) The concentrations of four proteins in the validation group using the ELISA method. (B) The prognostic value of four proteins for overall survival in the TCGA dataset. Table S1: The clinicopathological characteristics and treatment strategies of the enrolled patients based on the PD and non-PD groups. Table S2: Disease control rates and objective response rates in groups stratified by different models. Table S3: The clinicopathological characteristics and treatment strategies of the enrolled patients in the validation cohort.

Author Contributions: Conceptualization, Y.-L.W. and C.C.L.W.; methodology, C.C.L.W.; software, G.-L.J., L.-X.P. and M.-M.Z.; validation, G.-L.J., S.-R.W., M.-M.Z., Z.-H.C., H.-X.T. and L.-X.P.; formal analysis, G.-L.J., H.S., S.-R.W., S.-Y.M.L. and K.Y.; investigation, Y.-L.W.; resources, H.-Y.T., J.-J.Y., X.-C.Z. and Q.Z.; data curation, Y.-L.W. and C.C.L.W.; writing—original draft preparation, G.-L.J.; writing—review and editing, G.-L.J., C.C.L.W. and Y.-L.W.; visualization, G.-L.J.; supervision, Y.-L.W. and C.C.L.W.; project administration, Y.-L.W.; funding acquisition, Y.-L.W. and C.C.L.W. All authors have read and agreed to the published version of the manuscript.

Funding: This work was supported by the Key Lab System Project of Guangdong Science and Technology Department, Guangdong Provincial Key Lab of Translational Medicine in Lung Cancer (2017B030314120 to YL Wu), Guangdong Provincial People's Hospital Scientific Research Funds for Leading Medical Talents in Guangdong Province (KJ012019426 to YL Wu), Clinical Research Operating Fund of Central High Level Hospitals, Medical and Scientific Innovation Project of the Chinese Academy of Medical Science (2022-12M-1-004 to CCL.W.), Training Program of the Big Science Strategy Plan (2020YFE0202200 to CCL.W.), National Natural Science Foundation of China Grants (32150005 to CCL.W.), Ministry of Science and Technology of China (2022YFA0806000 to CCL.W.), and Research Funds from Health@InnoHK Program launched by Innovation Technology Commission of the Hong Kong Special Administrative Region.

Institutional Review Board Statement: The study was conducted in accordance with the Declaration of Helsinki, and was approved by the Institutional Review Board of Guangdong Provincial People's Hospital (approval no: 2013185H).

Informed Consent Statement: Informed consent was obtained from all patients involved in the study.

Data Availability Statement: Public database with RNA sequencing data could be obtained online. The proteomics data were uploaded in the supplementary material.

Acknowledgments: The authors are grateful for the help provided by Nan Zhang and Shuai-Xin Gao in mass spectrometry analysis and writing in mass spectrometry method.

Conflicts of Interest: Yi-Long Wu reports consulting and advisory services and declares speaker fees for Roche, AstraZeneca, Eli Lilly, Boehringer Ingelheim, Sanofi, Merck Sharp & Dohme, and Bristol Myers Squibb. All other authors declare that they have no conflict of interest.

References

1. Oxnard, G.R.; Hu, Y.; Mileham, K.F.; Husain, H.; Costa, D.B.; Tracy, P.; Feeney, N.; Sholl, L.M.; Dahlberg, S.E.; Redig, A.J.; et al. Assessment of Resistance Mechanisms and Clinical Implications in Patients with *EGFR* T790M-Positive Lung Cancer and Acquired Resistance to Osimertinib. *JAMA Oncol.* **2018**, *4*, 1527–1534. [CrossRef] [PubMed]
2. Go, H.; Jeon, Y.K.; Park, H.J.; Sung, S.W.; Seo, J.W.; Chung, D.H. High *MET* Gene Copy Number Leads to Shorter Survival in Patients with Non-Small Cell Lung Cancer. *J. Thorac. Oncol.* **2010**, *5*, 305–313. [CrossRef] [PubMed]
3. Engelman, J.A.; Zejnullahu, K.; Mitsudomi, T.; Song, Y.; Hyland, C.; Park, J.O.; Lindeman, N.; Gale, C.M.; Zhao, X.; Christensen, J.; et al. *MET* Amplification Leads to Gefitinib Resistance in Lung Cancer by Activating *ERBB3* Signaling. *Science* **2007**, *316*, 1039–1043. [CrossRef] [PubMed]
4. Guo, R.; Luo, J.; Chang, J.; Rekhtman, N.; Arcila, M.; Drilon, A. *MET*-Dependent Solid Tumours—Molecular Diagnosis and Targeted Therapy. *Nat. Rev. Clin. Oncol.* **2020**, *17*, 569–587. [CrossRef]

5. Schildhaus, H.U.; Schultheis, A.M.; Ruschoff, J.; Binot, E.; Merkelbach-Bruse, S.; Fassunke, J.; Schulte, W.; Ko, Y.D.; Schlesinger, A.; Bos, M.; et al. *MET* Amplification Status in Therapy-Naive Adeno- and Squamous Cell Carcinomas of the Lung. *Clin. Cancer Res.* **2015**, *21*, 907–915. [CrossRef]
6. Li, A.; Niu, F.Y.; Han, J.F.; Lou, N.N.; Yang, J.J.; Zhang, X.C.; Zhou, Q.; Xie, Z.; Su, J.; Zhao, N.; et al. Predictive and Prognostic Value of De Novo MET Expression in Patients with Advanced Non-Small-Cell Lung Cancer. *Lung Cancer* **2015**, *90*, 375–380. [CrossRef]
7. Reis, H.; Metzenmacher, M.; Goetz, M.; Savvidou, N.; Darwiche, K.; Aigner, C.; Herold, T.; Eberhardt, W.E.; Skiba, C.; Hense, J.; et al. MET Expression in Advanced Non-Small-Cell Lung Cancer: Effect on Clinical Outcomes of Chemotherapy, Targeted Therapy, and Immunotherapy. *Clin. Lung Cancer* **2018**, *19*, e441–e463. [CrossRef]
8. Wu, Y.L.; Zhang, L.; Kim, D.W.; Liu, X.; Lee, D.H.; Yang, J.C.; Ahn, M.J.; Vansteenkiste, J.F.; Su, W.C.; Felip, E.; et al. Phase Ib/II Study of Capmatinib (INC280) Plus Gefitinib after Failure of Epidermal Growth Factor Receptor (EGFR) Inhibitor Therapy in Patients with *EGFR*-Mutated, *MET* Factor-Dysregulated Non-Small-Cell Lung Cancer. *J. Clin. Oncol.* **2018**, *36*, 3101–3109. [CrossRef]
9. Wu, Y.-L.; Cheng, Y.; Zhou, J.; Lu, S.; Zhang, Y.; Zhao, J.; Kim, D.-W.; Soo, R.A.; Kim, S.-W.; Pan, H.; et al. Tepotinib Plus Gefitinib in Patients with *EGFR*-Mutant Non-Small-Cell Lung Cancer with MET Overexpression or *MET* Amplification and Acquired Resistance to Previous EGFR Inhibitor (INSIGHT Study): An Open-Label, Phase 1b/2, Multicentre, Randomised Trial. *Lancet Respir. Med.* **2020**, *8*, 1132–1143. [CrossRef]
10. Sequist, L.V.; Han, J.-Y.; Ahn, M.-J.; Cho, B.C.; Yu, H.; Kim, S.-W.; Yang, J.C.-H.; Lee, J.S.; Su, W.-C.; Kowalski, D.; et al. Osimertinib Plus Savolitinib in Patients with *EGFR* Mutation-Positive, *MET*-Amplified, Non-Small-Cell Lung Cancer after Progression on EGFR Tyrosine Kinase Inhibitors: Interim Results from a Multicentre, Open-Label, Phase 1b Study. *The Lancet Oncology* **2020**, *21*, 373–386. [CrossRef]
11. Cappuzzo, F.; Marchetti, A.; Skokan, M.; Rossi, E.; Gajapathy, S.; Felicioni, L.; Del Grammastro, M.; Sciarrotta, M.G.; Buttitta, F.; Incarbone, M.; et al. Increased *MET* Gene Copy Number Negatively Affects Survival of Surgically Resected Non-Small-Cell Lung Cancer Patients. *J. Clin. Oncol.* **2009**, *27*, 1667–1674. [CrossRef] [PubMed]
12. Noonan, S.A.; Berry, L.; Lu, X.; Gao, D.; Baron, A.E.; Chesnut, P.; Sheren, J.; Aisner, D.L.; Merrick, D.; Doebele, R.C.; et al. Identifying the Appropriate FISH Criteria for Defining *MET* Copy Number-Driven Lung Adenocarcinoma through Oncogene Overlap Analysis. *J. Thorac. Oncol.* **2016**, *11*, 1293–1304. [CrossRef] [PubMed]
13. Lai, G.G.Y.; Guo, R.; Drilon, A.; Shao Weng Tan, D. Refining Patient Selection of *MET*-Activated Non-Small Cell Lung Cancer through Biomarker Precision. *Cancer Treat. Rev.* **2022**, *110*, 102444. [CrossRef]
14. Guo, R.; Berry, L.D.; Aisner, D.L.; Sheren, J.; Boyle, T.; Bunn, P.A., Jr.; Johnson, B.E.; Kwiatkowski, D.J.; Drilon, A.; Sholl, L.M.; et al. Met Ihc Is a Poor Screen for *MET* Amplification or *MET* Exon 14 Mutations in Lung Adenocarcinomas: Data from a Tri-Institutional Cohort of the Lung Cancer Mutation Consortium. *J. Thorac. Oncol.* **2019**, *14*, 1666–1671. [CrossRef]
15. Yin, W.; Guo, M.; Tang, Z.; Toruner, G.A.; Cheng, J.; Medeiros, L.J.; Tang, G. MET Expression Level in Lung Adenocarcinoma Loosely Correlates with *MET* Copy Number Gain/Amplification and Is a Poor Predictor of Patient Outcome. *Cancers* **2022**, *14*, 2433. [CrossRef]
16. Zhang, B.; Whiteaker, J.R.; Hoofnagle, A.N.; Baird, G.S.; Rodland, K.D.; Paulovich, A.G. Clinical Potential of Mass Spectrometry-Based Proteogenomics. *Nat. Rev. Clin. Oncol.* **2019**, *16*, 256–268. [CrossRef]
17. Geyer, P.E.; Holdt, L.M.; Teupser, D.; Mann, M. Revisiting Biomarker Discovery by Plasma Proteomics. *Mol. Syst. Biol.* **2017**, *13*, 942. [CrossRef] [PubMed]
18. Gillette, M.A.; Satpathy, S.; Cao, S.; Dhanasekaran, S.M.; Vasaikar, S.V.; Krug, K.; Petralia, F.; Li, Y.; Liang, W.W.; Reva, B.; et al. Proteogenomic Characterization Reveals Therapeutic Vulnerabilities in Lung Adenocarcinoma. *Cell* **2020**, *182*, 200–225.e235. [CrossRef]
19. Eltahir, M.; Isaksson, J.; Mattsson, J.S.M.; Karre, K.; Botling, J.; Lord, M.; Mangsbo, S.M.; Micke, P. Plasma Proteomic Analysis in Non-Small Cell Lung Cancer Patients Treated with PD-1/PD-L1 Blockade. *Cancers* **2021**, *13*, 3116. [CrossRef]
20. Niu, L.; Thiele, M.; Geyer, P.E.; Rasmussen, D.N.; Webel, H.E.; Santos, A.; Gupta, R.; Meier, F.; Strauss, M.; Kjaergaard, M.; et al. Noninvasive Proteomic Biomarkers for Alcohol-Related Liver Disease. *Nat. Med.* **2022**, *28*, 1277–1287. [CrossRef]
21. Zhu, Y.; Aebersold, R.; Mann, M.; Guo, T. Snapshot: Clinical Proteomics. *Cell* **2021**, *184*, 4840. [CrossRef] [PubMed]
22. Mignard, X.; Ruppert, A.M.; Antoine, M.; Vasseur, J.; Girard, N.; Mazieres, J.; Moro-Sibilot, D.; Fallet, V.; Rabbe, N.; Thivolet-Bejui, F.; et al. C-MET Overexpression as a Poor Predictor of *MET* Amplifications or Exon 14 Mutations in Lung Sarcomatoid Carcinomas. *J. Thorac. Oncol.* **2018**, *13*, 1962–1967. [CrossRef] [PubMed]
23. Cancer Genome Atlas Research, N. Comprehensive Molecular Profiling of Lung Adenocarcinoma. *Nature* **2014**, *511*, 543–550. [CrossRef]
24. Peng, L.X.; Jie, G.L.; Li, A.N.; Liu, S.Y.; Sun, H.; Zheng, M.M.; Zhou, J.Y.; Zhang, J.T.; Zhang, X.C.; Zhou, Q.; et al. *MET* Amplification Identified by Next-Generation Sequencing and Its Clinical Relevance for Met Inhibitors. *Exp. Hematol. Oncol.* **2021**, *10*, 52. [CrossRef] [PubMed]
25. Solomon, J.P.; Yang, S.R.; Choudhury, N.J.; Ptashkin, R.N.; Eslamdoost, N.; Falcon, C.J.; Martin, A.; Plodkowski, A.; Wilhelm, C.; Shen, R.; et al. Bioinformatically-Expanded Next-Generation Sequencing Analysis Optimizes Identification of Therapeutically Relevant *MET* Copy Number Alterations in >50,000 Tumors. *Clin. Cancer Res.* **2022**, *28*, 4649–4659. [CrossRef] [PubMed]

26. Oscorbin, I.P.; Smertina, M.A.; Pronyaeva, K.A.; Voskoboev, M.E.; Boyarskikh, U.A.; Kechin, A.A.; Demidova, I.A.; Filipenko, M.L. Multiplex Droplet Digital PCR Assay for Detection of *MET* and *HER2* Genes Amplification in Non-Small Cell Lung Cancer. *Cancers* **2022**, *14*, 1458. [CrossRef] [PubMed]
27. Jorgensen, J.T.; Mollerup, J. Companion Diagnostics and Predictive Biomarkers for MET-Targeted Therapy in NSCLC. *Cancers* **2022**, *14*, 2150. [CrossRef] [PubMed]
28. Lai, G.G.Y.; Lim, T.H.; Lim, J.; Liew, P.J.R.; Kwang, X.L.; Nahar, R.; Aung, Z.W.; Takano, A.; Lee, Y.Y.; Lau, D.P.X.; et al. Clonal *MET* Amplification as a Determinant of Tyrosine Kinase Inhibitor Resistance in Epidermal Growth Factor Receptor-Mutant Non-Small-Cell Lung Cancer. *J. Clin. Oncol.* **2019**, *37*, 876–884. [CrossRef]
29. Fan, Y.; Sun, R.; Wang, Z.; Zhang, Y.; Xiao, X.; Liu, Y.; Xin, B.; Xiong, H.; Lu, D.; Ma, J. Detection of *MET* Amplification by Droplet Digital PCR in Peripheral Blood Samples of Non-Small Cell Lung Cancer. *J. Cancer Res. Clin. Oncol.* **2022**. [CrossRef]
30. Friedlaender, A.; Subbiah, V.; Russo, A.; Banna, G.L.; Malapelle, U.; Rolfo, C.; Addeo, A. *EGFR* and *HER2* Exon 20 Insertions in Solid Tumours: From Biology to Treatment. *Nat. Rev. Clin. Oncol.* **2022**, *19*, 51–69. [CrossRef]
31. Culy, C.R.; Faulds, D. Gefitinib. *Drugs* **2002**, *62*, 2237–2248. [CrossRef] [PubMed]
32. Yu, W.; Hurley, J.; Roberts, D.; Chakrabortty, S.K.; Enderle, D.; Noerholm, M.; Breakefield, X.O.; Skog, J.K. Exosome-Based Liquid Biopsies in Cancer: Opportunities and Challenges. *Ann. Oncol.* **2021**, *32*, 466–477. [CrossRef] [PubMed]
33. Ren, Z.; Spaargaren, M.; Pals, S.T. Syndecan-1 and Stromal Heparan Sulfate Proteoglycans: Key Moderators of Plasma Cell Biology and Myeloma Pathogenesis. *Blood* **2021**, *137*, 1713–1718. [CrossRef] [PubMed]
34. Pal, A.; Shinde, R.; Miralles, M.S.; Workman, P.; de Bono, J. Applications of Liquid Biopsy in the Pharmacological Audit Trail for Anticancer Drug Development. *Nat. Rev. Clin. Oncol.* **2021**, *18*, 454–467. [CrossRef]
35. Hu, S.; Ren, S.; Cai, Y.; Liu, J.; Han, Y.; Zhao, Y.; Yang, J.; Zhou, X.; Wang, X. Glycoprotein PTGDS Promotes Tumorigenesis of Diffuse Large B-Cell Lymphoma by MYH9-Mediated Regulation of Wnt-Beta-Catenin-STAT3 Signaling. *Cell Death Differ.* **2022**, *29*, 642–656. [CrossRef]
36. Lin, X.; Li, A.M.; Li, Y.H.; Luo, R.C.; Zou, Y.J.; Liu, Y.Y.; Liu, C.; Xie, Y.Y.; Zuo, S.; Liu, Z.; et al. Silencing MYH9 Blocks HBx-Induced GSK3beta Ubiquitination and Degradation to Inhibit Tumor Stemness in Hepatocellular Carcinoma. *Signal Transduct. Target. Ther.* **2020**, *5*, 13. [CrossRef]
37. Zimmermannova, O.; Doktorova, E.; Stuchly, J.; Kanderova, V.; Kuzilkova, D.; Strnad, H.; Starkova, J.; Alberich-Jorda, M.; Falkenburg, J.H.F.; Trka, J.; et al. An Activating Mutation of *GNB1* Is Associated with Resistance to Tyrosine Kinase Inhibitors in ETV6-ABL1-Positive Leukemia. *Oncogene* **2017**, *36*, 5985–5994. [CrossRef]
38. Jiang, T.; Zhou, B.; Li, Y.M.; Yang, Q.Y.; Tu, K.J.; Li, L.Y. ALOX12B Promotes Carcinogenesis in Cervical Cancer by Regulating the PI3K/ERK1 Signaling Pathway. *Oncol. Lett.* **2020**, *20*, 1360–1368. [CrossRef]
39. Liu, Z.; Li, L.; Li, X.; Hua, M.; Sun, H.; Zhang, S. Prediction and Prognostic Significance of ALOX12B and PACSIN1 Expression in Gastric Cancer by Genome-Wide RNA Expression and Methylation Analysis. *J. Gastrointest. Oncol.* **2021**, *12*, 2082–2092. [CrossRef]
40. Schekman, R. Peroxisomes: Another Branch of the Secretory Pathway? *Cell* **2005**, *122*, 1–2. [CrossRef]
41. Yamashita, S.; Hattori, N.; Fujii, S.; Yamaguchi, T.; Takahashi, M.; Hozumi, Y.; Kogawa, T.; El-Omar, O.; Liu, Y.Y.; Arai, N.; et al. Multi-Omics Analyses Identify *HSD17B4* Methylation-Silencing as a Predictive and Response Marker of *HER2*-Positive Breast Cancer to HER2-Directed Therapy. *Sci. Rep.* **2020**, *10*, 15530. [CrossRef] [PubMed]
42. Audet-Walsh, E.; Bellemare, J.; Lacombe, L.; Fradet, Y.; Fradet, V.; Douville, P.; Guillemette, C.; Levesque, E. The Impact of Germline Genetic Variations in Hydroxysteroid (17-Beta) Dehydrogenases on Prostate Cancer Outcomes after Prostatectomy. *Eur. Urol.* **2012**, *62*, 88–96. [CrossRef] [PubMed]

 *cancers*

Article

Identification of Novel MET Exon 14 Skipping Variants in Non-Small Cell Lung Cancer Patients: A Prototype Workflow Involving in Silico Prediction and RT-PCR

Riku Das, Maureen A. Jakubowski, Jessica Spildener and Yu-Wei Cheng *

Department of Laboratory Medicine, Robert J. Tomsich Pathology and Laboratory Medicine Institute, Cleveland Clinic, 9500 Euclid Avenue, Cleveland, OH 44195, USA
* Correspondence: chengy@ccf.org; Tel.: +1-216-445-0757; Fax: +1-216-445-0681

Highlights:

MET exon 14 skipping is an oncogenic targetable driver mutation in lung cancer.

- Two novel non-canonical splice site variants identified in MET genome.
- Predicted splicing strength using in silico splicing prediction tools.
- Tested routine cytological smear slides for RNA-based molecular diagnostics.
- RT-PCR and Sanger sequencing analysis confirmed MET exon 14 skipping.

Simple Summary: Non-small Cell Lung cancer (NSCLC) contributes to 85% of total lung cancer diagnoses in the United States. With the discovery of various targetable genetic markers and FDA approval of drugs against these markers, genetic testing has become a routine part of the diagnosis and staging process of NSCLC. MET gain of function mutations have been of particular interest as FDA has recently approved two MET inhibitors for the treatment of NSCLC patients with MET exon 14 skipping (METex14) mutations. However, an effective workflow for the classification of various METex14 mutations in the clinical testing laboratory has not been explored. In this report, we reveal two novel METex14 variants and propose a cost-effective and robust workflow for molecular diagnosis of MET variants contributing to exon 14 skipping with the use of readily available specimen sources.

Abstract: Background and aims: The MET exon 14 skipping (METex14) is an oncogenic driver mutation that provides a therapeutic opportunity in non-small cell lung cancer (NSCLCs) patients. This event often results from sequence changes at the MET canonical splicing sites. We characterize two novel non-canonical splicing site variants of MET that produce METex14. Materials and Methods: Two variants were identified in three advanced-stage NSCLC patients in a next-generation sequencing panel. The potential impact on splicing was predicted using in silico tools. METex14 mutation was confirmed using reverse transcription (RT)-PCR and a Sanger sequencing analysis on RNA extracted from stained cytology smears. Results: The interrogated MET (RefSeq ID NM_000245.3) variants include a single nucleotide substitution, c.3028+3A>T, in intron 14 and a deletion mutation, c.3012_3028del, in exon 14. The in silico prediction analysis exhibited reduced splicing strength in both variants compared with the MET normal transcript. The RT-PCR and subsequent Sanger sequencing analyses confirmed METex14 skipping in all three patients carrying these variants. Conclusion: This study reveals two non-canonical MET splice variants that cause exon 14 skipping, concurrently also proposes a clinical workflow for the classification of such non-canonical splicing site variants detected by routine DNA-based NGS test. It shows the usefulness of in silico prediction to identify potential METex14 driver mutation and exemplifies the opportunity of routine cytology slides for RNA-based testing.

Keywords: MET proto-oncogene; non-canonical splicing site; exon skipping; next-generation sequencing (NGS); non-small cell lung cancer (NSCLC); in silico prediction

Citation: Das, R.; Jakubowski, M.A.; Spildener, J.; Cheng, Y.-W. Identification of Novel MET Exon 14 Skipping Variants in Non-Small Cell Lung Cancer Patients: A Prototype Workflow Involving in Silico Prediction and RT-PCR. *Cancers* **2022**, *14*, 4814. https://doi.org/10.3390/cancers14194814

Academic Editor: David Wong

Received: 30 June 2022
Accepted: 22 September 2022
Published: 1 October 2022

Publisher's Note: MDPI stays neutral with regard to jurisdictional claims in published maps and institutional affiliations.

1. Introduction

Lung cancer is the leading cause of cancer death in the United States, with non-small cell lung cancer (NSCLC) contributing to 85% of total lung cancer diagnoses [1]. NSCLC patients with driver mutation who receive the appropriate targeted therapy have shown improved outcomes [2,3]. Among the actionable mutations for NSCLC treatment, mutations in EGFR, BRAF, KRAS, NTRK1/2/3, and ALK and ROS1 rearrangements are worth mentioning. Recently, the FDA has approved two drugs, capmatinib and tepotinib, for metastatic NSCLC with MET exon 14 skipping (METex14) mutation [3–7]. Upon treatment with MET tyrosin kinase inhibitor, patients with METex14 stage IV NSCLC survived longer. METex14 is observed in 3 to 4% of total NSCLC adenocarcinomas, the prevalence of which is greater or equal to some of the other oncogenic driver mutations: ROS1 (1–2%), NTRK1/2/3 (<1%), and BRAF (1–5%) [2,3,5,8–12].

MET proto-oncogene is located at chromosome 7q21-q31, which encodes for a receptor tyrosine kinase, c-Met, and is activated by ligand hepatocyte growth factor (HGF). Upon activation, MET phosphorylates its substrate and results in the activation of multiple signaling pathways (PI3K-AKT-mTOR, RAS-RAF-MEK-ERK, and FAK) leading to cell growth, proliferation, survival, adhesion, migration, and differentiation [13]. MET gain-of-function mutation has been recognized as a primary oncogenic driver that contributes to resistance towards many tyrosine kinase inhibitors in NSCLC treatment. Various MET gene alterations that lead to gain-of-function are sequence changes at MET exon 14 and flanking intronic regions, MET gene amplification, and MET gene fusions. Among them, METex14 is the most widely reported, 4–40% of which can occur concurrently with MET amplification [3,14,15]. However, other mechanisms of increased MET expression also play an important role in tumorigenesis driven by c-Met [16].

MET gene exon 14 encodes for a regulatory site in the juxtamembrane domain of c-Met protein. This site bears the binding site of Cbl, an E3 ubiquitin ligase, which leads to c-Met degradation upon binding [17]. Therefore, any alterations that cause exon 14 skipping leads to enhanced c-Met signaling and oncogenic transformation [18,19]. These alterations on the DNA level could be within the exon 14 (Y1003X or D1010X), in the intronic region surrounding the exon 14, or the total deletion of exon 14. Interestingly, the majority of these reported alterations are either partially deleted exon 14, or disruptions of the canonical splicing acceptor (AG) or donor (GT) sites of MET intron 13 and intron 14, respectively. However, the impact on METex14 caused by MET variants not involving the intron 14 canonical splicing donor site has seldom been addressed. With the approval of c-Met targeted drugs, identifying and accurately interpreting MET variants that increase c-Met signaling is of great targeted therapeutic importance. In this report, we describe a prototype workflow using in silico splicing prediction tools to identify MET variants of potential impact on the exon 14 splicing, followed by an RT-PCR and Sanger sequencing to confirm the splicing event, with a special focus on two novel MET variants located near the exon 14 and intron 14 juncture, but which do not disrupt the intron 14 canonical splicing site. Additionally, routine cytological smear slides were used to extract total RNA for the RT-PCR to determine the impact on METex14. Thereby, this study adds two novel variants to the growing list of METex14 variants [8,20,21] and demonstrates the utility of cytology slides as valuable sources for molecular diagnostic testing.

2. Methods

2.1. Sample Selection

With 3 years (2017 to 2019) of monitoring of the NSCLC specimens that were undergone in-house via the Cancer hotspot NGS test, we identified 20 cases of MET exon 14 and intron 14 genomic alterations. Out of the 20, we have identified three potential METex14 cases that do not involve canonical splicing sites. Two patients had novel variants identified in intron 14, c.3028+3A>T, and the third patient carried a variant in Exon 14, c.3012_3028del. These three NSCLC specimens were further investigated for the impact of MET exon 14 skipping at the RNA level. For the positive control, a patient's specimen (cytology slides) with MET exon 14 canonical splicing donor site mutation that causes MET exon 14 skipping was used. For the negative control, an RNA specimen from a patient's white blood cells without a history of NSCLC was used. The study was conducted according to the approved protocols of Cleveland Clinic's Institutional Review Board (IRB; 17–177 and 19–329).

2.2. Patient Samples, DNA and RNA Extraction

Genomic DNA was extracted from the bronchial fluid of NSCLC patients, which was preserved in PreservCyt solution using the Maxwell RSC Cell DNA purification kit according to the manufacturer's instruction (Promega, Madison, WI, USA). The quantity and quality of purified DNA were evaluated using Nanodrop and Qubit and stored at 4 °C until tested by Cancer hotspot NGS [22,23]. Direct smears were prepared from residual bronchial fluid, which were either diff-Quick or Papanicolaou (Pap)-stained. Selected diff-Quick and Pap-stained specimens were used for RNA extraction [24]. Total RNA was extracted using the Maxwell RSC RNA FFPE kit (Promega, Madison, WI, USA) from the smears to use in the RNA-based assay. The quantity and quality of total nucleic acid were evaluated using Nanodrop and Qubit and stored at −70 °C until tested.

2.3. Cancer Hotspot Panel Library Preparation, Sequencing, and Data Analysis

Cancer hotspot NGS library preparation was performed as described previously [22,23]. Briefly, 10 ng of genomic DNA and 207 PCR primers pairs (AmpliSeq Cancer Hotspot Panel v2.0 kit, Thermo Fisher Scientific, Waltham, MA, USA) were used for multiplex PCR to analyze approximately 2800 hotspot mutations in 50 genes. An oligonucleotide barcode was introduced into each sample to properly separate the sequencing reads of individual sample libraries. PCR amplicons were analyzed by Bioanalyzer 2100 for quality check and samples with >200 pM were pooled, followed by sequencing on the MiSeq instrument. The sequencing data were aligned to human genome build 19 (HG19/GRCh37) and variants in mutation hot spot regions in *BRAF, EGFR, ERBB2, KRAS,* and *MET* were identified using NextGENe Software (Soft Genetics, State College, PA, USA). The Integrative Genomics Viewer (IGV) was used to visually inspect the quality of read alignment and variant calls. A quality score of Q30 was used as filtering criteria to determine the sequence read quality. For a given sample, the minimum coverage requirement of targeted regions was 100×. Variants with variant allele frequencies (VAFs) as low as 2% may be identified using this method. The MET RefSeq transcript NM_000245.3 is used for variant data analysis and reporting.

2.4. In Silico Prediction

In silico splice tools, including SpliceSiteFinder-like, MaxEntScan, NNSplice, and GeneSplicer, were integrated in the Alamut Visual Plus (Version 1.3, SOPHiA GENETICS, Lausanne, Switzerland) for the prediction of the MET variant's impact on gene splicing. In the Alamut Visual Plus, impacts on gene splicing from individual tools are represented either with a vertical blue bar for 5′ donor sites or a vertical green bar for 3′ acceptor sites. Assigned scores, which are proportional to the heights of each bar, are indicators of splicing donor or acceptor signals that impact the splicing strength. Known constitutive signals are displayed as a small blue triangle for 5′ or a green triangle for 3′, close to the sequence letters.

2.5. RT-PCR and Sanger Sequencing

RNA specimens from the patients and negative control were reverse transcribed using the Ipsogen Reverse Transcription kit (Qiagen, Hilden, Germany) with random primers. The obtained cDNA was amplified using a forward primer specific to MET Exon 13, 5′-GCTGGTGTTGTCTCAATATCAA-3′ and a reverse primer specific to MET Exon 15, 5′-GGCATGAACCGTTCTGAGAT-3′. The PCR conditions are as follows: 95 °C for 3 min and 45 cycles of 95 °C for 30 s, 55 °C for 3 s, and 72 °C for 2 min. The PCR products were analyzed using Bioanalyzer and the splicing products were subjected to

Sanger sequencing. Sanger sequencing was performed using a modified protocol supplied by Applied Biosystems BigDye Terminator 1.1 and 3.1 Cycle sequencing kits. Fragments were then analyzed using Applied Biosystems 48-capillary 3730 Genomic Analyzer.

3. Results

3.1. Demographic and Clinical Characteristics of NSCLC Patients with Two Novel MET Variants

Three advanced-stage NSCLC patients with one exon 14 and the other intron 14 novel MET variants were identified from an in-house lung cancer NGS test. Patients' demography and clinical characteristics are shown in Table 1. Patient 1 and Patient 2 harbored the same MET single nucleotide variant c.3028+3A>T at the beginning of intron 14, with 24% and 37% allelic fractions, respectively. This variant was near but not at the canonical splice donor sequence (Figure 1A,B). Patient 3 carried another rare MET variant with a 17-nucleotide deletion at the 3′ end of the exon 14, c.3012_3028delAGCTACTTTTCCAGAAG, with 11% allelic fraction (Figure 1A,C). In all cases, variants were identified with very high numbers of sequencing coverages (Table 1). Patients 1 and 3 did not carry other actionable mutations in BRAF, EGFR, HER2, KRAS, and ALK rearrangement. Patient 2 had a mutation in KRAS (NM_004985.3 c.34G>T, p.Gly12Cys) with 10% allelic fraction.

Since the MET variants were identified near the exon–intron junction, we performed an in silico analysis for possible impact in splicing. Using the splicing prediction tool analysis, we have observed a drastic reduction in splicing strength at the MET intron 14 splicing donor site in both variants (c.3028+3A>T and c.3012_3028del), compared with the MET wild-type transcript (Figure 1B,C), suggesting the possibility of splicing alterations leading to the exon 14 skipping.

Table 1. Demographic and clinical characteristics.

Characteristics	Patient 1	Patient 2	Patient 3
Histology	Adenocarcinoma	Adenocarcinoma	Necrotic NSCLC
Tumor%	40	80	90
MET variant NG_008996.1 (NM_000245.3)	c.3028+3A>T VAF = 24% NGS read depth = 3761	c.3028+3A>T VAF = 37% NGS read depth = 10,184	c.3012_3028del VAF = 11% NGS read depth = 3518
Other activating mutations in hotspots of BRAF, EGFR, HER2, and KRAS	Negative	KRAS(NM_004985.3) c.34G>T (p.Gly12Cys)	Negative

Abbreviations, VAF = variant allele fraction.

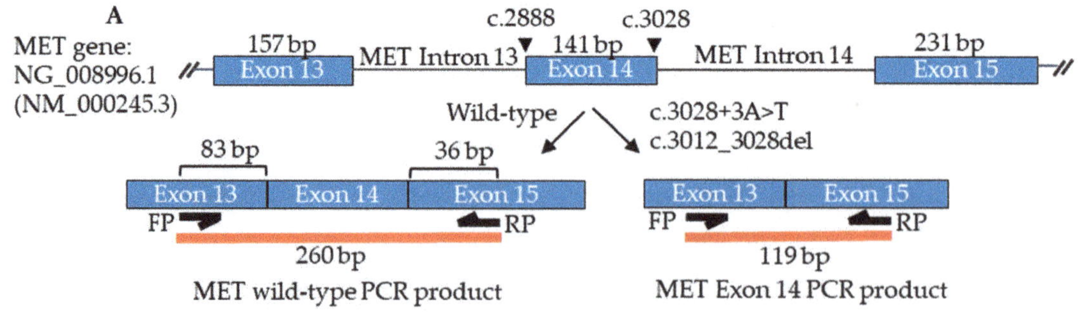

Figure 1. *Cont.*

B MET (NM_000245.3) c.3028+3A>T

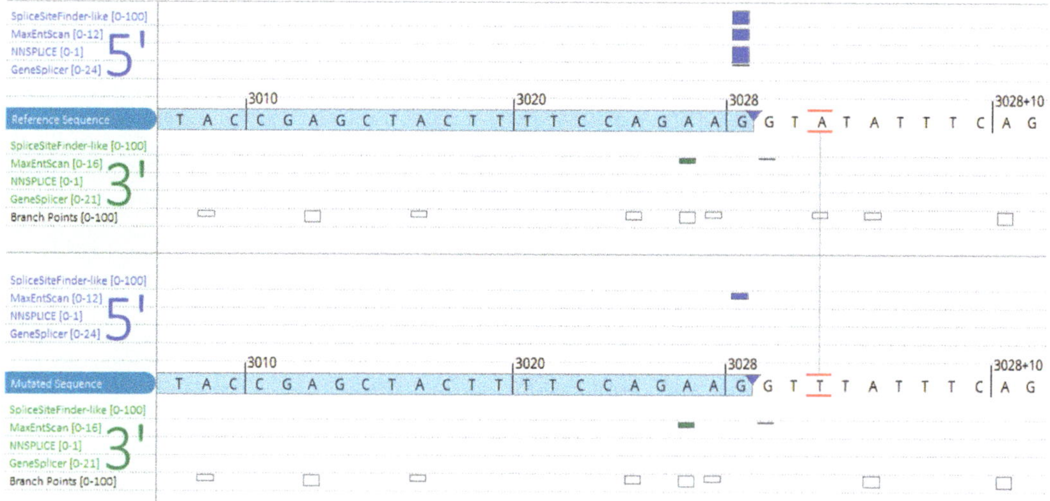

C MET(NM_000245.3) c.3012_3028del

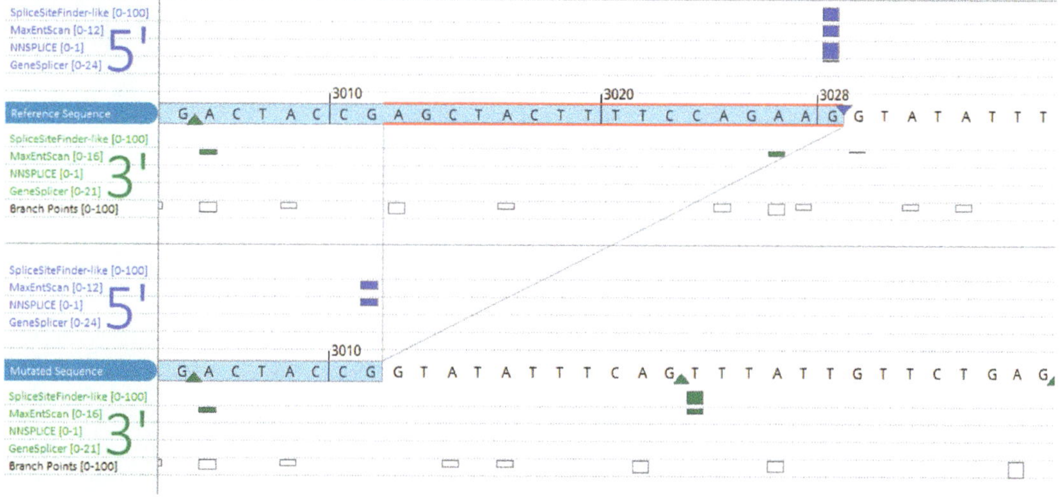

Figure 1. (**A**) Graphic representation of MET exon 13, 14, and 15 sizes, primers for RT-PCR binding sites, and predicted size of wild-type and MET exon 14 PCR products. (**B,C**) Alamut visual prediction of MET splicing with c.3028+3A>T and c.3012_3028del variants. With the use of four splicing predictors, donor prediction signals, shown with vertical blue bars (each bar corresponded to an individual splicing predictor) at the 5′ donor sites are reduced, suggesting that these variants alter splicing. Vertical green bars are for 3′ acceptor sites showing minimum or no change. Heights of the bars are proportional to splicing strength. Known constitutive signals are displayed as a small blue triangle for 5′ or a green triangle for 3′.

3.2. Confirmation of MET Exon 14 Mutation in Two Novel MET Variants

To provide functional evidence of these two MET variants causing a splicing defect and exon 14 skipping, an RT-PCR and Sanger sequencing analysis were performed. The positions of PCR primers and predicted amplicon sizes for MET wild-type (WT) and METex14 are shown in Figure 1A. The primers were designed—with forward binding to exon 13 and reverse to exon 15—and estimated to produce 260 bp wild-type (without exon 14 skipping) or 119 bp METex14 amplicons. RNA isolated from the diff-Quick smears were reverse transcribed to cDNA using random primers, followed by the amplification of cDNA with the gene-specific primers. As shown in Figure 2, the negative patient control produced a single fragment of approximately 260 bp in size, which matches with the calculated WT amplicon size. However, in addition to the WT PCR product, all three patients with MET variants and positive control produce a smaller fragment of roughly 119 bp in size, an expected METex14 product size. Of note, Patient 1 and Patient 2, as well as the positive control, showed a more robust amplification of the METex14 allele compared to the WT allele, whereas both alleles were somewhat equally amplified in Patient 3. A similar RT-PCR result of Patient 1 was also observed using RNA extracted from a Pap-stained slide (data not shown). Sequencing of the three patients' 119 bp PCR products revealed the splicing junction spanning the last nucleotide of exon 13 and the first nucleotide of exon 15 with the total omission of the exon 14 sequence (Figure 3A–C). Sequencing of the 260 bp fragment from the negative patient control indeed showed MET WT amplicon with the sequence spanning the entire exon 14 sequence and portions of exon 13 and 15 (data not shown). Altogether, these data suggest that the two novel MET variants (c.3028+3A>T and c.3012_3028del) identified in the lung cancer panel cause exon 14 skipping in the MET transcript.

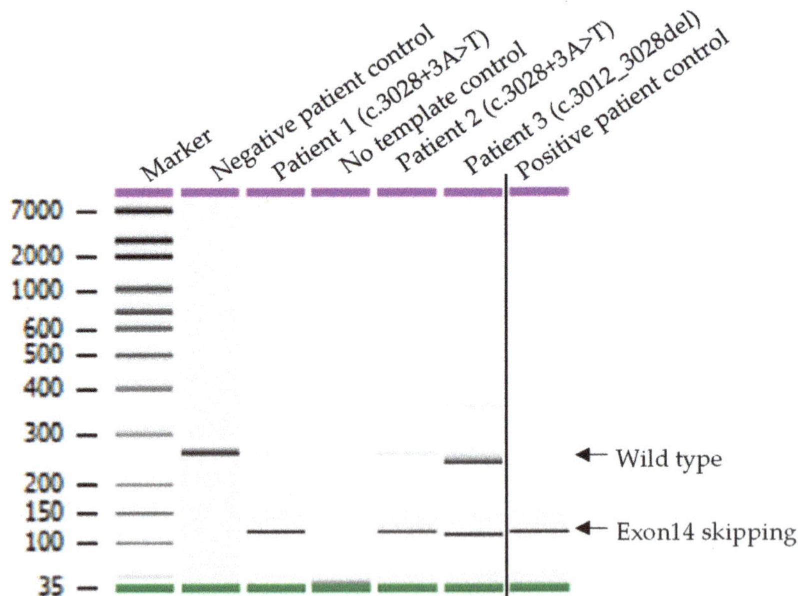

Figure 2. RT-PCR analysis on RNA from negative control and four patients with suspected (Patient 1 to 3) or known (positive control) METEx14 skipping variant, bronchial fine needle aspirate specimens. Gel picture from Bioanalyzer showing PCR products amplified using specific primers (MET_FP1 and MET_RP1). A fragment size of 260 indicates MET WT, and a fragment of 119 bp in size indicates MET exon 14 variant. No variant was identified in negative leukocytes.

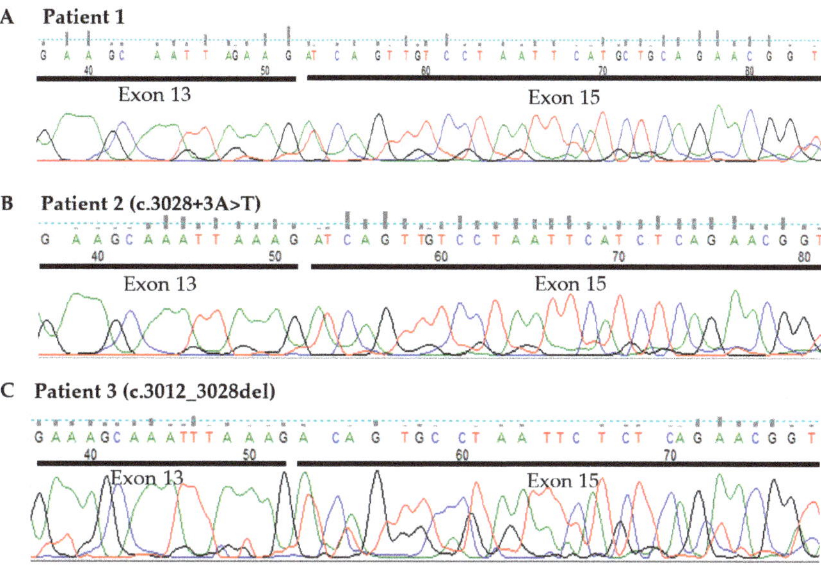

Figure 3. Sanger sequencing on PCR products obtained from Bioanalyzer polyacrylamide gel. Exon 13 and 15 sequences are indicated with black bars. (**A**) Negative patient control with 260 bp PCR product. (**B**) Patient 1 with MET c.3028+3A>T variant, 119 bp product. (**C**) Patient 1 with MET c.3028+3A>T variant, 119 bp product. Patient 3 with MET c.3012_3028del, 119 bp product.

4. Discussion

MET mutations that produce MET gain-of-function have been growing in interest among clinicians for their use as an actionable oncogenic therapeutic target for NSCLC patients. Clinical trial data, based on which the first MET-targeted therapy was approved in 2020, indicated that NSCLC patients with METex14 somatic mutation show better outcomes with longer survival [3,6,7,25–27]. In this report, for the benefit of NSCLC patient management, we demonstrate a cost-effective and robust workflow (Figure 4) to definitively determine MET variants that contribute to exon 14 skipping.

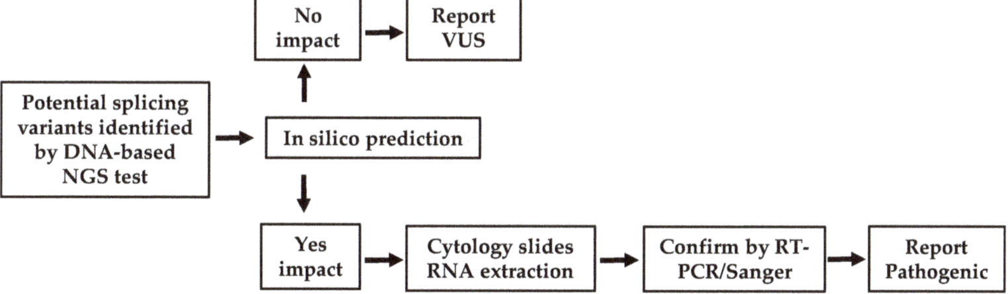

Figure 4. Model depicting a cost-effective clinical workflow to enable a potential genomic slicing variant screening process. When variants of uncertain significance around canonical splicing sites are identified in routine DNA-based NGS test, specimens can be assigned to a rapid in silico analysis to identify their impact in mRNA splicing. A variant can be reported as VUS if no impact is found in in silico prediction. If predictions suggest a significant impact in splicing, a specimen is assigned for an RNA work-up. The available cytology slides can be used to extract RNA, followed by RT-PCR and Sanger sequencing to convincingly identify splicing products.

It is well known that the canonical splice donor GT and acceptor AG dinucleotide sites are required for spliceosome interaction and subsequent splicing of the intronic sequences in pre-mRNA. Thus, in a molecular diagnostic laboratory, variants identified at the canonical splice sites are mostly classified as likely pathogenic (LP) or pathogenic due to the well-established biological impacts on gene splicing. Evidence also suggests that the immediate vicinity of 12–30 and 15–33 nucleotides surrounding the intronic donor and acceptor site, respectively, may contribute to the splicing efficiency by proving a preferential low folding strength [28]. In addition, splicing signals are also present in the exons, which are either called exonic splicing enhancer (ESE) to facilitate the splicing, or exonic splicing suppressor (ESS) to suppress splicing. These are located close to the splicing donor or acceptor sites and serve as binding sites of Ser/Arg-rich proteins (SR proteins) through their RNA-binding domain that help multiple steps of the splicing pathway, including the recruitment of spliceosome to the exon–intron junctions. The importance of these sites was previously widely validated in the mutational analysis experiments [29–32], as well as a computational method [33,34]. However, the impacts of non-canonical splice site variants on gene splicing remain investigational and rely on bioinformatic prediction tools to identify any potential candidates. Additionally, according to the Standards and Guidelines for the Interpretation of Sequence Variants issued by the American College of Medical Genetics and Genomics and the Association for Molecular Pathology, computational evidence predicting a deleterious effect is not sufficient to promote the identified variant from a variant of unknown significance (VUS) to the LP category without functional data [35].

To our knowledge, the presence of the c.3028+3A>T variant was not reported previously, as searched in the COSMIC, cBioPortal or in the population database, gnomAD. Rather, another variant was reported at the same nucleotide position (c.3028+3A>G) in a patient with pulmonary sarcomatoid carcinoma, which led to MET exon 14 skipping [36]. Consequently, when the MET c.3028+3A>T variant was identified in our clinical laboratory, it was classified as VUS because of the lack of direct evidence to meet the LP classification criteria, while reported with a caveat alongside the VUS classification, suggesting the likelihood of the c.3028+3A>T variant's contribution to MET exon 14 skipping. However, for the best practice of a molecular diagnostic laboratory, it is important to issue a report with a definitive test result as well as interpretation to avoid miscommunication between the testing laboratories and the caring clinicians. It is of interest to mention Patient 3, who carries the c.3012_3028del variant and is predicted to be a gain-of-function mutation. This patient has no other disease-causing variants in the hotspot regions of BRAF, EGFR, ERBB2, KRAS, and has no genomic rearrangements in the ALK and ROS1 genes. More importantly, the patient's condition was dramatically improved in just 8 weeks of treatment with crizotinib, even though this particular variant has not been reported in the somatic cancer databases. Our study now shows that it has a drastic impact on exon 14 splicing. Furthermore, a substitution mutation at c.3028G>A has shown the disruption of the splice donor site causing METex14 skipping [8]. The same study also demonstrated that the genomic deletion involving MET c.3028 and the canonical intron 14 splicing donor site (e.g., c.3010_3028+8del, c.3018_3028+8del, c.3020_3028+24del) accounts for 61% of MET exon 14 skipping mutations [8]. Altogether, our data and others indicate the importance of nucleotide position c.3028 and the surrounding sequences in regulating the MET exon 14 splicing event.

We have observed in the RT-PCR results that WT transcript levels are not proportional to the allelic fractions observed in the DNA-NGS analysis in Patients 1 and 2. WT transcripts were either near the detection limit (Patient 1) or at reduced levels (Patient 2), even when the variant allelic fractions were well below 50%. These observations are likely attributed to either (1) an inaccurate estimation of tumor cells in each specimen; (2) the non-quantification nature of the end-point PCR test; (3) the uneven distribution of the tumor and infiltrating stromal cells in the process of making various types of specimens for different downstream analyses. For the NGS analysis, paraffin blocks were used, whereas cell-smeared slides (Diff-Quick and pap-stained) were used for the RT-PCR assay. Even two smear slides made from the same specimen source will not have equal proportions of normal and malignant cells due to uncontrolled cell separation during the smearing preparation. A similar unproportioned transcript pattern was also seen in the previous reports of METex14 analysis in various RNA-based assays [8,21].

The ability of a diagnostic laboratory to determine the impact of a variant on gene splicing is essential. Our study warrants the importance of detecting actionable mutations with METex14 for NSCLC patients, including targets in MET exon 14 and surrounding introns. Additionally, METex14 detects better in an RNA-based NGS assay at a 4.2% rate compared to a 1.3% rate in a DNA-based NGS assay, which prompts clinicians to use a supplemental RNA-based panel [37,38]. However, the majority of molecular diagnostics laboratories use DNA-based NGS tests as a routine method to identify METex14 variants, which may not provide the proper functional evidence of exon 14 splicing. Here, we present a workflow (Figure 4) that facilitates the variant triage process to determine those potential candidates that require RT-PCR confirmation of the splicing products. The combination of in silico prediction, RT-PCR, and Sanger sequencing can be readily adopted to a laboratory standard operating procedure as a routine practice. It is worth noting that although fresh and frozen tissues are often the sample of choice for RNA-based techniques, there is a growing demand for the use of cytology samples that are already processed and stained for downstream molecular testing. The possibility of utilizing cytological slides in RNA-based diagnostic methods was previously validated using smeared and Giemsa or Diff-Quik stained slides [39]. In our study, cytology slides for the corresponding three patients were retrieved, and tissues from these slides were used for RNA extraction and the RT-PCR. The successful outcome of our procedure further affirms the possibility of incorporating cytology slides when other tissue sources are scarce in patients for the benefit of targeted lung cancer therapy.

Author Contributions: Conceptualization, Supervision & Formal analysis: Y.-W.C., Investigation: M.A.J., J.S. and R.D., Visualization and Writing: R.D., Review & Editing: Y.-W.C., R.D., M.A.J. and J.S. All authors have read and agreed to the published version of the manuscript.

Funding: This study was in part supported by the research fund of the Laboratory Medicine Department of the Robert J. Tomsich Pathology and Laboratory Medicine Institute, Cleveland Clinic.

Institutional Review Board Statement: The study was conducted according to the approved protocols of Cleveland Clinic's Institutional Review Board (IRB; 17–177 and 19–329).

Informed Consent Statement: Patient consent was waived due to the retrospective nature of the study.

Data Availability Statement: Data supporting the reported results can be obtained from the corresponding author.

Acknowledgments: The Sanger sequencing analysis was carried out by the Genomics Core Facility of the CWRU School of Medicine's Genetics and Genome Sciences Department.

Conflicts of Interest: The author declare no conflict of interest.

References

1. Siegel, R.L.; Miller, K.D.; Jemal, A. Cancer statistics, 2020. *CA Cancer J. Clin.* **2020**, *70*, 7–30. [CrossRef] [PubMed]
2. Hirsch, F.R.; Zaric, B.; Rabea, A.; Thongprasert, S.; Lertprasertsuke, N.; Dalurzo, M.L.; Varella-Garcia, M. Biomarker Testing for Personalized Therapy in Lung Cancer in Low- and Middle-Income Countries. *Am. Soc. Clin. Oncol. Educ. Book* **2017**, *37*, 403–408. [CrossRef] [PubMed]
3. Frampton, G.M.; Ali, S.M.; Rosenzweig, M.; Chmielecki, J.; Lu, X.; Bauer, T.M.; Akimov, M.; Bufill, J.A.; Lee, C.; Jentz, D.; et al. Activation of MET via diverse exon 14 splicing alterations occurs in multiple tumor types and confers clinical sensitivity to MET inhibitors. *Cancer Discov.* **2015**, *5*, 850–859. [CrossRef] [PubMed]
4. Yoda, S.; Dagogo-Jack, I.; Hata, A.N. Targeting oncogenic drivers in lung cancer: Recent progress, current challenges and future opportunities. *Pharmacol. Ther.* **2019**, *193*, 20–30. [CrossRef]
5. Vuong, H.G.; Ho, A.T.N.; Altibi, A.M.A.; Nakazawa, T.; Katoh, R.; Kondo, T. Clinicopathological implications of MET exon 14 mutations in non-small cell lung cancer—A systematic review and meta-analysis. *Lung Cancer* **2018**, *123*, 76–82. [CrossRef]

6. NIH. National Cancer Institute: Drugs Approved for Lung Cancer. 2021. Available online: https://www.cancer.gov/about-cancer/treatment/drugs/lung (accessed on 7 February 2022).

7. TABRECTA (Capmatinib) Tablets: How TABRECTA May Help. 2021. Available online: https://www.us.tabrecta.com/met-exon-14-skipping-mutation-nsclc/about-tabrecta/how-tabrecta-may-help/ (accessed on 7 February 2022).

8. Awad, M.M.; Oxnard, G.R.; Jackman, D.M.; Savukoski, D.O.; Hall, D.; Shivdasani, P.; Heng, J.C.; Dahlberg, S.E.; Jänne, P.A.; Verma, S.; et al. MET Exon 14 Mutations in Non-Small-Cell Lung Cancer Are Associated With Advanced Age and Stage-Dependent MET Genomic Amplification and c-Met Overexpression. *J. Clin. Oncol.* **2016**, *34*, 721–730. [CrossRef]

9. Lindeman, N.I.; Cagle, P.T.; Aisner, D.L.; Arcila, M.E.; Beasley, M.B.; Bernicker, E.H.; Colasacco, C.; Dacic, S.; Hirsch, F.R.; Kerr, K.; et al. Updated Molecular Testing Guideline for the Selection of Lung Cancer Patients for Treatment With Targeted Tyrosine Kinase Inhibitors: Guideline From the College of American Pathologists, the International Association for the Study of Lung Cancer, and the Association for Molecular Pathology. *J. Thorac. Oncol.* **2018**, *13*, 323–358. [PubMed]

10. Barlesi, F.; Mazieres, J.; Merlio, J.P.; Debieuvre, D.; Mosser, J.; Lena, H.; Ouafik, L.H.; Besse, B.; Rouquette, I.; Westeel, V.; et al. Routine molecular profiling of patients with advanced non-small-cell lung cancer: Results of a 1-year nationwide programme of the French Cooperative Thoracic Intergroup (IFCT). *Lancet* **2016**, *387*, 1415–1426. [CrossRef]

11. Mitsudomi, T.; Suda, K.; Yatabe, Y. Surgery for NSCLC in the era of personalized medicine. *Nat. Rev. Clin. Oncol.* **2013**, *10*, 235–244. [CrossRef]

12. Xu, Z.; Li, H.; Dong, Y.; Cheng, P.; Luo, F.; Fu, S.; Gao, M.; Kong, L.; Che, N. Incidence and PD-L1 Expression of MET 14 Skipping in Chinese Population: A Non-Selective NSCLC Cohort Study Using RNA-Based Sequencing. *OncoTargets Ther.* **2020**, *13*, 6245–6253. [CrossRef]

13. Comoglio, P.M.; Trusolino, L.; Boccaccio, C. Known and novel roles of the MET oncogene in cancer: A coherent approach to targeted therapy. *Nat. Rev. Cancer* **2018**, *18*, 341–358. [CrossRef] [PubMed]

14. Schmidt, L.; Duh, F.M.; Chen, F.; Kishida, T.; Glenn, G.; Choyke, P.; Scherer, S.W.; Zhuang, Z.; Lubensky, I.; Dean, M.; et al. Germline and somatic mutations in the tyrosine kinase domain of the MET proto-oncogene in papillary renal carcinomas. *Nat. Genet.* **1997**, *16*, 68–73. [CrossRef] [PubMed]

15. Di Renzo, M.F.; Olivero, M.; Martone, T.; Maffe, A.; Maggiora, P.; Stefani, A.D.; Valente, G.; Giordano, S.; Cortesina, G.; Comoglio, P.M. Somatic mutations of the MET oncogene are selected during metastatic spread of human HNSC carcinomas. *Oncogene* **2000**, *19*, 1547–1555. [CrossRef] [PubMed]

16. Lu, X.; Peled, N.; Greer, J.; Wu, W.; Choi, P.; Berger, A.H.; Wong, S.; Jen, K.-Y.; Seo, Y.; Hann, B.; et al. MET Exon 14 Mutation Encodes an Actionable Therapeutic Target in Lung Adenocarcinoma. *Cancer Res.* **2017**, *77*, 4498–4505. [CrossRef]

17. Awad, M.M. Impaired c-Met Receptor Degradation Mediated by MET Exon 14 Mutations in Non-Small-Cell Lung Cancer. *J. Clin. Oncol. Off. J. Am. Soc. Clin. Oncol.* **2016**, *34*, 879–881. [CrossRef]

18. Peschard, P.; Fournier, T.M.; Lamorte, L.; Naujokas, M.A.; Band, H.; Langdon, W.Y.; Park, M. Mutation of the c-Cbl TKB domain binding site on the Met receptor tyrosine kinase converts it into a transforming protein. *Mol. Cell* **2001**, *8*, 995–1004. [CrossRef]

19. Kong-Beltran, M.; Seshagiri, S.; Zha, J.; Zhu, W.; Bhawe, K.; Mendoza, N.; Holcomb, T.; Pujara, K.; Stinson, J.; Fu, L.; et al. Somatic mutations lead to an oncogenic deletion of met in lung cancer. *Cancer Res.* **2006**, *66*, 283–289. [CrossRef]

20. Socinski, M.A.; Pennell, N.A.; Davies, K.D. MET Exon 14 Skipping Mutations in Non-Small-Cell Lung Cancer: An Overview of Biology, Clinical Outcomes, and Testing Considerations. *JCO Precis. Oncol.* **2021**, *5*, 653–663. [CrossRef]

21. Shi, M.; Ma, J.; Feng, M.; Liang, L.; Chen, H.; Wang, T.; Xie, Z. Novel MET exon 14 skipping analogs characterized in non-small cell lung cancer patients: A case study. *Cancer Genet.* **2021**, *256–257*, 62–67. [CrossRef]

22. Reynolds, J.P.; Zhou, Y.; Jakubowski, M.A.; Wang, Z.; Brainard, J.A.; Klein, R.D.; Farver, C.F.; Almeida, F.A.; Cheng, Y.-W. Next-generation sequencing of liquid-based cytology non-small cell lung cancer samples. *Cancer Cytopathol.* **2017**, *125*, 178–187. [CrossRef]

23. Cheng, Y.W.; Stefaniuk, C.; Jakubowski, M.A. Real-time PCR and targeted next-generation sequencing in the detection of low level EGFR mutations: Instructive case analyses. *Respir. Med. Case Rep.* **2019**, *28*, 100901. [CrossRef] [PubMed]

24. Zhou, W.; Geiersbach, K.; Chadwick, B. Rapid removal of cytology slide coverslips for DNA and RNA isolation. *J. Am. Soc. Cytopathol.* **2017**, *6*, 24–27. [CrossRef] [PubMed]

25. Jenkins, R.W.; Oxnard, G.R.; Elkin, S.; Sullivan, E.K.; Carter, J.L.; Barbie, D.A. Response to Crizotinib in a Patient With Lung Adenocarcinoma Harboring a MET Splice Site Mutation. *Clin. Lung Cancer* **2015**, *16*, e101–e104. [CrossRef] [PubMed]

26. Mendenhall, M.A.; Goldman, J.W. MET-Mutated NSCLC with Major Response to Crizotinib. *J. Thorac. Oncol. Off. Publ. Int. Assoc. Study Lung Cancer* **2015**, *10*, e33–e34. [CrossRef] [PubMed]

27. Paik, P.K.; Drilon, A.; Fan, P.D.; Yu, H.; Rekhtman, N.; Ginsberg, M.S.; Borsu, L.; Schultz, N.; Berger, M.F.; Rudin, C.M.; et al. Response to MET inhibitors in patients with stage IV lung adenocarcinomas harboring MET mutations causing exon 14 skipping. *Cancer Discov.* **2015**, *5*, 842–849. [CrossRef] [PubMed]

28. Zafrir, Z.; Tuller, T. Nucleotide sequence composition adjacent to intronic splice sites improves splicing efficiency via its effect on pre-mRNA local folding in fungi. *RNA* **2015**, *21*, 1704–1718. [CrossRef]

29. Fairbrother, W.G.; Yeo, G.W.; Yeh, R.; Goldstein, P.; Mawson, M.; Sharp, P.A.; Burge, C.B. RESCUE-ESE identifies candidate exonic splicing enhancers in vertebrate exons. *Nucleic Acids Res.* **2004**, *32*, W187–W190. [CrossRef]

30. Liu, H.X.; Zhang, M.; Krainer, A.R. Identification of functional exonic splicing enhancer motifs recognized by individual SR proteins. *Genes Dev.* **1998**, *12*, 1998–2012. [CrossRef]

31. Schaal, T.D.; Maniatis, T. Selection and characterization of pre-mRNA splicing enhancers: Identification of novel SR protein-specific enhancer sequences. *Mol. Cell. Biol.* **1999**, *19*, 1705–1719. [CrossRef]
32. Coulter, L.R.; Landree, M.A.; Cooper, T.A. Identification of a new class of exonic splicing enhancers by in vivo selection. *Mol. Cell. Biol.* **1997**, *17*, 2143–2150. [CrossRef]
33. Zhang, X.H.; Kangsamaksin, T.; Chao, M.S.; Banerjee, J.K.; Chasin, L.A. Exon inclusion is dependent on predictable exonic splicing enhancers. *Mol. Cell. Biol.* **2005**, *25*, 7323–7332. [CrossRef] [PubMed]
34. Zhang, X.H.; Leslie, C.S.; Chasin, L.A. Computational searches for splicing signals. *Methods* **2005**, *37*, 292–305. [CrossRef] [PubMed]
35. Richards, S.; Aziz, N.; Bale, S.; Bick, D.; Das, S.; Gastier-Foster, J.; Grody, W.W.; Hegde, M.; Lyon, E.; Spector, E.; et al. Standards and guidelines for the interpretation of sequence variants: A joint consensus recommendation of the American College of Medical Genetics and Genomics and the Association for Molecular Pathology. *Genet. Med.* **2015**, *17*, 405–424. [CrossRef] [PubMed]
36. Liu, X.; Jia, Y.; Stoopler, M.B.; Shen, Y.; Cheng, H.; Chen, J.; Mansukhani, M.; Koul, S.; Halmos, B.; Borczuk, A.C. Next-Generation Sequencing of Pulmonary Sarcomatoid Carcinoma Reveals High Frequency of Actionable MET Gene Mutations. *J. Clin. Oncol. Off. J. Am. Soc. Clin. Oncol.* **2016**, *34*, 794–802. [CrossRef] [PubMed]
37. Davies, K.D.; Lomboy, A.; Lawrence, C.A.; Yourshaw, M.; Bocsi, G.T.; Camidge, D.R.; Aisner, D.L. DNA-Based versus RNA-Based Detection of MET Exon 14 Skipping Events in Lung Cancer. *J. Thorac. Oncol.* **2019**, *14*, 737–741. [CrossRef] [PubMed]
38. Jurkiewicz, M.; Saqi, A.; Mansukhani, M.M.; Hodel, V.; Krull, A.; Shu, C.A.; Fernandes, H.D. Efficacy of DNA versus RNA NGS-based Methods in MET Exon 14 skipping mutation detection. *J. Clin. Oncol.* **2020**, *38*, 9036. [CrossRef]
39. Gentien, D.; Piqueret-Stephan, L.; Henry, E.; Albaud, B.; Rapinat, A.; Koscielny, S.; Scoazec, J.-Y.; Vielh, P. Digital Multiplexed Gene Expression Analysis of mRNA and miRNA from Routinely Processed and Stained Cytological Smears: A Proof-of-Principle Study. *Acta Cytol.* **2021**, *65*, 88–98. [CrossRef]

Review

Targeting *MET* Amplification: Opportunities and Obstacles in Therapeutic Approaches

Yuichi Kumaki [1,*], **Goshi Oda** [1] **and Sadakatsu Ikeda** [2,3,*]

[1] Department of Specialized Surgery, Tokyo Medical and Dental University, Tokyo 113-8519, Japan; odasrg2@tmd.ac.jp

[2] Center for Innovative Cancer Treatment, Tokyo Medical and Dental University, Tokyo 113-8519, Japan

[3] Moores Cancer Center, University of California San Diego, La Jolla, CA 92037, USA

[*] Correspondence: kumaki.srg2@tmd.ac.jp (Y.K.); ikeda.canc@tmd.ac.jp (S.I.)

Simple Summary: The *MET* gene is crucial for cell growth and has shown promise as a cancer treatment target. However, distinguishing between focal amplification and polysomy, different types of gene multiplication, is challenging. Accurate differentiation requires techniques such as in situ hybridization (ISH) or next generation sequencing (NGS). As the effectiveness of MET inhibitors can vary, careful patient selection and defining the perfect amplification threshold are critical. Future studies should focus on determining optimal therapy combinations and innovating new treatments targeting *MET* amplification.

Abstract: The *MET* gene plays a vital role in cellular proliferation, earning it recognition as a principal oncogene. Therapies that target *MET* amplification have demonstrated promising results both in preclinical models and in specific clinical cases. A significant obstacle to these therapies is the ability to distinguish between focal amplification and polysomy, a task for which simple *MET* copy number measurement proves insufficient. To effectively differentiate between the two, it is crucial to utilize comparative measures, including in situ hybridization (ISH) with the centromere or next generation sequencing (NGS) with adjacent genes. Despite the promising potential of *MET* amplification treatment, the judicious selection of patients is paramount to maximize therapeutic efficacy. The effectiveness of MET inhibitors can fluctuate depending on the extent of *MET* amplification. Future research must seek to establish the ideal threshold value for *MET* amplification, identify the most efficacious combination therapies, and innovate new targeted treatments for patients exhibiting *MET* amplification.

Keywords: *MET* amplification; polysomy; aneuploidy; acquired resistance; MET inhibitor

Citation: Kumaki, Y.; Oda, G.; Ikeda, S. Targeting *MET* Amplification: Opportunities and Obstacles in Therapeutic Approaches. *Cancers* **2023**, *15*, 4552. https://doi.org/10.3390/cancers15184552

Academic Editors: Aamir Ahmad, Jan Trøst Jørgensen and Jens Mollerup

Received: 21 June 2023
Revised: 1 September 2023
Accepted: 11 September 2023
Published: 14 September 2023

1. Introduction

The *MET* gene, also referred to as c-MET, encodes a receptor tyrosine kinase (RTK) known as MET. This gene has garnered significant attention in the oncology field due to its pivotal role in tumorigenesis and metastasis [1]. The aberrant activation of MET signaling has been implicated in the development and progression of several malignancies, including non-small cell lung cancer (NSCLC) [2], gastric cancer [3,4], colorectal cancer [5], papillary renal cell carcinoma (PRCC) [6], hepatocellular carcinoma [7], and breast cancer [8,9].

Multiple genetic alterations within *MET* have been identified as oncogenic drivers in cancer. One such alteration is exon 14 skipping, which results in an aberrant MET protein lacking a critical regulatory domain [10]. The event leads to constitutive activation of the MET signaling pathway and contributes to tumorigenesis, particularly in NSCLC [11]. Notably, studies have revealed that approximately 3% of NSCLC patients exhibit *MET* exon 14 skipping, highlighting its clinical relevance [10].

In addition to exon 14 skipping, gene amplification and single nucleotide variants have been implicated in MET-driven oncogenesis. *MET* amplification is characterized by an increased copy number of the *MET* gene, leading to elevated MET protein levels and hyperactivation of downstream signaling cascades, including the RAS-ERK/MAPK, PI3K-AKT-mTOR, or PLCgamma-PKC pathways [12–15]. This phenomenon has been observed in various cancer types and holds promise as a therapeutic target [16,17].

The identification of specific *MET* alterations has paved the way for targeted therapies in cancer treatment. For instance, in the case of NSCLC patients with *MET* exon 14 skipping, the development of MET tyrosine kinase inhibitors (TKIs), such as capmatinib and tepotinib, has shown remarkable efficacy in clinical trials [18–20]. These TKIs selectively inhibit the activated MET kinase, thereby suppressing aberrant MET signaling and impeding tumor growth.

Furthermore, *MET* amplification has emerged as an intriguing therapeutic target. Targeted therapies designed to counteract *MET* amplification have shown promise in preclinical studies and a subset of clinical studies [18,21]. However, challenges remain, particularly regarding the development of effective inhibitors that can overcome resistance mechanisms associated with *MET* amplification [22–26].

In light of the significance of *MET* alterations in cancer, this review aims to comprehensively summarize the measurement, frequency, and therapeutic implications of *MET* amplification. We seek to elucidate the underlying mechanisms driving *MET* amplification and explore its potential as a therapeutic target. Additionally, we will discuss the challenges and prospects associated with targeting *MET* amplification, including the emergence of resistance mechanisms.

2. Gene Amplification and Protein Overexpression

Gene amplification is a prevalent genetic alteration observed in cancer [27–29]. It usually causes protein overexpression by enhancing levels of the products encoded by the amplified gene [27]. Although gene amplification and protein overexpression are distinct phenomena, they are often associated with each other [27,28]. In general, protein overexpression of RTKs can have oncogenic effects by increasing local receptor concentration, leading to auto-dimerization of receptors, and subsequent hyperactivation of downstream signaling pathways [27]. For instance, amplification of the *HER2* (*ERBB2*) gene on chromosome 17 and the overexpression of HER2 protein play a crucial role in the pathogenesis of HER2-positive breast cancer [30]. Fluorescent in situ hybridization (FISH) is a method that detects *ERRB2* gene amplification, while HER2 immunochemistry (IHC) identifies overexpression of HER2 protein. These two assays are commonly used and exhibit a strong correlation, serving as biomarkers for anti-HER2 therapy [31,32]. In the clinical setting of breast cancer, HER2 status determination often involves measuring HER2 protein expression using IHC initially, followed by FISH when the IHC result is equivocal (2+). Targeted therapies specially designed for HER2-positive breast cancer have demonstrated remarkable success, making them the most effective treatment in the realm of personalized medicine [33].

However, the situation regarding MET differs significantly from HER2. Although the method of measurement and threshold for *MET* amplification is controversial, *MET* amplification has been reported in a number of cancer types. It occurs 1% to 6% in NSCLC patients [34–37], 1% to 10% in gastric cancers [3,4,38,39], 1% to 4% in colorectal cancers [40,41], 3% to 13% in PRCCs [42], and 8% in breast cancers [43].

It has been observed that MET protein expression assessed by IHC does not always align with *MET* amplification [44–46]. Therefore, it is crucial to recognize that the patient populations identified by *MET* amplification techniques such as FISH and next generation sequencing (NGS) may differ substantially from those identified using IHC for treatment selection. The underlying reasons for the discordance between MET protein expression and gene amplification are not yet fully understood, but intra-tumor heterogeneity has been suggested as a contributing factor [46,47]. Previous studies investigating targeted therapies for MET based on IHC measurements have yielded inconsistent results [48–50].

Consequently, IHC may not be the appropriate approach to identifying the population that would benefit from MET-targeted therapy.

3. Focal Gene Amplification and Polysomy

Gene copy number gain in cancer can manifest as either focal gene amplification or polysomy (or aneuploidy of chromosome) [51]. Focal amplification refers to the specific gain of gene copies in the specific gene, while polysomy involves an overall increase in chromosome copy number [51]. In the case of HER2, FISH is employed to measure gene copy number (GCN) and the ratio of *ERRB2* to centromere enumerator probe (CEP)17. However, it is crucial to consider the presence of polysomy when interpreting FISH results [52]. Polysomy has been identified as the primary cause of equivocal HER2 FISH results, and its association with the efficacy or prognosis of trastuzumab treatment remains unclear [53,54]. Consequently, the selection of appropriate candidates for anti-HER2 therapy requires careful consideration [55].

Similarly, in the context of MET, focal amplification of the *MET* gene entails a specific gain of gene copies, while polysomy involves an increase in chromosome 7 copy number, often accompanied by co-amplification of adjacent genes such as *CDK6* and *BRAF* [56] (Figure 1). It has been observed that polysomy does not exhibit a favorable response to treatment with MET inhibitors alone [57]. Currently, there is a lack of standardized methods, including the choice of diagnostic tools and the establishment of thresholds, to effectively define focal *MET* amplification as a therapeutic target [56,58].

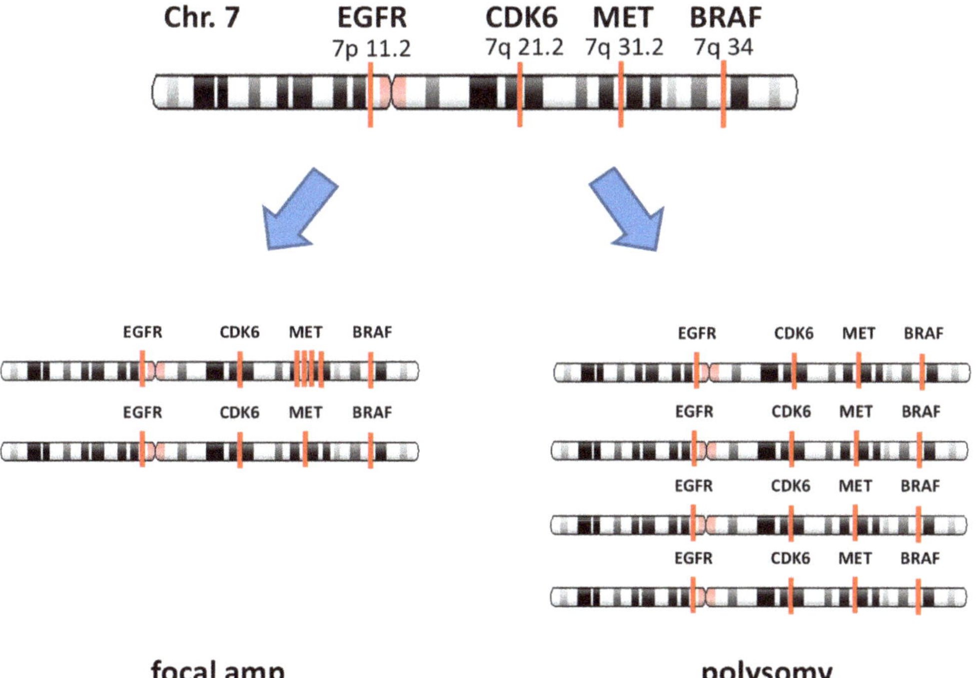

Figure 1. Schematic representation of focal gene amplification and polysomy involving the *MET* gene and chromosome 7 copy number alterations.

4. *MET* Amplification Detection by FISH

Various methods, such as FISH and NGS, have been employed to detect *MET* amplification [33]. Among these methods, FISH has been regarded as the gold standard for measuring *MET* amplification [59–61]. In previous studies, *MET* amplification was defined based on the *MET*/CEP7 ratio, using FISH analysis, and it was found to occur in approximately 5% of patients with NSCLC or gastric adenocarcinoma, with a cut-off value of *MET*/CEP > 2.2 [16]. The correlation between FISH and IHC is under investigation.

However, a phase I study reported that patients with *MET* amplification, defined by a cut-off of *MET*/CEP $\geq$ 2.0, did not exhibit a favorable response to therapy with a MET inhibitor [62]. This raises concerns regarding the accurate selection of the target population for *MET* amplification-directed treatment. Further investigations utilizing FISH measurements have revealed that only cases with high *MET*/CEP ratios, such as *MET*/CEP > 5, exhibit a favorable response to MET inhibitors [63–65]. Guo et al. conducted a comprehensive analysis of the relationship between the definition of *MET* amplification by FISH and the response rate to targeted therapy in previous clinical trials. They consistently observed the most favorable outcomes in NSCLCs characterized by high-level *MET* amplification [57].

It is worth noting that the efficacy of MET inhibitors alone may be diminished when *MET* amplification coexists with other driver mutations. In such cases, combination therapies may be necessary to achieve optimal treatment outcomes.

5. *MET* Amplification Detection by NGS

MET amplification detection using NGS has gained popularity in clinical settings, offering a comprehensive analysis of genomic alterations. However, its efficacy as a tool for identifying optimal biomarkers to guide MET therapy remains a topic of debate. Several studies have indicated that NGS may not be as effective as FISH in predicting response to MET inhibitors [66,67]. For instance, a comparative analysis between NGS-based gene copy number (GCN) assessment and FISH revealed that NGS alone is not sufficient for predicting MET inhibitor response [68]. One potential limitation of NGS is its inability to accurately distinguish between polysomy and focal amplification, which may impact the identification of suitable therapeutic targets. This might stem from inaccurate interpretation of copy number gain result. Without recognizing the potential polysomy, the inaccurate interpretation of copy number gain might occur. Strategies to address this challenge are still under investigation, and a recent study proposed a method to differentiate polysomy from focal *MET* amplification by co-examining the amplification of genes on chromosome 7, such as *BRAF* and *CDK6* [56].

Our group analyzed the data from 1025 patients with advanced solid tumors using a non-invasive cfDNA NGS panel known as Guardant360 [56]. An algorithm to define focal amplification was developed. Focal amplification was defined as *MET* amplification without polysomy or an increase in the chromosome copy number itself. The Guardant360 test examined four genes located on chromosome 7: *EGFR* in 7p11.2, *CDK6* in 7q21.2, *MET* in 7q31.2, and *BRAF* in 7q34. The *MET* gene was classified as focally amplified if there were no co-amplification of adjacent genes, such as *CDK6* or *BRAF*. Conversely, *MET* non-focal amplification was defined as a *MET* copy number increase associated with polysomy, in which *MET* copy number increased together with either *CDK6* and/or *BRAF*.

Several examples were given to illustrate this: a sample with only *MET* amplification would be classified as focal. If a sample had co-amplification of *MET* and *EGFR* without *CDK6* amplification, it would be categorized as focal amplification, given that polysomy could not occur without increasing the copy numbers of all three genes together. The algorithm to describe focal amplification was defined as follows:

(a) *MET* copy number $\geq$ 2.2.
(b) *MET* is amplified without co-amplification of *CDK6* and *BRAF*. Co-amplification status was defined as "increased together" when the copy number of the other gene (*CDK6* or *BRAF*) $\geq$ 2.2, and the difference with *MET* amplification is within $+/-0.5$.
(c) *MET* amplification that satisfies both (a) and (b) is defined as focal.

Results of analyzing patient cohort consisting with the testing cohort (291 patients), and validation cohort (734 patients) showed *MET* alterations related to abnormal signaling in approximately 10.7% (110 patients) of the entire patient population across nine different cancer types, most notably non-small cell and small cell lung cancers, gastroesophageal cancer, and prostate adenocarcinoma. *MET* alterations were found in 37 out of 291 patients. Among these, 24 had amplifications, 5 had exon 14 skipping, and 13 had single nucleotide variants (SNVs). Co-alterations, such as amplification, SNVs, were found in four samples.

Of the 24 *MET* amplifications, about 30% (7/24) were classified as focal amplification. The *MET* copy number was significantly higher with focal amplification compared to non-focal amplification (polysomy). In a validation cohort, focal *MET* amplification was detected in 4.2% of patients. Overall, the rate of focal amplification was 3.7% (=38/1025) across all patients.

This study showed that this approach can distinguish focal from non-focal *MET* amplification using comprehensive genomic profiling with NGS in patients with advanced cancer. The study also suggests that only 30% of all *MET* amplifications detected by the NGS are focal, and these are associated with a higher plasma *MET* copy number.

Nevertheless, it is pertinent to acknowledge a limitation inherent in this study, namely the selective focus on a restricted gene set localized within chromosome 7, encompassing *MET, EGFR, BRAF,* and *CDK6*. Inclusion of a broader array of genes or single nucleotide polymorphisms (SNPs) has the potential to substantially enhance the precision and fidelity of detecting focal amplification events. Furthermore, the absence of a well-defined correlation between *EGFR* and *MET* necessitates a more comprehensive exposition.

Although clinical investigations are warranted, NGS holds promise as a valuable tool for *MET* amplification analysis due to its widespread use in clinical practice and the potential to simultaneously examine other driver genes, such as *EGFR*. The integration of NGS with established methods such as FISH has been proposed in HER2-positive cases to capture those missed by the current IHC and FISH-based approaches [69]. By leveraging the strengths of both techniques, NGS-based detection may enhance the sensitivity and accuracy of *MET* amplification identification, leading to improved patient stratification and personalized treatment decisions.

To fully exploit the potential of NGS in *MET* amplification assessment, further research is needed to establish standardized guidelines for interpreting NGS results, particularly in distinguishing between polysomy and focal amplification. Additionally, the development of bioinformatics algorithms and computational tools specific to *MET* amplification analysis will aid in the accurate classification of NGS data. Efforts to validate the clinical utility of NGS-based detection of focal amplification and its correlation with treatment response are crucial steps toward integrating NGS into routine clinical practice for precision cancer care.

6. *MET* Amplification as an Acquired Resistance Mechanism

MET amplification has emerged as a significant mechanism of acquired resistance in various targeted therapies, particularly in *EGFR*-mutant NSCLCs. *MET* amplification is known to stimulate signaling pathways such as RAS-ERK/MAPK, STAT, and PI3K/AKT downstream of EGFR, leading to resistance against EGFR TKIs [70,71]. This resistance mechanism has been observed across all generations of EGFR TKIs [72–74].

Clinical studies have shed light on the prevalence of *MET* amplification as a resistance mechanism in NSCLC. Coleman et al. conducted a comprehensive analysis of key clinical trials in NSCLC and reported that *MET* amplification was identified as a mechanism of resistance in 7–15% of patients who experienced treatment failure with first-line osimertinib and in 10–22% following second-line osimertinib [26]. These findings highlight the clinical relevance of *MET* amplification in acquired resistance scenarios.

Moreover, it has been recognized that *MET* amplification may contribute to treatment resistance not only in *EGFR*-positive NSCLCs but also in other oncogenic gene-positive NSCLCs, such as those with *ALK* fusion [75], *RET*-fusion [76,77], *ROS1*-fusion [78], *NTRK*-fusion [79], or *KRAS* [80]. Collectively, it is estimated that approximately 15% of NSCLC tumors harboring *EGFR, KRAS, ALK* fusion, and *RET* fusion alterations exhibit *MET* amplification [26]. This underscores the importance of considering *MET* amplification as a potential resistance mechanism in the management of various oncogenic driver alterations.

Understanding the prevalence and clinical implications of *MET* amplification in acquired resistance scenarios is crucial for optimizing treatment strategies. Efforts are underway to develop effective therapeutic approaches targeting *MET* amplification to overcome acquired resistance. These include the investigation of combination therapies involving MET inhibitors with other targeted agents or immunotherapies to enhance treatment efficacy and prevent the emergence of resistance. Additionally, ongoing research aims to elucidate the underlying mechanisms and molecular interactions associated with *MET* amplification-mediated resistance, which will inform the development of novel therapeutic strategies for patients with *MET*-amplified NSCLC.

7. Treatment Option Targeting for MET Amplification

7.1. Monotherapy

As previously discussed, initial studies have indicated limited effectiveness of MET inhibitors in patients with *MET* amplification ($MET/CEP \geq 2.0$) [62]. However, subsequent research has demonstrated that the efficacy of MET inhibitors improves with higher levels of amplification [18,81–88].

Table 1 summarizes the clinical trials investigating monotherapy targeting *MET* amplification. For instance, the PROFILE 1001 study examined the activity of crizotinib, an ALK/ROS1/MET tyrosine kinase inhibitor (TKI), in NSCLC patients categorized into high ($MET/CEP \geq 4$), medium ($4 > MET/CEP > 2.2$), or low ($2.2 \geq MET/CEP \geq 1.8$) amplification groups [84]. The high-amplification group exhibited the highest objective response rate (ORR) of 38.1% and median progression-free survival (mPFS) of 6.7 months.

Table 1. *MET* amplification targeting therapy trials (monotherapy).

Drug	Cancer Type	Study	*MET* Amplification Criteria	Clinical Outcome	
SAR125844	Solid tumors	Phase I (n = 72) Angevin et al., 2017 [81]	GCN > 4 and $MET/CEP \geq 2.0$ by FISH (or IHC 2+/3+)	ORR 17% (5/29)	
AMG337	Gastric cancers	Phase II (n = 60) Van Cutsem et al., 2019 [82]	$MET/CEP \geq 2.0$ by FISH	ORR 18% (8/45)	
	Solid tumors	Phase I (n = 111) Hong et al., 2019 [83]	$MET/CEP \geq 2.0$ by FISH	$4 > MET/CEP$	ORR 0% (0/2)
				$MET/CEP \geq 4$	ORR 60% (6/10)
crizotinib	NSCLC	Phase I (n = 38) [PROFILE 1001] Camidge et al., 2021 [84]	$MET/CEP \geq 1.8$ by FISH	$2.2 \geq MET/CEP \geq 1.8$	ORR 33% (1/3) mPFS 1.8 months
				$4.0 > MET/CEP > 2.2$	ORR 14% (2/14) mPFS 1.9 months
				$MET/CEP \geq 4.0$	ORR 38% (8/21) mPFS 6.7 months
		Phase II (n = 26) [METROS] Landi et al., 2019 [85]	$MET/CEP > 2.2$ by FISH	$5.0 > MET/CEP > 2.2$	ORR 36% (5/14)
				$MET/CEP \geq 5.0$	ORR 0% (0/2)
		Phase II (n = 25) [AcSe] Moro-Sibilot et al., 2019 [86]	GCN $\geq$ 6 by FISH	ORR 16% (4/25) mPFS 3.2 months	

Table 1. *Cont.*

Drug	Cancer Type	Study	*MET* Amplification Criteria	Clinical Outcome	
capmatinib	NSCLC	Phase I (n = 44) Schuler et al., 2020 [87]	GCN $\geq$ 5 or *MET*/CEP $\geq$ 2.0 by FISH (or IHC 2+/3+ or H-score $\geq$ 150)	4 > GCN	ORR 0% (0/17)
				6 > GCN $\geq$ 4	ORR 17% (2/12)
				GCN $\geq$ 6	ORR 47% (7/15) mPFS 9.3 months
		Phase II (n = 195) [GEOMETRY mono-1] Wolf et al., 2020 [18]	determined by FISH, NGS	4 > GCN	ORR 7% (2/30) mPFS 3.6 months
				GCN 4 or 5	ORR 9% (5/54) mPFS 2.7 months
				GCN 6 to 9	ORR 12% (5/42) mPFS 2.7 months
				GCN $\geq$ 10	ORR 29% (20/69) mPFS 4.1 months
	Hepatocellular carcinoma	Phase II (n = 30) Qin et al., 2019 [88]	GCN $\geq$ 5 or *MET*/CEP $\geq$ 2.0 by FISH (or IHC 2+/3+)	ORR 10% (3/30)	

ORR; objective response rate, mPFS; median progression free survival.

Another study evaluated the impact of capmatinib in NSCLC patients with *MET* amplification, revealing ORR rates of 29% (GCN $\geq$ 10), 12% (GCN 6 to 9), 9% (GCN 4 or 5), and 7% (GCN < 4), indicating an increased therapeutic effect with higher degrees of *MET* amplification (GEOMETRY mono-1 study, phase II) [18]. Several other MET-targeted TKIs have also been assessed in relation to the degree of *MET* amplification, consistently showing improved outcomes in the higher amplification groups (Table 1).

However, it is worth noting that the cutoff criteria for defining amplification varied across trials, and establishing standardized criteria for identifying optimal treatment targets remains a future challenge.

7.2. Combination Therapy

Combination therapy involving primary oncogene TKIs and MET TKIs has emerged as a rational strategy for addressing acquired resistance resulting from *MET* amplification. In recent years, significant advancements have been made in targeting both MET and EGFR, leading to the initiation of several clinical trials investigating the combination of MET inhibitors and EGFR inhibitors for acquired resistance in NSCLC. Table 2 provides an overview of combination therapy trials targeting *MET* amplification in *EGFR*-mutant NSCLC.

Table 2. *MET* amplification combination therapy trials for *EGFR*-mutant NSCLC.

Drug	Study	Patient/*MET* Amplification Criteria	Clinical Outcome	
capmatinib/gefitinib	Phase II (n = 100) Wu et al., 2018 [89]	acquired resistance to EGFR-TKI GCN $\geq$ 4 by FISH (or IHC 3+)	4 > GCN	ORR 12% (5/41) mPFS 3.9 months
			6 > GCN $\geq$ 4	ORR 22% (4/18) mPFS 5.4 months
			GCN $\geq$ 6	ORR 47% (17/36) mPFS 5.5 months

Table 2. *Cont.*

Drug	Study	Patient/*MET* Amplification Criteria	Clinical Outcome	
tepotinib/gefitinib	Phase II (n = 12) [INSIGHIT] Wu et al., 2020 [90]	acquired resistance to EGFR-TKI and T790 negative GCN ≥ 5 or *MET*/CEP ≥ 2.0 by FISH (or IHC 2+/3+)	ORR 67% (8/12) mPFS 16.6mo (vs. 4.2 months with chemotherapy, HR = 0.13, 90% CI = 0.04–0.43) OS 37.3mo (vs. 13.1 months with chemotherapy, HR = 0.08, 90% CI = 0.01–0.51)	
savolitinib/osimertinib	Phase Ib (n = 174) [TATTON] Sequist et al., 2020 [91]	acquired resistance to EGFR-TKI GCN ≥ 5 or *MET*/CEP ≥ 2.0 by FISH or GCN ≥ 5 by NGS	Cohort B1 *; after 3rd gen. EGFR-TKI and T790 negative	ORR 30% (21/69) mPFS 5.4 months
			Cohort B2 *; after 1st/2nd gen. EGFR-TKI and T790 negative	ORR 65% (33/51) mPFS 9.0 months
			Cohort B3 *; after 1st/2nd gen. EGFR-TKI and T790 positive	ORR 67% (12/18) mPFS 11.0 months
			Cohort D **; after 1st/2nd gen. EGFR-TKI and T790 negative	ORR 64% (23/36) mPFS 9.1 months
savolitinib/gefitinib	Phase I (n = 57) Yang et al., 2021 [92]	acquired resistance to EGFR-TKI and T790 negative GCN ≥ 5 or *MET*/CEP ≥ 2.0 by FISH	ORR 52% (12/23)	
Telisotuzumab vedotin/erlotinib	Phase Ib (n = 28) *** Camidge et al., 2023 [50]	acquired resistance to EGFR-TKI (H-score ≥ 150 by IHC)	ORR 32% (9/28) mPFS 5.9 months	

* savolitinib 600/300 mg plus osimertinib 80 mg. ** savolitinib 300 mg plus osimertinib 80 mg. *** including 4 *MET* amplified. ORR; objective response rate, mPFS; median progression free survival, OS; overall survival.

Capmatinib and tepotinib, both FDA-approved for NSCLC with *MET* exon14 skipping [18,19], have demonstrated positive outcomes when combined with gefitinib in pretreated NSCLC cases harboring both *EGFR* mutations and *MET* amplification [89,90]. In a phase II study, the combination of capmatinib and gefitinib in patients with acquired resistance to EGFR-TKI and *MET* amplification (GCN ≥ 4 by FISH) or MET overexpression (IHC 3+) showed improved activity, particularly in the high *MET* amplification group (GCN ≥ 6; ORR 47% and mPFS 5.5 months) [89]. Another phase II study, known as the INSIGHT study, investigated the combination of tepotinib and gefitinib in patients with acquired resistance to EGFR-TKI, *MET* amplification (GCN ≥ 5 or *MET*/CEP ≥ 2.0 by FISH), or MET overexpression (IHC 2+/3+). The ORR in this study was 67%, with a median PFS of 16.6 months compared to 4.2 months with chemotherapy (HR = 0.13, 90% CI = 0.04–0.43). Additionally, the overall survival was 37.3 months vs. 13.1 months with chemotherapy (HR = 0.08, 90% CI = 0.01–0.51) [90].

Savolitinib, a potent and selective MET TKI, in combination with osimertinib, was evaluated in the TATTON study for its efficacy in pretreated *EGFR* mutation-positive lung cancers with *MET* amplification [91]. In patients who experienced resistance after first- or second-generation EGFR-TKIs with *MET* amplification (GCN ≥ 5 or *MET*/CEP ≥ 2.0 by FISH or GCN ≥ 5 by NGS), the ORR ranged from 64% to 67%. For patients with resistance after third-generation EGFR-TKIs and *MET* amplification, the objective response rate was 30%. A phase I study examining the combination of savolitinib and gefitinib in patients with acquired resistance to EGFR-TKI and *MET* amplification (GCN ≥ 5 or *MET*/CEP ≥ 2.0 by FISH) reported an objective response rate of 52% [92]. It is important to note that the criteria for defining *MET* amplification varied across these studies. Furthermore, resistance-related amplification is often considered a subclonal population, which may warrant a lower threshold for defining the degree of amplification compared to de novo amplification [57,58]. Nevertheless, studies have consistently demonstrated that higher degrees of amplification correspond to improved treatment efficacy [57].

Ongoing investigations are exploring the use of antibody-based therapies targeting MET. Telisotuzumab vedotin, an antibody-drug conjugate targeting MET, exhibited promising antitumor activity and acceptable toxicity in NSCLC cases with acquired resistance to EGFR-TKI and MET overexpression (H-score $\geq$ 150 by IHC). In a study involving patients with MET overexpression (including four patients with *MET* amplification out of a total of 28 patients), ORR was 32%, and mPFS was 5.9 months [50].

7.3. Ongoing Study

Several ongoing clinical trials are currently investigating treatment strategies targeting *MET* amplification. These trials aim to further evaluate the efficacy and safety of different therapeutic approaches. One notable trial is a phase III study evaluating the combination of savolitinib and osimertinib for acquired resistance in NSCLC (ClinicalTrials.gov Identifier: NCT05015608). This trial seeks to assess the potential benefits of combining a MET inhibitor (savolitinib) with a third-generation EGFR inhibitor (osimertinib) in patients who have developed resistance to previous treatments.

In addition, a phase II trial is underway to investigate the efficacy of savolitinib in combination with osimertinib for acquired resistance in NSCLC (ClinicalTrials.gov Identifier: NCT03778229). This study aims to provide further insights into the potential synergistic effects of these targeted therapies in overcoming resistance mechanisms, specifically in patients with *MET* amplification.

Furthermore, a phase II trial is evaluating the combination of savolitinib and durvalumab, an immune checkpoint inhibitor, for advanced gastric cancer (ClinicalTrials.gov Identifier: NCT05620628). This trial explores the potential of combining MET inhibition with immune checkpoint blockade to enhance therapeutic outcomes in this patient population.

Lastly, a phase I trial is investigating amivantamab, a human bispecific antibody targeting both EGFR and MET, in advanced NSCLC (ClinicalTrials.gov Identifier: NCT02609776). This study aims to assess the safety and efficacy of this novel antibody therapy in patients with advanced NSCLC harboring *EGFR* mutations and *MET* amplification.

These ongoing trials represent important steps in advancing our understanding of *MET* amplification as a therapeutic target and may contribute to the development of more effective treatment strategies for patients with acquired resistance in various cancer types. The results from these studies will provide valuable insights into the clinical utility of targeting *MET* amplification and may guide future treatment decisions in precision oncology.

8. Conclusions

The treatment of *MET* amplification represents a promising avenue, particularly in combination with EGFR-TKI therapy for pretreated NSCLC patients. However, the appropriate selection of patients is crucial for maximizing treatment efficacy. The underlying mechanisms contribute to the varying effectiveness of MET inhibitors based on the degree of *MET* amplification. Nevertheless, there remains a lack of consensus regarding the specific threshold or cut-off value for *MET* amplification across different studies.

Further research is needed to determine the optimal cutoff value for *MET* amplification, to identify the best combination therapies, and to develop new targeted therapies for patients with *MET* amplification.

Author Contributions: Conceptualization, Y.K. and S.I.; Manuscript writing, Y.K.; Reviewing of the manuscript, G.O. and S.I.; Supervision, S.I.; Project administration, S.I. All authors have read and agreed to the published version of the manuscript.

Funding: This research received no external funding.

Institutional Review Board Statement: Not applicable.

Informed Consent Statement: Not applicable.

Data Availability Statement: The data presented in this study are available in this article.

Conflicts of Interest: The authors declare no conflict of interest.

References

1. De Bono, J.S.; Yap, T.A. c-MET: An exciting new target for anticancer therapy. *Ther. Adv. Med. Oncol.* **2011**, *3* (Suppl. S1), S3–S5. [CrossRef]
2. Van Der Steen, N.; Pauwels, P.; Gil-Bazo, I.; Castañon, E.; Raez, L.; Cappuzzo, F.; Rolfo, C. cMET in NSCLC: Can We Cut off the Head of the Hydra? From the Pathway to the Resistance. *Cancers* **2015**, *7*, 556–573. [CrossRef] [PubMed]
3. Peng, Z.; Zhu, Y.; Wang, Q.; Gao, J.; Li, Y.; Ge, S.; Shen, L. Prognostic significance of MET amplification and expression in gastric cancer: A systematic review with meta-analysis. *PLoS ONE* **2014**, *9*, e84502. [CrossRef] [PubMed]
4. Kawakami, H.; Okamoto, I. MET-targeted therapy for gastric cancer: The importance of a biomarker-based strategy. *Gastric Cancer* **2016**, *19*, 687–695. [CrossRef] [PubMed]
5. Parizadeh, S.M.; Jafarzadeh-Esfehani, R.; Fazilat-Panah, D.; Hassanian, S.M.; Shahidsales, S.; Khazaei, M.; Parizadeh, S.M.R.; Ghayour-Mobarhan, M.; Ferns, G.A.; Avan, A. The potential therapeutic and prognostic impacts of the c-MET/HGF signaling pathway in colorectal cancer. *IUBMB Life* **2019**, *71*, 802–811. [CrossRef] [PubMed]
6. Linehan, W.M.; Spellman, P.T.; Ricketts, C.J.; Creighton, C.J.; Fei, S.S.; Davis, C.; Wheeler, D.A.; Murray, B.A.; Schmidt, L.; Vocke, C.D.; et al. Comprehensive Molecular Characterization of Papillary Renal-Cell Carcinoma. *N. Engl. J. Med.* **2016**, *374*, 135–145.
7. Qi, X.S.; Guo, X.Z.; Han, G.H.; Li, H.Y.; Chen, J. MET inhibitors for treatment of advanced hepatocellular carcinoma: A review. *World J. Gastroenterol.* **2015**, *21*, 5445–5453. [CrossRef]
8. Gastaldi, S.; Comoglio, P.M.; Trusolino, L. The Met oncogene and basal-like breast cancer: Another culprit to watch out for? *Breast Cancer Res.* **2010**, *12*, 208. [CrossRef]
9. Ho-Yen, C.M.; Jones, J.L.; Kermorgant, S. The clinical and functional significance of c-Met in breast cancer: A review. *Breast Cancer Res.* **2015**, *17*, 52. [CrossRef]
10. Frampton, G.M.; Ali, S.M.; Rosenzweig, M.; Chmielecki, J.; Lu, X.; Bauer, T.M.; Akimov, M.; Bufill, J.A.; Lee, C.; Jentz, D.; et al. Activation of MET via diverse exon 14 splicing alterations occurs in multiple tumor types and confers clinical sensitivity to MET inhibitors. *Cancer Discov.* **2015**, *5*, 850–859. [CrossRef]
11. Schrock, A.B.; Frampton, G.M.; Suh, J.; Chalmers, Z.R.; Rosenzweig, M.; Erlich, R.L.; Halmos, B.; Goldman, J.; Forde, P.; Leuenberger, K.; et al. Characterization of 298 Patients with Lung Cancer Harboring MET Exon 14 Skipping Alterations. *J. Thorac. Oncol.* **2016**, *11*, 1493–1502. [CrossRef]
12. Organ, S.L.; Tsao, M.S. An overview of the c-MET signaling pathway. *Ther. Adv. Med. Oncol.* **2011**, *3* (Suppl. S1), S7–S19. [CrossRef] [PubMed]
13. Cecchi, F.; Rabe, D.C.; Bottaro, D. Targeting the HGF/Met signaling pathway in cancer therapy. *Expert Opin. Ther. Targets* **2012**, *16*, 553–572. [CrossRef] [PubMed]
14. Garajová, I.; Giovannetti, E.; Biasco, G.; Peters, G.J. c-Met as a Target for Personalized Therapy. *Transl. Oncogenomics* **2015**, *7* (Suppl. S1), 13–31. [PubMed]
15. De Silva, D.M.; Roy, A.; Kato, T.; Cecchi, F.; Lee, Y.H.; Matsumoto, K.; Bottaro, D.P. Targeting the hepatocyte growth factor/Met pathway in cancer. *Biochem. Soc. Trans.* **2017**, *45*, 855–870. [CrossRef]
16. Kawakami, H.; Okamoto, I.; Okamoto, W.; Tanizaki, J.; Nakagawa, K.; Nishio, K. Targeting MET Amplification as a New Oncogenic Driver. *Cancers* **2014**, *6*, 1540–1552. [CrossRef]
17. Liu, L.; Kalyani, F.S.; Yang, H.; Zhou, C.; Xiong, Y.; Zhu, S.; Yang, N.; Qu, J. Prognosis and Concurrent Genomic Alterations in Patients with Advanced NSCLC Harboring MET Amplification or MET Exon 14 Skipping Mutation Treated with MET Inhibitor: A Retrospective Study. *Front. Oncol.* **2021**, *11*, 649766. [CrossRef]
18. Wolf, J.; Seto, T.; Han, J.Y.; Reguart, N.; Garon, E.B.; Groen, H.J.M.; Tan, D.S.W.; Hida, T.; de Jonge, M.; Orlov, S.V.; et al. Capmatinib in MET Exon 14-Mutated or MET-Amplified Non-Small-Cell Lung Cancer. *N. Engl. J. Med.* **2020**, *383*, 944–957. [CrossRef]
19. Paik, P.K.; Felip, E.; Veillon, R.; Sakai, H.; Cortot, A.B.; Garassino, M.C.; Mazieres, J.; Viteri, S.; Senellart, H.; Van Meerbeeck, J.; et al. Tepotinib in Non-Small-Cell Lung Cancer with MET Exon 14 Skipping Mutations. *N. Engl. J. Med.* **2020**, *383*, 931–943. [CrossRef]
20. Mazieres, J.; Paik, P.K.; Garassino, M.C.; Le, X.; Sakai, H.; Veillon, R.; Smit, E.F.; Cortot, A.B.; Raskin, J.; Viteri, S.; et al. Tepotinib Treatment in Patients with MET Exon 14-Skipping Non-Small Cell Lung Cancer: Long-term Follow-up of the VISION Phase 2 Nonrandomized Clinical Trial. *JAMA Oncol.* **2023**. [CrossRef]
21. Kawakami, H.; Okamoto, I.; Arao, T.; Okamoto, W.; Matsumoto, K.; Taniguchi, H.; Kuwata, K.; Yamaguchi, H.; Nishio, K.; Nakagawa, K.; et al. MET amplification as a potential therapeutic target in gastric cancer. *Oncotarget* **2013**, *4*, 9–17. [CrossRef] [PubMed]
22. Botting, G.M.; Rastogi, I.; Chhabra, G.; Nlend, M.; Puri, N. Mechanism of Resistance and Novel Targets Mediating Resistance to EGFR and c-Met Tyrosine Kinase Inhibitors in Non-Small Cell Lung Cancer. *PLoS ONE* **2015**, *10*, e0136155. [CrossRef] [PubMed]
23. Bean, J.; Brennan, C.; Shih, J.Y.; Riely, G.; Viale, A.; Wang, L.; Chitale, D.; Motoi, N.; Szoke, J.; Broderick, S.; et al. MET amplification occurs with or without T790M mutations in EGFR mutant lung tumors with acquired resistance to gefitinib or erlotinib. *Proc. Natl. Acad. Sci. USA* **2007**, *104*, 20932–20937. [CrossRef] [PubMed]

24. Engelman, J.A.; Zejnullahu, K.; Mitsudomi, T.; Song, Y.; Hyland, C.; Park, J.O.; Lindeman, N.; Gale, C.M.; Zhao, X.; Christensen, J.; et al. MET amplification leads to gefitinib resistance in lung cancer by activating ERBB3 signaling. *Science* **2007**, *316*, 1039–1043. [CrossRef] [PubMed]

25. Wang, Q.; Yang, S.; Wang, K.; Sun, S.Y. MET inhibitors for targeted therapy of EGFR TKI-resistant lung cancer. *J. Hematol. Oncol.* **2019**, *12*, 63. [CrossRef]

26. Coleman, N.; Hong, L.; Zhang, J.; Heymach, J.; Hong, D.; Le, X. Beyond epidermal growth factor receptor: MET amplification as a general resistance driver to targeted therapy in oncogene-driven non-small-cell lung cancer. *ESMO Open* **2021**, *6*, 100319. [CrossRef]

27. Schwab, M. Oncogene amplification in solid tumors. *Semin. Cancer Biol.* **1999**, *9*, 319–325. [CrossRef]

28. Albertson, D.G. Gene amplification in cancer. *Trends Genet.* **2006**, *22*, 447–455. [CrossRef]

29. Matsui, A.; Ihara, T.; Suda, H.; Mikami, H.; Semba, K. Gene amplification: Mechanisms and involvement in cancer. *Biomol. Concepts* **2013**, *4*, 567–582. [CrossRef]

30. Krishnamurti, U.; Silverman, J.F. HER2 in breast cancer: A review and update. *Adv. Anat. Pathol.* **2014**, *21*, 100–107. [CrossRef]

31. Pauletti, G.; Godolphin, W.; Press, M.F.; Slamon, D.J. Detection and quantitation of HER-2/neu gene amplification in human breast cancer archival material using fluorescence in situ hybridization. *Oncogene* **1996**, *13*, 63–72. [PubMed]

32. Kobayashi, M.; Ooi, A.; Oda, Y.; Nakanishi, I. Protein overexpression and gene amplification of c-erbB-2 in breast carcinomas: A comparative study of immunohistochemistry and fluorescence in situ hybridization of formalin-fixed, paraffin-embedded tissues. *Hum. Pathol.* **2002**, *33*, 21–28. [CrossRef]

33. Kreutzfeldt, J.; Rozeboom, B.; Dey, N.; De, P. The trastuzumab era: Current and upcoming targeted HER2+ breast cancer therapies. *Am. J. Cancer Res.* **2020**, *10*, 1045–1067. [PubMed]

34. Onozato, R.; Kosaka, T.; Kuwano, H.; Sekido, Y.; Yatabe, Y.; Mitsudomi, T. Activation of MET by Gene Amplification or by Splice Mutations Deleting the Juxtamembrane Domain in Primary Resected Lung Cancers. *J. Thorac. Oncol.* **2009**, *4*, 5–11. [CrossRef] [PubMed]

35. Okuda, K.; Sasaki, H.; Yukiue, H.; Yano, M.; Fujii, Y. Met gene copy number predicts the prognosis for completely resected non-small cell lung cancer. *Cancer Sci.* **2008**, *99*, 2280–2285. [CrossRef]

36. Cappuzzo, F.; Marchetti, A.; Skokan, M.; Rossi, E.; Gajapathy, S.; Felicioni, L.; Del Grammastro, M.; Sciarrotta, M.G.; Buttitta, F.; Incarbone, M.; et al. Increased MET Gene Copy Number Negatively Affects Survival of Surgically Resected Non–Small-Cell Lung Cancer Patients. *J. Clin. Oncol.* **2009**, *27*, 1667–1674. [CrossRef]

37. Schildhaus, H.-U.; Schultheis, A.M.; Rüschoff, J.; Binot, E.; Merkelbach-Bruse, S.; Fassunke, J.; Schulte, W.; Ko, Y.D.; Schlesinger, A.; Bos, M.; et al. MET Amplification Status in Therapy-Naïve Adeno- and Squamous Cell Carcinomas of the Lung. *Clin. Cancer Res.* **2015**, *21*, 907–915. [CrossRef]

38. Graziano, F.; Galluccio, N.; Lorenzini, P.; Ruzzo, A.; Canestrari, E.; D'Emidio, S.; Catalano, V.; Sisti, V.; Ligorio, C.; Andreoni, F.; et al. Genetic activation of the MET pathway and prognosis of patients with high-risk, radically resected gastric cancer. *J. Clin. Oncol.* **2011**, *29*, 4789–4795. [CrossRef]

39. Peng, Z.; Li, Z.; Gao, J.; Lu, M.; Gong, J.; Tang, E.T.; Oliner, K.S.; Hei, Y.J.; Zhou, H.; Shen, L. Tumor MET Expression and Gene Amplification in Chinese Patients with Locally Advanced or Metastatic Gastric or Gastroesophageal Junction Cancer. *Mol. Cancer Ther.* **2015**, *14*, 2634–2641. [CrossRef]

40. Raghav, K.; Morris, V.; Tang, C.; Morelli, P.; Amin, H.M.; Chen, K.; Manyam, G.C.; Broom, B.; Overman, M.J.; Shaw, K.; et al. MET amplification in metastatic colorectal cancer: An acquired response to EGFR inhibition, not a de novo phenomenon. *Oncotarget* **2016**, *7*, 54627–54631. [CrossRef]

41. Zhang, M.; Li, G.; Sun, X.; Ni, S.; Tan, C.; Xu, M.; Huang, D.; Ren, F.; Li, D.; Wei, P.; et al. MET amplification, expression, and exon 14 mutations in colorectal adenocarcinoma. *Hum. Pathol.* **2018**, *77*, 108–115. [CrossRef] [PubMed]

42. Pal, S.K.; Ali, S.M.; Yakirevich, E.; Geynisman, D.M.; Karam, J.A.; Elvin, J.A.; Frampton, G.M.; Huang, X.; Lin, D.I.; Rosenzweig, M.; et al. Characterization of Clinical Cases of Advanced Papillary Renal Cell Carcinoma via Comprehensive Genomic Profiling. *Eur. Urol.* **2018**, *73*, 71–78. [CrossRef]

43. Gonzalez-Angulo, A.M.; Chen, H.; Karuturi, M.S.; Chavez-MacGregor, M.; Tsavachidis, S.; Meric-Bernstam, F.; Do, K.A.; Hortobagyi, G.N.; Thompson, P.A.; Mills, G.B.; et al. Frequency of mesenchymal-epithelial transition factor gene (MET) and the catalytic subunit of phosphoinositide-3-kinase (PIK3CA) copy number elevation and correlation with outcome in patients with early stage breast cancer. *Cancer* **2013**, *119*, 7–15. [CrossRef] [PubMed]

44. Gunia, S.; Erbersdobler, A.; Hakenberg, O.W.; Koch, S.; May, M. C-MET is expressed in the majority of penile squamous cell carcinomas and correlates with polysomy-7 but is not associated with MET oncogene amplification, pertinent histopathologic parameters, or with cancer-specific survival. *Pathol. Res. Pr.* **2013**, *209*, 215–220. [CrossRef] [PubMed]

45. Mignard, X.; Ruppert, A.M.; Antoine, M.; Vasseur, J.; Girard, N.; Mazières, J.; Moro-Sibilot, D.; Fallet, V.; Rabbe, N.; Thivolet-Bejui, F.; et al. c-MET Overexpression as a Poor Predictor of MET Amplifications or Exon 14 Mutations in Lung Sarcomatoid Carcinomas. *J. Thorac. Oncol.* **2018**, *13*, 1962–1967. [CrossRef] [PubMed]

46. Guo, R.; Berry, L.D.; Aisner, D.L.; Sheren, J.; Boyle, T.; Bunn, P.A.; Johnson, B.E.; Kwiatkowski, D.J.; Drilon, A.; Sholl, L.M.; et al. MET IHC Is a Poor Screen for MET Amplification or MET Exon 14 Mutations in Lung Adenocarcinomas: Data from a Tri-Institutional Cohort of the Lung Cancer Mutation Consortium. *J. Thorac. Oncol.* **2019**, *14*, 1666–1671. [CrossRef]

47. Lai, G.G.Y.; Lim, T.H.; Lim, J.; Liew, P.J.R.; Kwang, X.L.; Nahar, R.; Aung, Z.W.; Takano, A.; Lee, Y.Y.; Lau, D.P.X.; et al. Clonal MET Amplification as a Determinant of Tyrosine Kinase Inhibitor Resistance in Epidermal Growth Factor Receptor-Mutant Non-Small-Cell Lung Cancer. *J. Clin. Oncol.* **2019**, *37*, 876–884. [CrossRef]

48. Spigel, D.R.; Edelman, M.J.; O'Byrne, K.; Paz-Ares, L.; Mocci, S.; Phan, S.; Shames, D.S.; Smith, D.; Yu, W.; Paton, V.E.; et al. Results from the Phase III Randomized Trial of Onartuzumab Plus Erlotinib Versus Erlotinib in Previously Treated Stage IIIB or IV Non-Small-Cell Lung Cancer: METLung. *J. Clin. Oncol.* **2017**, *35*, 412–420. [CrossRef]

49. Neal, J.W.; Dahlberg, S.E.; Wakelee, H.A.; Aisner, S.C.; Bowden, M.; Huang, Y.; Carbone, D.P.; Gerstner, G.J.; Lerner, R.E.; Rubin, J.L.; et al. Erlotinib, cabozantinib, or erlotinib plus cabozantinib as second-line or third-line treatment of patients with EGFR wild-type advanced non-small-cell lung cancer (ECOG-ACRIN 1512): A randomised, controlled, open-label, multicentre, phase 2 trial. *Lancet Oncol.* **2016**, *17*, 1661–1671. [CrossRef]

50. Camidge, D.R.; Barlesi, F.; Goldman, J.W.; Morgensztern, D.; Heist, R.; Vokes, E.; Spira, A.; Angevin, E.; Su, W.C.; Hong, D.S.; et al. Phase Ib Study of Telisotuzumab Vedotin in Combination with Erlotinib in Patients with c-Met Protein-Expressing Non-Small-Cell Lung Cancer. *J. Clin. Oncol.* **2023**, *41*, 1105–1115. [CrossRef]

51. Orsaria, M.; Khelifa, S.; Buza, N.; Kamath, A.; Hui, P. Chromosome 17 polysomy: Correlation with histological parameters and HER2NEU gene amplification. *J. Clin. Pathol.* **2013**, *66*, 1070–1075. [CrossRef] [PubMed]

52. Vanden Bempt, I.; Van Loo, P.; Drijkoningen, M.; Neven, P.; Smeets, A.; Christiaens, M.R.; Paridaens, R.; De Wolf-Peeters, C. Polysomy 17 in breast cancer: Clinicopathologic significance and impact on HER-2 testing. *J. Clin. Oncol.* **2008**, *26*, 4869–4874. [CrossRef] [PubMed]

53. Yosepovich, A.; Avivi, C.; Bar, J.; Polak-Charcon, S.; Mardoukh, C.; Barshack, I. Breast cancer HER2 equivocal cases: Is there an alternative to FISH testing? A pilot study using two different antibodies sequentially. *Isr. Med. Assoc. J.* **2010**, *12*, 353–356. [PubMed]

54. Kong, H.; Bai, Q.; Li, A.; Zhou, X.; Yang, W. Characteristics of HER2-negative breast cancers with FISH-equivocal status according to 2018 ASCO/CAP guideline. *Diagn. Pathol.* **2022**, *17*, 5. [CrossRef]

55. Lim, T.H.; Lim, A.S.; Thike, A.A.; Tien, S.L.; Tan, P.H. Implications of the Updated 2013 American Society of Clinical Oncology/College of American Pathologists Guideline Recommendations on Human Epidermal Growth Factor Receptor 2 Gene Testing Using Immunohistochemistry and Fluorescence In Situ Hybridization for Breast Cancer. *Arch. Pathol. Lab. Med.* **2016**, *140*, 140–147.

56. Kumaki, Y.; Olsen, S.; Suenaga, M.; Nakagawa, T.; Uetake, H.; Ikeda, S. Comprehensive Genomic Profiling of Circulating Cell-Free DNA Distinguishes Focal MET Amplification from Aneuploidy in Diverse Advanced Cancers. *Curr. Oncol.* **2021**, *28*, 3717–3728. [CrossRef]

57. Guo, R.; Luo, J.; Chang, J.; Rekhtman, N.; Arcila, M.; Drilon, A. MET-dependent solid tumours—Molecular diagnosis and targeted therapy. *Nat. Rev. Clin. Oncol.* **2020**, *17*, 569–587. [CrossRef]

58. Camidge, D.R.; Davies, K.D. MET Copy Number as a Secondary Driver of Epidermal Growth Factor Receptor Tyrosine Kinase Inhibitor Resistance in EGFR-Mutant Non-Small-Cell Lung Cancer. *J. Clin. Oncol.* **2019**, *37*, 855–857. [CrossRef]

59. Jørgensen, J.T.; Mollerup, J. Companion Diagnostics and Predictive Biomarkers for MET-Targeted Therapy in NSCLC. *Cancers* **2022**, *14*, 2150. [CrossRef]

60. Castiglione, R.; Alidousty, C.; Holz, B.; Duerbaum, N.; Wittersheim, M.; Binot, E.; Merkelbach-Bruse, S.; Friedrichs, N.; Dettmer, M.S.; Bosse, A.; et al. MET-FISH Evaluation Algorithm: Proposal of a Simplified Method. *J. Cancer Sci. Clin. Ther.* **2022**, *6*, 411–427. [CrossRef]

61. Chrzanowska, N.M.; Kowalewski, J.; Lewandowska, M.A. Use of Fluorescence In Situ Hybridization (FISH) in Diagnosis and Tailored Therapies in Solid Tumors. *Molecules* **2020**, *25*, 1864. [CrossRef] [PubMed]

62. Jardim, D.L.F.; Tang, C.; Gagliato, D.e.M.; Falchook, G.S.; Hess, K.; Janku, F.; Fu, S.; Wheler, J.J.; Zinner, R.G.; Naing, A.; et al. Analysis of 1115 Patients Tested for MET Amplification and Therapy Response in the MD Anderson Phase I Clinic. *Clin. Cancer Res.* **2014**, *20*, 6336–6345. [CrossRef] [PubMed]

63. Tong, J.H.; Yeung, S.F.; Chan, A.W.; Chung, L.Y.; Chau, S.L.; Lung, R.W.; Tong, C.Y.; Chow, C.; Tin, E.K.; Yu, Y.H.; et al. MET Amplification and Exon 14 Splice Site Mutation Define Unique Molecular Subgroups of Non-Small Cell Lung Carcinoma with Poor Prognosis. *Clin. Cancer Res.* **2016**, *22*, 3048–3056. [CrossRef] [PubMed]

64. Noonan, S.A.; Berry, L.; Lu, X.; Gao, D.; Barón, A.E.; Chesnut, P.; Sheren, J.; Aisner, D.L.; Merrick, D.; Doebele, R.C.; et al. Identifying the Appropriate FISH Criteria for Defining MET Copy Number-Driven Lung Adenocarcinoma through Oncogene Overlap Analysis. *J. Thorac. Oncol.* **2016**, *11*, 1293–1304. [CrossRef] [PubMed]

65. Caparica, R.; Yen, C.T.; Coudry, R.; Ou, S.I.; Varella-Garcia, M.; Camidge, D.R.; de Castro, G. Responses to Crizotinib Can Occur in High-Level MET-Amplified Non-Small Cell Lung Cancer Independent of MET Exon 14 Alterations. *J. Thorac. Oncol.* **2017**, *12*, 141–144. [CrossRef]

66. Li, J.; Wang, Y.; Zhang, B.; Xu, J.; Cao, S.; Zhong, H. Characteristics and response to crizotinib in lung cancer patients with MET amplification detected by next-generation sequencing. *Lung Cancer* **2020**, *149*, 17–22. [CrossRef]

67. Peng, L.X.; Jie, G.L.; Li, A.N.; Liu, S.Y.; Sun, H.; Zheng, M.M.; Zhou, J.Y.; Zhang, J.T.; Zhang, X.C.; Zhou, Q.; et al. MET amplification identified by next-generation sequencing and its clinical relevance for MET inhibitors. *Exp. Hematol. Oncol.* **2021**, *10*, 52. [CrossRef]

68. Schubart, C.; Stöhr, R.; Tögel, L.; Fuchs, F.; Sirbu, H.; Seitz, G.; Seggewiss-Bernhardt, R.; Leistner, R.; Sterlacci, W.; Vieth, M.; et al. MET Amplification in Non-Small Cell Lung Cancer (NSCLC)-A Consecutive Evaluation Using Next-Generation Sequencing (NGS) in a Real-World Setting. *Cancers* **2021**, *13*, 5023. [CrossRef]

69. Morsberger, L.; Pallavajjala, A.; Long, P.; Hardy, M.; Park, R.; Parish, R.; Nozari, A.; Zou, Y.S. HER2 amplification by next-generation sequencing to identify HER2-positive invasive breast cancer with negative HER2 immunohistochemistry. *Cancer Cell Int.* **2022**, *22*, 350. [CrossRef]

70. Turke, A.B.; Zejnullahu, K.; Wu, Y.L.; Song, Y.; Dias-Santagata, D.; Lifshits, E.; Toschi, L.; Rogers, A.; Mok, T.; Sequist, L.; et al. Preexistence and clonal selection of MET amplification in EGFR mutant NSCLC. *Cancer Cell* **2010**, *17*, 77–88. [CrossRef]

71. Drilon, A.; Cappuzzo, F.; Ou, S.I.; Camidge, D.R. Targeting MET in Lung Cancer: Will Expectations Finally Be MET? *J. Thorac. Oncol.* **2017**, *12*, 15–26. [CrossRef] [PubMed]

72. Benedettini, E.; Sholl, L.M.; Peyton, M.; Reilly, J.; Ware, C.; Davis, L.; Vena, N.; Bailey, D.; Yeap, B.Y.; Fiorentino, M.; et al. Met activation in non-small cell lung cancer is associated with de novo resistance to EGFR inhibitors and the development of brain metastasis. *Am. J. Pathol.* **2010**, *177*, 415–423. [CrossRef] [PubMed]

73. Remon, J.; Morán, T.; Majem, M.; Reguart, N.; Dalmau, E.; Márquez-Medina, D.; Lianes, P. Acquired resistance to epidermal growth factor receptor tyrosine kinase inhibitors in EGFR-mutant non-small cell lung cancer: A new era begins. *Cancer Treat. Rev.* **2014**, *40*, 93–101. [CrossRef] [PubMed]

74. Ou, S.I.; Agarwal, N.; Ali, S.M. High MET amplification level as a resistance mechanism to osimertinib (AZD9291) in a patient that symptomatically responded to crizotinib treatment post-osimertinib progression. *Lung Cancer* **2016**, *98*, 59–61. [CrossRef]

75. Dagogo-Jack, I.; Yoda, S.; Lennerz, J.K.; Langenbucher, A.; Lin, J.J.; Rooney, M.M.; Prutisto-Chang, K.; Oh, A.; Adams, N.A.; Yeap, B.Y.; et al. MET Alterations Are a Recurring and Actionable Resistance Mechanism in ALK-Positive Lung Cancer. *Clin. Cancer Res.* **2020**, *26*, 2535–2545. [CrossRef]

76. Lin, J.J.; Liu, S.V.; McCoach, C.E.; Zhu, V.W.; Tan, A.C.; Yoda, S.; Peterson, J.; Do, A.; Prutisto-Chang, K.; Dagogo-Jack, I.; et al. Mechanisms of resistance to selective RET tyrosine kinase inhibitors in RET fusion-positive non-small-cell lung cancer. *Ann. Oncol.* **2020**, *31*, 1725–1733. [CrossRef]

77. Rosen, E.Y.; Johnson, M.L.; Clifford, S.E.; Somwar, R.; Kherani, J.F.; Son, J.; Bertram, A.A.; Davare, M.A.; Gladstone, E.; Ivanova, E.V.; et al. Overcoming MET-Dependent Resistance to Selective RET Inhibition in Patients with RET Fusion-Positive Lung Cancer by Combining Selpercatinib with Crizotinib. *Clin. Cancer Res.* **2021**, *27*, 34–42. [CrossRef] [PubMed]

78. Tyler, L.C.; Le, A.T.; Chen, N.; Nijmeh, H.; Bao, L.; Wilson, T.R.; Chen, D.; Simmons, B.; Turner, K.M.; Perusse, D.; et al. MET gene amplification is a mechanism of resistance to entrectinib in ROS1+ NSCLC. *Thorac. Cancer* **2022**, *13*, 3032–3041. [CrossRef]

79. Cocco, E.; Schram, A.M.; Kulick, A.; Misale, S.; Won, H.H.; Yaeger, R.; Razavi, P.; Ptashkin, R.; Hechtman, J.F.; Toska, E.; et al. Resistance to TRK inhibition mediated by convergent MAPK pathway activation. *Nat. Med.* **2019**, *25*, 1422–1427. [CrossRef]

80. Awad, M.M.; Liu, S.; Rybkin, I.I.; Arbour, K.C.; Dilly, J.; Zhu, V.W.; Johnson, M.L.; Heist, R.S.; Patil, T.; Riely, G.J.; et al. Acquired Resistance to KRAS(G12C) Inhibition in Cancer. *N. Engl. J. Med.* **2021**, *384*, 2382–2393. [CrossRef]

81. Angevin, E.; Spitaleri, G.; Rodon, J.; Dotti, K.; Isambert, N.; Salvagni, S.; Moreno, V.; Assadourian, S.; Gomez, C.; Harnois, M.; et al. A first-in-human phase I study of SAR125844, a selective MET tyrosine kinase inhibitor, in patients with advanced solid tumours with MET amplification. *Eur. J. Cancer* **2017**, *87*, 131–139. [CrossRef] [PubMed]

82. Van Cutsem, E.; Karaszewska, B.; Kang, Y.K.; Chung, H.C.; Shankaran, V.; Siena, S.; Go, N.F.; Yang, H.; Schupp, M.; Cunningham, D. A Multicenter Phase II Study of AMG 337 in Patients with MET-Amplified Gastric/Gastroesophageal Junction/Esophageal Adenocarcinoma and Other MET-Amplified Solid Tumors. *Clin. Cancer Res.* **2019**, *25*, 2414–2423. [CrossRef] [PubMed]

83. Hong, D.S.; LoRusso, P.; Hamid, O.; Janku, F.; Kittaneh, M.; Catenacci, D.V.T.; Chan, E.; Bekaii-Saab, T.; Gadgeel, S.M.; Loberg, R.D.; et al. Phase I Study of AMG 337, a Highly Selective Small-molecule MET Inhibitor, in Patients with Advanced Solid Tumors. *Clin. Cancer Res.* **2019**, *25*, 2403–2413. [CrossRef] [PubMed]

84. Camidge, D.R.; Otterson, G.A.; Clark, J.W.; Ignatius Ou, S.H.; Weiss, J.; Ades, S.; Shapiro, G.I.; Socinski, M.A.; Murphy, D.A.; Conte, U.; et al. Crizotinib in Patients with MET-Amplified NSCLC. *J. Thorac. Oncol.* **2021**, *16*, 1017–1029. [CrossRef]

85. Landi, L.; Chiari, R.; Tiseo, M.; D'Incà, F.; Dazzi, C.; Chella, A.; Delmonte, A.; Bonanno, L.; Giannarelli, D.; Cortinovis, D.L.; et al. Crizotinib in MET-Deregulated or ROS1-Rearranged Pretreated Non-Small Cell Lung Cancer (METROS): A Phase II, Prospective, Multicenter, Two-Arms Trial. *Clin. Cancer Res.* **2019**, *25*, 7312–7319. [CrossRef]

86. Moro-Sibilot, D.; Cozic, N.; Pérol, M.; Mazières, J.; Otto, J.; Souquet, P.J.; Bahleda, R.; Wislez, M.; Zalcman, G.; Guibert, S.D.; et al. Crizotinib in c-MET- or ROS1-positive NSCLC: Results of the AcSé phase II trial. *Ann. Oncol.* **2019**, *30*, 1985–1991. [CrossRef]

87. Schuler, M.; Berardi, R.; Lim, W.T.; de Jonge, M.; Bauer, T.M.; Azaro, A.; Gottfried, M.; Han, J.Y.; Lee, D.H.; Wollner, M.; et al. Molecular correlates of response to capmatinib in advanced non-small-cell lung cancer: Clinical and biomarker results from a phase I trial. *Ann. Oncol.* **2020**, *31*, 789–797. [CrossRef]

88. Qin, S.; Chan, S.L.; Sukeepaisarnjaroen, W.; Han, G.; Choo, S.P.; Sriuranpong, V.; Pan, H.; Yau, T.; Guo, Y.; Chen, M.; et al. A phase II study of the efficacy and safety of the MET inhibitor capmatinib (INC280) in patients with advanced hepatocellular carcinoma. *Ther. Adv. Med. Oncol.* **2019**, *11*, 1758835919889001. [CrossRef]

89. Wu, Y.L.; Zhang, L.; Kim, D.W.; Liu, X.; Lee, D.H.; Yang, J.C.; Ahn, M.J.; Vansteenkiste, J.F.; Su, W.C.; Felip, E.; et al. Phase Ib/II Study of Capmatinib (INC280) Plus Gefitinib After Failure of Epidermal Growth Factor Receptor (EGFR) Inhibitor Therapy in Patients with EGFR-Mutated, MET Factor-Dysregulated Non-Small-Cell Lung Cancer. *J. Clin. Oncol.* **2018**, *36*, 3101–3109. [CrossRef]

90. Wu, Y.L.; Cheng, Y.; Zhou, J.; Lu, S.; Zhang, Y.; Zhao, J.; Kim, D.W.; Soo, R.A.; Kim, S.W.; Pan, H.; et al. Tepotinib plus gefitinib in patients with EGFR-mutant non-small-cell lung cancer with MET overexpression or MET amplification and acquired resistance to previous EGFR inhibitor (INSIGHT study): An open-label, phase 1b/2, multicentre, randomised trial. *Lancet Respir. Med.* **2020**, *8*, 1132–1143. [CrossRef]

91. Sequist, L.V.; Han, J.Y.; Ahn, M.J.; Cho, B.C.; Yu, H.; Kim, S.W.; Yang, J.C.; Lee, J.S.; Su, W.C.; Kowalski, D.; et al. Osimertinib plus savolitinib in patients with EGFR mutation-positive, MET-amplified, non-small-cell lung cancer after progression on EGFR tyrosine kinase inhibitors: Interim results from a multicentre, open-label, phase 1b study. *Lancet Oncol.* **2020**, *21*, 373–386. [CrossRef] [PubMed]

92. Yang, J.J.; Fang, J.; Shu, Y.Q.; Chang, J.H.; Chen, G.Y.; He, J.X.; Li, W.; Liu, X.Q.; Yang, N.; Zhou, C.; et al. A phase Ib study of the highly selective MET-TKI savolitinib plus gefitinib in patients with EGFR-mutated, MET-amplified advanced non-small-cell lung cancer. *Investig. New Drugs* **2021**, *39*, 477–487. [CrossRef] [PubMed]

Review

Companion Diagnostics and Predictive Biomarkers for MET-Targeted Therapy in NSCLC

Jan Trøst Jørgensen [1,*] and Jens Mollerup [2]

[1] Department: Medical Sciences, Dx-Rx Institute, Baunevaenget 76, 3480 Fredensborg, Denmark
[2] Pathology Division, Agilent Technologies Denmark ApS, Produktionsvej 42, 2600 Glostrup, Denmark
*   Correspondence: jan.trost@dx-rx.dk

Simple Summary: MET is a receptor tyrosine kinase encoded by the *MET* proto-oncogene that has a significant role in cancer cell progression. Several drugs targeting MET are under development for the treatment of different cancers, including non-small-cell lung cancer (NSCLC). However, until now, relatively few of these drugs have shown sufficient clinical activity and obtained regulatory approval. One of the reasons for this could be the lack of effective biomarkers to select the right patients for treatment. In a number of clinical trials, different biomarkers have been studied, but so far, *MET* exon 14 skipping mutation is the only one that has shown sufficient predictive properties. Another interesting biomarker is *MET* amplification detected by fluorescence in situ hybridization (FISH), which has shown promising results in the treatment of patients with NSCLC. Future clinical research will show whether *MET* amplification by FISH is an effective predictive biomarker for MET-targeted therapy.

Abstract: Dysregulation of the MET tyrosine kinase receptor is a known oncogenic driver, and multiple genetic alterations can lead to a clinically relevant oncogenesis. Currently, a number of drugs targeting MET are under development as potential therapeutics for different cancer indications, including non-small cell lung cancer (NSCLC). However, relatively few of these drugs have shown sufficient clinical activity and obtained regulatory approval. One of the reasons for this could be the lack of effective predictive biomarkers to select the right patient populations for treatment. So far, capmatinib is the only MET-targeted drug approved with a companion diagnostic (CDx) assay, which is indicated for the treatment of metastatic NSCLC in patients having a mutation resulting in *MET* exon 14 skipping. An alternative predictive biomarker for MET therapy is *MET* amplification, which has been identified as a resistance mechanism in patients with *EGFR*-mutated NSCLC. Results obtained from different clinical trials seem to indicate that the *MET*/CEP7 ratio detected by FISH possesses the best predictive properties, likely because this method excludes *MET* amplification caused by polysomy. In this article, the concept of CDx assays will be discussed, with a focus on the currently FDA-approved MET targeted therapies for the treatment of NSCLC.

Keywords: MET; NSCLC; *MET* exon 14 skipping mutation; *MET* amplification; NGS; FISH; crizotinib; capmatinib; tepotinib; amivantamab

Citation: Jørgensen, J.T.; Mollerup, J. Companion Diagnostics and Predictive Biomarkers for MET-Targeted Therapy in NSCLC. *Cancers* **2022**, *14*, 2150. https://doi.org/10.3390/cancers14092150

Academic Editor: Andrea Cavazzoni

Received: 17 March 2022
Accepted: 22 April 2022
Published: 26 April 2022

1. Introduction

For more than 20 years, companion diagnostics (CDx) and predictive biomarkers have had a significant impact on the development of a number of targeted hematological and oncological drugs as well as their subsequent use in the clinic. The first drug to have a CDx linked to its use was the monoclonal antibody trastuzumab (Herceptin, Roche/Genentech, Basel, Switzerland/ San Francisco, CA, USA), which was approved for the treatment of HER2-positive metastatic breast cancer together with the immunohistochemical assay (IHC) HercepTest™ (Dako/Agilent Technologies, Glostrup, Denmark) in 1998 [1,2]. The HercepTest™ assay became the first ever CDx to obtain regulatory approval by the Food and Drug Administration (FDA), and, since then, the number of drug–CDx combinations have increased considerably. By the end of 2021, this number was close to 50 [3]. For most of these targeted hematological and oncological drugs, the CDx assay plays an important role in selecting patients who are likely to respond, and, without such assays, most drugs will lose their value.

MET was originally discovered as the transforming gene in a chemically transformed cell line derived from human osteosarcoma [4]. Since then, it has been established that *MET* is a proto-oncogene on chromosome 7q31, and it encodes a transmembrane receptor with intrinsic tyrosine kinase activity. The receptor tyrosine kinase is also called c-Met or hepatocyte growth factor receptor (HGFR) after its ligand, hepatocyte growth factor (HGF) [4–6]. Dysregulation of the MET tyrosine kinase is a known oncogenic driver; however, compared to most other proto-oncogenes, the *MET* gene is special, as different genomic states such as amplification, mutation, and rearrangement can lead to a clinically relevant oncogenesis [7]. Currently, several small-molecular inhibitors and antibody-based drugs targeting MET are under development as potential therapeutics for different cancer indications, but so far, only a few of these have obtained regulatory approval and reached the clinic [8]. One of the reasons for this might be the challenges in finding the right predictive biomarkers to guide the use of these drugs. Until now, capmatinib (Tabrecta, Novartis, Basel, Switzerland) is the only MET inhibitor that has an FDA-approved CDx linked to its use. Capmatinib can be used for the treatment of patients with metastatic non-small-cell lung cancer (NSCLC), when tumors have a mutation that leads to *MET* exon 14 skipping (*MET*ex14) [9,10]. In this article, the concept of CDx assays and predictive biomarkers will be discussed, with a focus on the current FDA-approved MET inhibitors for the treatment of NSCLC.

2. Companion Diagnostics

The successful development of trastuzumab for HER2-positive metastatic breast cancer opened the way for the drug-diagnostic co-development model, in which a predictive biomarker test is developed along with the drug. The use of a biomarker-guided clinical enrichment strategy often leads to an increased power of the individual clinical trial and a higher likelihood of a positive outcome [2]. For trastuzumab, this strategy was of immense importance, since without a CDx assay to enrich the clinical trial population with HER2-positive patients, the clinical development program would likely have failed [11]. A few years after the regulatory approval of trastuzumab, alternative sample size calculations for the phase III trial that led to the initial approval of trastuzumab for HER2-positive metastatic breast cancer was published by Richard Simon of the US National Cancer Institute in *Clinical Cancer Research* [12]. In this trial, an enrichment strategy was used, and here, 469 HER2-positive patients were included and randomized to receive trastuzumab plus chemotherapy or chemotherapy alone [1]. One of the alternative sample size calculations was made for an all-comers trial design, where no testing for HER2 positivity was performed, and this calculation showed that the number of patients to be included would have been 8050, to demonstrate the same statistically significant difference between the two arms, as in the original phase III trial. This corresponds to 17.2 times more patients and demonstrates the importance of the clinical enrichment trial design for the development of trastuzumab and for many other cancer drugs developed for different cancer indications over the past 20 years [3,12].

In 2014, the FDA issued the first guideline on In Vitro Companion Diagnostic Devices in which they officially defined a CDx assay [13]. This definition states that a CDx is an assay that provides information that is essential for the safe and effective use of a corresponding therapeutic product. In relation to this definition, it was noted that an inadequate performance of a CDx assay can have severe therapeutic consequences for the individual patient, as erroneous results could lead to the withholding of appropriate therapy or the administration of an inappropriate therapy. Consequently, the FDA classifies CDx assays as high-risk Class III devices, which requires the submission of substantial documentation for both the analytical and clinical performance before the assay can be approved and used in the clinic. Most CDx assays are developed using the prospective drug-diagnostic co-development model, so both drug and diagnostic can obtain simultaneous regulatory approval [3]. Furthermore, the use of a CDx must be included in both the labeling for the drug and the diagnostic, including the labeling of any subsequent generic equivalents of the drug. This emphasizes the importance of the CDx assay, and testing must be performed before prescribing the drug to the patient [13].

Several other countries worldwide, inducing the European Union (EU), have introduced similar definitions of a CDx assay as the FDA, and have tightened documentation requirements and regulatory approval procedures. In 2017, the European Parliament passed new regulations on In Vitro Diagnostic Medical Devices (IVDR) that will have a great impact on the development and use of CDx assays in European countries, as CE-IVD marking based on a self-declaration will no longer be possible [14]. The new IVDR was supposed to come into force in May 2022, but due to the extraordinary circumstances mainly caused by the COVID-19 pandemic, the European Commission has partly proposed to postpone the effective date for CDx assays to May 2026 [15].

3. MET-Targeted Therapy and NSCLC

Within different cancers, *MET* dysregulations can serve as primary drivers in promoting tumor growth, invasion, angiogenesis, and metastasis. NSCLC *MET* dysregulations in the form of *MET* amplification (*MET*amp) have also been shown to act as secondary drivers that can mediate resistance to targeted therapy for other oncogenes such as epidermal growth factor receptor (*EGFR*) mutations [7,8,16]. *MET* dysregulations have been found in 5% to 26% of NSCLC patients following treatment with an EGFR tyrosine kinase inhibitor (TKI) [17]. The first MET inhibitor to obtain FDA approval was the multi-kinase inhibitor crizotinib (Xalkori, Pfizer, New York, NY, USA), which was approved for the treatment of patients with *ALK*-rearranged NSCLC more than 10 years ago [18]. Subsequently, a few other drugs targeting MET have likewise been approved for various NSCLC indications (Figure 1). These drugs and their CDx assays are listed in Table 1 and described in more detail below. In addition, the target sites in relation to MET domains are schematically illustrated in Figure 1.

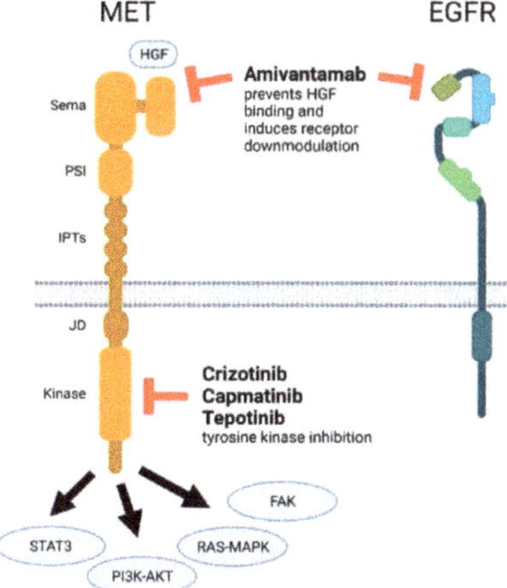

Figure 1. Target regions of MET inhibitors. Cartoon overview of the intra- and extracellular-domain structure of MET and the sites of inhibitor binding to MET and EGFR. Four major signaling pathways involved in MET signaling are indicated. Abbreviations: hepatocyte growth factor (HGF), Semaphorin (Sema), plexin-semaphorin-integrin (PSI), integrin-plexin-transcription factor (IPTs), juxtamembrane (JD), and epidermal growth factor receptor (EGFR).

Table 1. FDA-approved MET targeted drugs and their CDx assays. Only capmatinib and tepotinib are approved for a MET-specific indication [9,10,18–20].

Drug	Drug Class	Approved Indication(s)		FDA Approved CDx Assay(s)
Crizotinib	Small molecule inhibitor	Treatment of patients with metastatic NSCLC whose tumors are ALK or ROS1-positive as detected by an FDA-approved test	ALK	FoundationOne CDx VENTANA ALK (D5F3) CDx Assay Vysis ALK Break Apart FISH Probe Kit ROS1
			ROS1	Oncomine Dx Target Test
			MET	No approved CDx available
Capmatinib	Small molecule inhibitor	Treatment of adult patients with NSCLC whose tumors have a mutation that leads to *MET* exon 14 skipping as detected by an FDA-approved test	MET	FoundationOne CDx
Tepotinib	Small molecule inhibitor	Treatment of adult patients with metastatic NSCLC harboring *MET* exon 14 skipping alterations	MET	No approved CDx available
Amivantamab	Bispecific antibody	Treatment of adult patients with locally advanced or metastatic NSCLC with *EGFR* exon 20 insertion mutations, as detected by an FDA-approved test, whose disease has progressed on or after platinum-based chemotherapy	*EGFR* exon 20 insertion	Guardant360® CDx
			MET	No approved CDx available

CDx = Companion Diagnostic; ALK = Anaplastic Lymphoma Kinase; ROS1 = ROS proto-oncogene 1; MET = Mesenchymal Epithelial Transition Factor; EGFR = Epidermal Growth Factor Receptor; *MET*ex14 = *MET* exon 14 skipping mutation; NSCLC = Non-Small Cell Lung Cancer; FDA = Food and Drug Administration.

3.1. Crizotinib

Crizotinib is a small molecule inhibitor of receptor tyrosine kinases including ALK, MET, and ROS1, and belongs to the class Ia MET inhibitors [8,18]. Different in vitro studies have demonstrated a concentration-dependent inhibition by crizotinib of tyrosine phosphorylation mediated by ALK, ROS1, and MET in different tumor cell line-based assays. Furthermore, crizotinib has been used in vivo to show antitumor activity in mice having tumor xenografts that expressed MET or on the fusion proteins EML4-ALK or NPM-ALK [18].

The first indication in which crizotinib demonstrated important clinical activities was in metastatic NSCLC patients with *ALK*-rearrangement, which led to a regulatory approval of the drug by the FDA in 2011 [18,21,22]. In 2016, this indication was expanded to include NSCLC patients with *ROS1* rearrangement [18,23]. For the selection of patients for treatment with crizotinib, the FDA has approved CDx assays for the detection of both *ALK*- and *ROS1*-rearrangements [10]. Today, crizotinib must be regarded as a well-established therapy in NSCLC patients with *ALK* or *ROS1* rearrangement, but when it comes to patients with MET dysregulations, the documentation is less convincing, and a regulatory approval for this indication has not yet been granted.

A few clinical trials have shown varying activity of crizotinib in NSCLC patients with *MET*amp or an *MET*ex14 mutation, with objective response rates (ORR) in the range of 12% to 32% depending on the type of *MET* dysregulation [24–26]. In these trials, the *MET*ex14 mutation was detected by next-generation sequencing (NGS) and *MET*amp by fluorescence in situ hybridization (FISH). Different cut-off values were applied with regard to the detection of *MET*amp by FISH. In one of the trials, the cut-off value was based on a *MET*/CEP7 ratio > 2.2, and in another trial, it was based on a *MET* gene copy number (GCN) ≥ 6. The results of the different reported MET trials with crizotinib are shown in Table 2.

Table 2. Predictive Biomarkers and Companion Diagnostics for the FDA-approved MET Targeted Drugs in NSCLC.

Drug	Publication [Reference]	Method	Biomarker/CDx	N	Objective Response Rate
Crizotinib	Moro-Sibilot D et al. [24]	FISH	MET GCN $\geq$ 6	25	16%
		NGS	METex14	25	12%
	Landi L et al. [25]	FISH	MET/CEP7 > 2.2	16	31%
		NGS	METex14	10	20%
	Drilon A et al. [26]	NGS	METex14	65	32%
Capmatinib	Schuler M et al. [27]	FISH	MET GCN < 4	17	6%
			$4 \leq MET$ GCN < 6	12	25%
			MET GCN $\geq$ 6	15	47%
			MET/CEP7 $\geq$ 2.0	9	44%
			MET/CEP7 < 2.0	32	22%
		IHC	MET IHC2+	14	14%
			MET IHC3+	37	27%
	Wu YL et al. [17]	FISH	MET GCN < 4	41	12%
			$4 \leq MET$ GCN < 6	18	22%
			MET GCN $\geq$ 6	36	47%
		IHC	MET IHC2+	16	19%
			MET IHC3+	78	32%
	Wolf J et al. [28]	NGS	METex14	69 (Previous treated)	41%
			METex14	28 (Treatment naïve)	64%
		NGS	MET GCN < 4	30 (Previous treated)	7%
			MET GCN 4 or 5	54 (Previous treated)	9%
			MET GCN 6–9	42 (Previous treated)	12%
			MET GCN $\geq$ 10	69 (Previous treated)	28%
			MET GCN $\geq$ 10	15 (Treatment naïve)	40%
Tepotinib	Wu YL et al. [29]	IHC	MET IHC3+	19	68%
		FISH	MET/CEP7 $\geq$ 2.0	12	67%
	Paik PK et al. [30]	NGS	METex14	99	46%

CDx = Companion Diagnostic; FISH = Fluorescence In Situ Hybridization; NGS = Next-Generation Sequencing; IHC = Immunohistochemistry; MET = Mesenchymal Epithelial Transition Factor; GCN = Gene Copy Number; METex14 = MET exon 14 skipping mutation; CEP7 = Centromere 7.

3.2. Capmatinib

Capmatinib is a small molecule kinase inhibitor that belongs to the MET-class Ib inhibitors [8]. In murine tumor xenograft models derived from human lung tumors, capmatinib has been shown to inhibit tumor growth driven by METex14 mutation or METamp [8,9]. Capmatinib exerts its activity by inhibiting the MET phosphorylation triggered by the binding of HGF or by METamp, as well as MET-mediated phosphorylation of the different downstream signaling proteins, which results in the impaired proliferation and survival of the MET-dependent tumor cells [9].

In a phase I trial, a possible biomarker enrichment strategy for capmatinib was investigated. Here, 55 pretreated metastatic NSCLC patients with MET dysregulation were treated with capmatinib as monotherapy [27]. Several different approaches were used to test for MET dysregulation, including overexpression as IHC2+ or IHC3+ in $\geq$50% of the tumor cells by IHC and MET GCN $\geq$ 5 or MET/CEP7 ratio $\geq$ 2.0 by FISH. In addition, the

*MET*ex14 mutation was detected in a small subset of patient samples using NGS. Overall, for all enrolled patients, an ORR of 20% was observed. For the subgroup of patients with a *MET* GCN $\geq$ 6 (n = 15), an ORR of 47% was shown. For the group of patients with MET overexpression as IHC3+ (n = 37), the ORR was 27%. *MET*ex14 mutation was detected in four patients, and all responded to treatment with capmatinib. For more results from the phase I trial, please see Table 2. Based on the results from this explorative biomarker trial, it was concluded that capmatinib showed meaningful clinical activity in pretreated metastatic NSCLC patients with either *MET* GCN $\geq$ 6 or an *MET*ex14 mutation. When it comes to MET overexpression by IHC, this method was not considered a reliable predictive biomarker for the efficacy of capmatinib [27].

Another phase Ib/II trial also showed clinical activity of capmatinib in MET-dysregulated NSCLC patients with acquired EFGR-TKI resistance [17]. In this trial, 161 patients with MET dysregulation were selected based on *MET*amp by FISH or MET overexpression by IHC and were treated with capmatinib plus gefitinib (Iressa, AstraZeneca). For the phase Ib part, the patient selection criteria were either *MET* GCN $\geq$ 5 and/or a *MET*/CEP7 ratio $\geq$ 2.0 or MET overexpression as IHC2+ or IHC3+ in $\geq$ 50% of the tumor cells. For the phase II part, the selection criteria were initially defined as *MET* GCN $\geq$ 5 or MET IHC2+ or IHC3+ overexpression in $\geq$ 50% of tumor cells. However, in a protocol amendment, these criteria were revised to MET IHC2+ or IHC3+ plus *MET* GCN $\geq$ 5; these were subsequently changed once more to MET IHC3+ or *MET* GCN $\geq$ 4. Across the phase Ib/II trial and the different MET or *MET* selection criteria, the observed ORR was 27%. In a post-hoc subgroup analysis, an ORR of 47% was shown for phase II patients with *MET* GCN > 6 (n = 36). For patients with MET IHC3+ (n = 78), the ORR was 32%, and for the MET IHC2+ group (n = 16), the ORR was 19%. For more results from the phase Ib/II trial, please see Table 2. Overall, the post-hoc subgroup analysis showed that *MET* FISH using a cut-off value of *MET* GCN > 6 was more accurate compared to MET IHC in predicting the response in NSCLC patients receiving a combined treatment of capmatinib and gefitinib [17].

In a prospective, open-labeled, multiple-cohort phase 2 trial (GENOMETRY mono-1), capmatinib was further investigated in metastatic NSCLC patients with an *MET*ex14 mutation or *MET*amp [28]. The patients in this trial were assigned to different cohorts based on *MET* status and previous treatment. *MET*ex14 mutation was initially determined by a qualitative real-time reverse-transcription PCR (RT-PCR) assay. Detection of *MET*amp, as GCN, was initially determined by FISH. Subsequently, the *MET*ex14 mutation and *MET* GCN were retrospectively retested based on the baseline tissue samples from the trial using the NGS FoundationOne CDx assay (Foundation Medicine). A total of 364 patients were assigned to different study cohorts. For the group with the *MET*ex14 mutation, an ORR of 68% was obtained in the treatment-naïve patients (n = 28) compared to the previously treated patients (n = 69) with an ORR of 41%. In the patients with *MET*amp and a GCN $\geq$ 10, the ORR was 40% in the treatment-naïve (n = 15) and 29% in those previously treated (n = 69). For the patient cohorts previously treated with *MET* GCN < 10, the activity of capmatinib seemed to be limited, with ORR in the range of 7 to 12% [28]. For more results from the phase II trial, please see Table 2. In 2020, based on data from 97 patients with a *MET*ex14 mutation in the GENOMETRY mono-1 trial, capmatinib obtained FDA approval [9], and, along with this approval, the NGS FoundationOne CDx assay was approved as the CDx for the detection of the *MET*ex14 mutation in NSCLC patients who may benefit from treatment with capmatinib [10,31].

3.3. Tepotinib

Tepotinib (Tepmetko, Merck/EMD Serono; Darmstadt, Germany/Rockland, MA, USA) is a small molecule-class Ib inhibitor that targets MET by inhibiting HGF-dependent and -independent MET phosphorylation as well as the MET-dependent downstream signaling pathways [8]. In vitro and in vivo studies have shown that tepotinib inhibited the growth of MET-dysregulated tumor cells, and mice implanted with tumor cells expressing oncogenic active MET had a reduced formation of metastases [19].

In a phase Ib/II trial, the clinical activity of tepotinib was demonstrated in metastatic NSCLC patients with MET dysregulation who had developed a resistance to EGFR-TKI [29]. A total of 73 patients, 18 in the phase Ib part and 55 in the phase II part, were enrolled. In the phase II part, the patients were randomized to receive either tepotinib plus gefitinib or chemotherapy. MET dysregulation was defined as MET IHC2+ or IHC3+ by IHC or *MET*amp as GCN $\geq$ 5 or *MET*/CEP7 ratio $\geq$ 2.0 by FISH. In the phase 1b part, four of the seven patients with MET IHC3+ responded, which was similar for four of the six patients with *MET*amp. In the phase 2 part of the trial, 13 of the 19 MET IHC3+ patients responded (ORR 68%), which was similar to the *MET*amp patients, with 8 of 12 patients responding (ORR 67%). For the chemotherapy group, the ORR was 33% for MET IHC3+ and 43% for *MET*amp patients [29]. Based on the results of this trial, it was concluded that the combination of tepotinib and gefitinib showed similar activity in MET IHC3+ and *MET*amp patients.

In another phase II trial (VISION), tepotinib was investigated in patients with metastatic NSCLC who harbored a *MET*ex14 mutation [30]. Furthermore, the response to tepotinib was analyzed according to whether the presence of the *MET*ex14 mutation was detected from a tissue biopsy or from plasma as circulating tumor DNA (ctDNA). For the tumor tissue biopsies, the *MET*ex14 mutation was assessed using the NGS Oncomine Focus Assay (Thermo Fisher Scientific, Waltham, MA, USA), and for the liquid biopsy, the ctDNA was analyzed using another NGS assay, the Guardant360 (Guardant Health, Redwood City, MA, USA). Testing by both biopsy methods was not a requirement for inclusion in the trial. A total of 152 patients were enrolled in the trial, and in the combined biopsy group (n = 99), the ORR was 46%. For the 66 patients with liquid biopsies, the ORR was 48%, and it was 50% for the 60 patients with tissue biopsies. In 2021, based on clinical data from the VISION trial, tepotinib obtained FDA approval for the treatment of patients with metastatic NSCLC harboring the *MET*ex14 mutation [19]. Somewhat surprisingly, tepotinib was approved without the Oncomine Focus Assay and/or the Guardant360 as CDx for the detection of *MET*ex14 mutations. The full prescribing information for tepotinib emphasizes that an FDA-approved test for the detection of *MET*ex14 mutations in NSCLC for selecting patients for treatment is not available. Furthermore, it is stated that testing for the presence of *MET*ex14 mutations in plasma specimens is recommended only in patients in whom a tumor biopsy cannot be obtained. If a *MET*ex14 mutation is not detected in a plasma specimen, the possibility of a tumor biopsy should be reconsidered [19]. Information related to the use of liquid biopsies is the same as the FDA has given for other CDx assays using similar technologies [10]. The reason for this is that a negative result from a liquid biopsy does not exclude a potential oncogenic driver (here, the *MET*ex14 mutation), because some tumors do not shed a sufficient amount of DNA into the circulation to be detected by this method [32]

3.4. Amivantamab

Amivantamab (Rybrevant, Jansen Biotech, Horsham, PA, USA) is a bispecific antibody targeting the EGFR and MET [33]. In vitro and in vivo studies have shown that amivantamab is able to disrupt the EGFR and MET signaling functions by ligand blocking and receptor degradation. Furthermore, amivantamab has the ability to induce trogocytosis and engage immune effector cells to eliminate EGFR and MET-presenting tumor cells through antibody-dependent cellular cytotoxicity [20,33].

In a multicohort open-labeled phase I trial (CHRYSALIS) in patients with metastatic NSCLC, amivantamab was investigated in different molecular-defined subgroups, including patients with *EGFR* exon 20 insertion mutation, *MET*ex14 mutation, and *MET*amp. So far, only the results from the *EGFR* exon 20 insertion mutation cohort (n = 81) have been reported, and here, amivantamab given as monotherapy showed an ORR of 40% [34]. The clinical outcome data from this cohort of the CHRYSALIS trial led to an FDA approval of amivantamab for the treatment of adult patients with locally advanced NSCLC with the *EGFR* exon 20 insertion mutation, whose disease has progressed on or after platinum-based chemotherapy [20]. Together with the approval of amivantamab, the Guardant360 assay was approved as CDx for the detection of *EGFR* exon 20 insertions [10]. However, it will be interesting to study the results from the other cohorts of the trial, patients with the *MET*ex14 mutation and *MET*amp, and to see if they can support an expansion of the current indication for amivantamab.

4. Companion Diagnostics and Predictive Biomarkers for MET-Targeted Therapy

Taking into consideration the relatively large number of MET inhibitors that have been or are under development for different indications, it is disappointing to see that only a small proportion have obtained regulatory approval and subsequently reached the clinic. Besides being investigated for treatment of NSCLC, MET inhibitors are under clinical development for different indications, such as gastric and gastroesophageal cancer, hepatocellular carcinoma, and renal cell carcinoma [7,8,35]. These investigational drugs cover both small molecule inhibitors, mono- and bispecific antibodies, as well as antibody–drug conjugates targeting MET [35–37].

One of the reasons for the relatively low success rate of MET-targeted therapy in NSCLC might be the lack of predictive biomarkers with sufficient accuracy. For the FDA-approved drugs, Table 2 lists the biomarkers that were used to select patients for MET-targeted therapy in different clinical trials. These biomarkers include overexpression by IHC, GCN, and *MET*/CEP7 ratios by FISH, GCN by NGS, and *MET*ex14 mutation by NGS. The use of MET overexpression, either MET IHC2+ or IHC3+, has given inconsistent results, and the data in Table 2 shows ORR ranging from 14% to 68% depending on the level of overexpression, with the best response obtained in patients with MET IHC3+ tumors [17,27,29].

The data for *MET*amp in Table 2 also shows a variation with ORR ranging from 16% to 67% for patients with *MET* GCN > 6 or a *MET*/CEP7 ratio ≥ 2.0. For a *MET*/CEP7 ratio > 2.0, the ORR following treatment with MET-targeted therapy ranges from 33% to 67%, whereas *MET* GCN > 6 predicts ORR outcomes in the range of 16% to 67%. *MET* gene copy number gains can occur through both polysomy and amplification, but it seems that true amplification is more likely to lead to oncogenic addiction, which might explain the possibly better predictive properties of *MET*/CEP7 over *MET* GCN [7,16]. NGS was used as a method for the detection of *MET*amp in patients with NSCLC, but results from several comparative studies with FISH have shown poor reliability in detecting the various levels of *MET*amp [38–40]. Based on the results from one of these studies comparing clinical outcome data, the authors concluded that *MET*amp identified by FISH remains the optimal biomarker to identify suitable candidates for MET-TKI therapy [39]. One possible problem in relation to the use of NGS-based assays is likely the lack of control for CEP7, whereby a detected increase in the *MET* gene copy number could be the result of polysomy rather than a true *MET*amp [16].

For NSCLC patients with identified oncogenic drivers, the use of targeted therapy has significantly improved treatment outcome, with high response rates and improved progression-free survival. However, resistance to this type of therapy will be developed sooner or later, and here, *MET*amp seems to play a central role [7,8,16]. For treatment with EGFR inhibitors, *MET*amp has been established as a mechanism of acquired resistance, and evidence is accumulating that this could also occur in NSCLC with targeted therapies related to *ALK-*, *RET-*, and *ROS1*-rearrangements. For patients with *MET*amp who have developed resistance to EGFR inhibitors, the combination with a MET inhibitor seems to overcome the resistance [16,17,29]. In this perspective, it is important to clarify whether NGS can be used as a reliable platform for the detection of *MET*amp in patients with NSCLC, or whether the recommendation should be to use a FISH assay going forward.

In Table 2, the data for *MET*ex14 mutation detected by NGS also shows variability regarding the ORR observed. However, if data originate from larger populations, it seems to show a higher degree of consistency with ORR in the range of 32% to 64% [26,28,30]. In general, when comparing the data in Table 2, it is important to have in mind that these data originate from different patient populations and line of therapy as well as different MET-targeted drugs. Furthermore, several of the patient populations listed are relatively small, which is why one should be very careful not to draw firm conclusions based on the presented data.

So far, the FoundationOne CDx is the only assay linked to a MET inhibitor that has obtained FDA approval [10,31]. This CDx assay is used with capmatinib, and the clinical validation with respect to the detection of the *MET*ex14 mutation was performed based on samples and clinical data from the GENOMETRY mono-1 trial [28]. Here, a clinical bridging study was performed to show both analytical and clinical agreement between the enrollment assay and the FoundationOne CDx assay. In the GENOMETRY mono-1 trial, the patients were enrolled based on test results from a *MET*ex14 mutation RT-PCR assay. To establish concordance between the two assays, both cohorts of treatment-naïve and previously treated patients were analyzed. Based on the assay results, positive percent agreement (PPA), negative percent agreement (NPA), and overall agreement (OPA) were calculated. For the previously treated group of patients, PPA was 96.8%, NPA was 100%, and OPA was 99.1%. For the group of treatment-naïve patients, PPA, NPA, and OPA were all 100% and, thereby, a complete concordance between the two assays was achieved and likewise with respect to the clinical outcome [31].

MET-targeted therapy has also been investigated outside NSCLC, and here, it is more or less the same type of predictive biomarkers that have been used for patient selection. In a phase II trial, the experimental small molecule MET inhibitor AMG 337 was investigated in patients with gastric/gastroesophageal junction/esophageal adenocarcinoma, and a FISH assay with $MET/CEP7 \geq 2.0$ as a cut-off was used [41,42]. Furthermore, the small molecular inhibitors tivantinib and tepotinib have been investigated in patients with hepatocellular carcinoma using MET IHC2+ and IHC3+ overexpression for enrollment in the clinical trials [43,44]. Despite these ongoing development activities, none of the different MET inhibitors have obtained regulatory approval for indications other than NSCLC so far.

5. Conclusions

Despite intensive research and development efforts, relatively few MET inhibitors have shown sufficient clinical activity. One of the reasons for this could be the lack of effective predictive biomarkers to select the right patient population for treatment. So far, capmatinib is the only MET inhibitor that has been approved with a CDx assay. In 2020, capmatinib obtained FDA approval for the treatment of patients with metastatic NSCLC whose tumors harbor a *MET*ex14 mutation. Likewise, in different clinical trials, the *MET*ex14 mutation has also shown predictive properties for drugs such as tepotinib and crizotinib in patients with metastatic NSCLC. Another candidate biomarker is *MET* amplification, which plays an important role in the development of achieved resistance to EGFR inhibitors. Results obtained from different clinical trials indicate that determination of the $MET/CEP7$ ratio by FISH possesses the best predictive property, likely because this approach excludes *MET* amplification caused by polysomy. However, further clinical research will have to show whether $MET/CEP7$ by FISH is an effective predictive biomarker for MET-targeted therapy.

Author Contributions: Conceptualization, J.T.J.; Methodology, J.T.J. and J.M.; Investigation, J.T.J. and J.M.; Data Curation, J.T.J. and J.M.; Writing—Original Draft Preparation, J.T.J. and J.M.; Writing—Review and Editing, J.T.J. and J.M.; Supervision, J.T.J. All authors have read and agreed to the published version of the manuscript.

Funding: This research received no external funding.

Conflicts of Interest: Jan Trøst Jørgensen is an employee of the Dx-Rx Institute and has worked as a consultant for Agilent Technologies, Alligator Biosciences, Argenx, Azanta, Biovica, Euro Diagnostica, Leo Pharma, and Oncology Venture, and has given lectures at meetings sponsored by AstraZeneca, Merck Sharp & Dohme, and Roche. Jens Mollerup is an employee of Agilent Technologies Denmark ApS and a shareholder of Agilent Technologies Inc.

References

1. Slamon, D.J.; Leyland-Jones, B.; Shak, S.; Fuchs, H.; Paton, V.; Bajamonde, A.; Fleming, T.; Eiermann, W.; Wolter, J.; Pegram, M.; et al. Use of chemotherapy plus a monoclonal antibody against HER2 for metastatic breast cancer that overexpresses HER2. *N. Engl. J. Med.* **2001**, *344*, 783–792. [CrossRef]
2. Jørgensen, J.T.; Winther, H.; Askaa, J.; Andresen, L.; Olsen, D.; Mollerup, J. A Companion Diagnostic with Significant Clinical Impact in Treatment of Breast and Gastric Cancer. *J. Front. Oncol.* **2021**, *11*, 676939. [CrossRef]
3. Jørgensen, J.T. Oncology Drug-Companion Diagnostic Combinations. *Cancer Treat. Res. Commun.* **2021**, *29*, 100492. [CrossRef] [PubMed]
4. Cooper, C.S.; Park, M.; Blair, D.G.; Tainsky, M.A.; Huebner, K.; Croce, C.M.; Vande Woude, G. Molecular cloning of a new transforming gene from a chemically transformed human cell line. *Nature* **1984**, *311*, 29–33. [CrossRef] [PubMed]
5. Naldini, L.; Vigna, E.; Narsimhan, R.P.; Gaudino, G.; Zarnegar, R.; Michalopoulos, G.K.; Comoglio, P.M. Hepatocyte growth factor (HGF) stimulates the tyrosine kinase activity of the receptor encoded by the proto-oncogene c-MET. *Oncogene* **1991**, *6*, 501–504.
6. Matsumoto, K.; Umitsu, M.; De Silva, D.M.; Roy, A.; Bottaro, D.P. Hepatocyte growth factor/MET in cancer progression and biomarker discovery. *Cancer Sci.* **2017**, *108*, 296–307. [CrossRef]
7. Guo, R.; Luo, J.; Chang, J.; Rekhtman, N.; Arcila, M.; Drilon, A. MET-dependent solid tumours-molecular diagnosis and targeted therapy. *Nat. Rev. Clin. Oncol.* **2020**, *17*, 569–587. [CrossRef]
8. Recondo, G.; Che, J.; Jänne, P.A.; Awad, M.M. Targeting MET Dysregulation in Cancer. *Cancer Discov.* **2020**, *10*, 922–934. [CrossRef] [PubMed]
9. FDA. Capmatinib (Tabrecta) Full Prescribing Information. Revised: 5/2020. Available online: https://www.accessdata.fda.gov/drugsatfda_docs/label/2020/213591s000lbl.pdf (accessed on 3 December 2021).
10. FDA. List of Cleared or Approved Companion Diagnostic Devices (In Vitro and Imaging Tools). Update 1 December 2021. Available online: https://www.fda.gov/medical-devices/in-vitro-diagnostics/list-cleared-or-approved-companion-diagnostic-devices-in-vitro-and-imaging-tools (accessed on 2 December 2021).
11. Hortobagyi, G.N. Opportunities and challenges in the development of targeted therapies. *Semin. Oncol.* **2004**, *31*, 21–27. [CrossRef]
12. Simon, R.; Maitournam, A. Evaluating the efficiency of targeted designs for randomized clinical trials. *Clin. Cancer Res.* **2004**, *10*, 6759–6763, Erratum in *Clin. Cancer Res.* **2006**, *12*, 3229. [CrossRef]
13. FDA. Guidance for Industry and Food and Drug Administration Staff. In Vitro Companion Diagnostic Devices. 2014. Available online: https://www.fda.gov/media/81309/download (accessed on 6 December 2021).
14. Regulation (EU) 2017/746 of the European Parliament and of the Council of 5 April 2017 on In Vitro Diagnostic Medical Devices and Repealing Directive 98/79/EC and Commission Decision. 2010/227/EU. Available online: https://eur-lex.europa.eu/legal-content/EN/TXT/PDF/?uri=CELEX:32017R0746&from=EN (accessed on 6 December 2021).
15. Regulation 2021/0323 of the European Parliament and of the Council of 14 October 2021. Transitional Provisions for Certain In Vitro Diagnostic Medical Devices and Deferred Application of Requirements for In-House Devices. Available online: https://eur-lex.europa.eu/legal-content/EN/TXT/PDF/?uri=CELEX:52021PC0627&from=EN (accessed on 6 December 2021).
16. Coleman, N.; Hong, L.; Zhang, J.; Heymach, J.; Hong, D.; Le, X. Beyond epidermal growth factor receptor: MET amplification as a general resistance driver to targeted therapy in oncogene-driven non-small-cell lung cancer. *ESMO Open* **2021**, *6*, 100319. [CrossRef] [PubMed]
17. Wu, Y.-L.; Zhang, L.; Kim, D.-W.; Liu, X.; Lee, D.H.; Yang, J.C.-H.; Ahn, M.-J.; Vansteenkiste, J.F.; Su, W.-C.; Felip, E.; et al. Phase Ib/II Study of Capmatinib (INC280) Plus Gefitinib After Failure of Epidermal Growth Factor Receptor (EGFR) Inhibitor Therapy in Patients With EGFR-Mutated, MET Factor-Dysregulated Non-Small-Cell Lung Cancer. *J. Clin. Oncol.* **2018**, *36*, 3101–3109. [CrossRef]
18. FDA. Crizotinib (Xalkori) Full Prescribing Information. Revised 9/2021. Available online: https://www.accessdata.fda.gov/drugsatfda_docs/label/2021/202570s031lbl.pdf (accessed on 8 December 2021).
19. FDA. Tepotinib (Tepmetko) Full Prescribing Information. Revised 2/2021. Available online: https://www.accessdata.fda.gov/drugsatfda_docs/label/2021/214096s000lbl.pdf (accessed on 14 December 2021).
20. FDA. Amivantamab (Rybrevant) Full Prescribing Information. Revised 5/2021. Available online: https://www.accessdata.fda.gov/drugsatfda_docs/label/2021/761210s000lbl.pdf (accessed on 18 December 2021).
21. Kwak, E.L.; Bang, Y.-J.; Camidge, D.R.; Shaw, A.T.; Solomon, B.; Maki, R.G.; Ou, S.-H.I.; Dezube, B.J.; Jänne, P.A.; Costa, D.; et al. Anaplastic lymphoma kinase inhibition in non-small-cell lung cancer. *N. Engl. J. Med.* **2010**, *363*, 1693–1703. [CrossRef] [PubMed]
22. Shaw, A.T.; Kim, D.W.; Nakagawa, K.; Seto, T.; Crinó, L.; Ahn, M.J.; De Pas, T.; Besse, B.; Solomon, B.J.; Blackhall, F.; et al. Crizotinib versus chemotherapy in advanced ALK-positive lung cancer. *N. Engl. J. Med.* **2013**, *368*, 2385–2394. [CrossRef]
23. Shaw, A.T.; Ou, S.H.; Bang, Y.J.; Camidge, D.R.; Solomon, B.J.; Salgia, R.; Riely, G.J.; Varella-Garcia, M.; Shapiro, G.I.; Costa, D.B.; et al. Crizotinib in ROS1-rearranged non-small-cell lung cancer. *N. Engl. J. Med.* **2014**, *371*, 1963–1971. [CrossRef]
24. Moro-Sibilot, D.; Cozic, N.; Pérol, M.; Mazières, J.; Otto, J.; Souquet, P.J.; Bahleda, R.; Wislez, M.; Zalcman, G.; Guibert, S.D.; et al. Crizotinib in c-MET- or ROS1-positive NSCLC: Results of the AcSé phase II trial. *Ann. Oncol.* **2019**, *30*, 1985–1991. [CrossRef] [PubMed]
25. Landi, L.; Chiari, R.; Tiseo, M.; D'Incà, F.; Dazzi, C.; Chella, A.; Delmonte, A.; Bonanno, L.; Giannarelli, D.; Cortinovis, D.; et al. Crizotinib in MET-Deregulated or ROS1-Rearranged Pretreated Non-Small Cell Lung Cancer (METROS): A Phase II, Prospective, Multicenter, Two-Arms Trial. *Clin. Cancer Res.* **2019**, *25*, 7312–7319. [CrossRef]

26. Drilon, A.; Clark, J.W.; Weiss, J.; Ou, S.I.; Camidge, D.R.; Solomon, B.J.; Otterson, G.A.; Villaruz, L.C.; Riely, G.J.; Heist, R.S.; et al. Antitumor activity of crizotinib in lung cancers harboring a MET exon 14 alteration. *Nat. Med.* **2020**, *26*, 47–51. [CrossRef]

27. Schuler, M.; Berardi, R.; Lim, W.T.; de Jonge, M.; Bauer, T.M.; Azaro, A.; Gottfried, M.; Han, J.Y.; Lee, D.H.; Wollner, M.; et al. Molecular correlates of response to capmatinib in advanced non-small-cell lung cancer: Clinical and biomarker results from a phase I trial. *Ann. Oncol.* **2020**, *31*, 789–797. [CrossRef]

28. Wolf, J.; Seto, T.; Han, J.Y.; Reguart, N.; Garon, E.B.; Groen, H.J.M.; Tan, D.S.W.; Hida, T.; de Jonge, M.; Orlov, S.V.; et al. Capmatinib in MET Exon 14-Mutated or MET-Amplified Non-Small-Cell Lung Cancer. *N. Engl. J. Med.* **2020**, *383*, 944–957. [CrossRef]

29. Wu, Y.L.; Cheng, Y.; Zhou, J.; Lu, S.; Zhang, Y.; Zhao, J.; Kim, D.W.; Soo, R.A.; Kim, S.W.; Pan, H.; et al. Tepotinib plus gefitinib in patients with EGFR-mutant non-small-cell lung cancer with MET overexpression or MET amplification and acquired resistance to previous EGFR inhibitor (INSIGHT study): An open-label, phase 1b/2, multicentre, randomised trial. *Lancet Respir. Med.* **2020**, *8*, 1132–1143. [CrossRef]

30. Paik, P.K.; Felip, E.; Veillon, R.; Sakai, H.; Cortot, A.B.; Garassino, M.C.; Mazieres, J.; Viteri, S.; Senellart, H.; Van Meerbeeck, J.; et al. Tepotinib in Non-Small-Cell Lung Cancer with MET Exon 14 Skipping Mutations. *N. Engl. J. Med.* **2020**, *383*, 931–943. [CrossRef] [PubMed]

31. FDA. Summary of Safety and Effectiveness Data. FoundationOne CDx. 2020. Available online: https://www.accessdata.fda.gov/cdrh_docs/pdf17/P170019S011B.pdf (accessed on 14 December 2021).

32. Socinski, M.A.; Pennell, N.A.; Davies, K.D. MET Exon 14 Skipping Mutations in Non-Small-Cell Lung Cancer: An Overview of Biology, Clinical Outcomes, and Testing Considerations. *JCO Precis. Oncol.* **2021**, *5*, 653–663. [CrossRef]

33. Brazel, D.; Nagasaka, M. Spotlight on Amivantamab (JNJ-61186372) for EGFR Exon 20 Insertions Positive Non-Small Cell Lung Cancer. *Lung Cancer* **2021**, *12*, 133–138. [CrossRef] [PubMed]

34. Park, K.; Haura, E.B.; Leighl, N.B.; Mitchell, P.; Shu, C.A.; Girard, N.; Viteri, S.; Han, J.Y.; Kim, S.W.; Lee, C.K.; et al. Amivantamab in EGFR Exon 20 Insertion-Mutated Non-Small-Cell Lung Cancer Progressing on Platinum Chemotherapy: Initial Results from the CHRYSALIS Phase I Study. *J. Clin. Oncol.* **2021**, *39*, 3391–3402. [CrossRef] [PubMed]

35. Yao, H.P.; Hudson, R.; Wang, M.H. Progress and challenge in development of biotherapeutics targeting MET receptor for treatment of advanced cancer. *Biochim. Biophys. Acta Rev. Cancer* **2020**, *1874*, 188425. [CrossRef] [PubMed]

36. Garon, E.B.; Brodrick, P. Targeted Therapy Approaches for MET Abnormalities in Non-Small Cell Lung Cancer. *Drugs* **2021**, *81*, 547–554. [CrossRef]

37. Spigel, D.R.; Edelman, M.J.; O'Byrne, K.; Paz-Ares, L.; Mocci, S.; Phan, S.; Shames, D.S.; Smith, D.; Yu, W.; Paton, V.E.; et al. Results from the Phase III Randomized Trial of Onartuzumab Plus Erlotinib Versus Erlotinib in Previously Treated Stage IIIB or IV Non-Small-Cell Lung Cancer: METLung. *J. Clin. Oncol.* **2017**, *35*, 412–420. [CrossRef]

38. Schmitt, C.; Schulz, A.A.; Winkelmann, R.; Smith, K.; Wild, P.J.; Demes, M. Comparison of MET gene amplification analysis by next-generation sequencing and fluorescence in situ hybridization. *Oncotarget* **2021**, *12*, 2273–2282. [CrossRef]

39. Peng, L.X.; Jie, G.L.; Li, A.N.; Liu, S.Y.; Sun, H.; Zheng, M.M.; Zhou, J.Y.; Zhang, J.T.; Zhang, X.C.; Zhou, Q.; et al. MET amplification identified by next-generation sequencing and its clinical relevance for MET inhibitors. *Exp. Hematol. Oncol.* **2021**, *10*, 52. [CrossRef]

40. Schubart, C.; Stöhr, R.; Tögel, L.; Fuchs, F.; Sirbu, H.; Seitz, G.; Seggewiss-Bernhardt, R.; Leistner, R.; Sterlacci, W.; Vieth, M.; et al. MET Amplification in Non-Small Cell Lung Cancer (NSCLC)-A Consecutive Evaluation Using Next-Generation Sequencing (NGS) in a Real-World Setting. *Cancers* **2021**, *13*, 5023. [CrossRef]

41. Van Cutsem, E.; Karaszewska, B.; Kang, Y.K.; Chung, H.C.; Shankaran, V.; Siena, S.; Go, N.F.; Yang, H.; Schupp, M.; Cunningham, D. A Multicenter Phase II Study of AMG 337 in Patients with MET-Amplified Gastric/Gastroesophageal Junction/Esophageal Adenocarcinoma and Other MET-Amplified Solid Tumors. *Clin. Cancer Res.* **2019**, *25*, 2414–2423. [CrossRef] [PubMed]

42. Jørgensen, J.T.; Mollerup, J.; Yang, H.; Go, N.; Nielsen, K.B. MET deletion is a frequent event in gastric/gastroesophageal junction/esophageal cancer: A cross-sectional analysis of gene status and signal distribution in 1580 patients. *Ann. Transl. Med.* **2021**, *9*, 225. [CrossRef] [PubMed]

43. Kudo, M.; Morimoto, M.; Moriguchi, M.; Izumi, N.; Takayama, T.; Yoshiji, H.; Hino, K.; Oikawa, T.; Chiba, T.; Motomura, K.; et al. A randomized, double-blind, placebo-controlled, phase 3 study of tivantinib in Japanese patients with MET-high hepatocellular carcinoma. *Cancer Sci.* **2020**, *111*, 3759–3769. [CrossRef]

44. Decaens, T.; Barone, C.; Assenat, E.; Wermke, M.; Fasolo, A.; Merle, P.; Blanc, J.F.; Grando, V.; Iacobellis, A.; Villa, E.; et al. Phase 1b/2 trial of tepotinib in sorafenib pretreated advanced hepatocellular carcinoma with MET overexpression. *Br. J. Cancer* **2021**, *125*, 190–199. [CrossRef] [PubMed]

 *cancers*

Review

The Role of MET in Resistance to EGFR Inhibition in NSCLC: A Review of Mechanisms and Treatment Implications

Susan L. Feldt and Christine M. Bestvina *

Department of Medicine, University of Chicago, Chicago, IL 60637, USA; sfeldt@uchicagomedicine.org
* Correspondence: cbestvina@medicine.bsd.uchicago.edu; Tel.: +1-773-702-4627

Simple Summary: In non-small cell lung cancer that has spread to other locations in the body, identifying genetic abnormalities in a patient's cancer has allowed for the development of targeted treatments. For cancers that have genetic changes in *epidermal growth factor receptor (EGFR)*, treatments such as osimertinib have allowed patients to live longer. However, these cancers do eventually continue to grow after targeted therapy. In this paper, we aimed to summarize the current research identifying changes in the *mesenchymal-epithelial transition (MET)* gene that occur after EGFR-targeted therapy, allowing the cancer to become resistant. We summarized current medications that target MET, and early findings from trials that used medications targeting both EGFR and MET together. Targeting both mechanisms at the same time could be a promising new treatment strategy, and larger trials studying these treatments in combination are currently ongoing to understand the potential benefit to patients.

Abstract: Utilizing targeted therapy against activating mutations has opened a new era of treatment paradigms for patients with advanced non-small cell lung cancer (NSCLC). For patients with *epidermal growth factor (EGFR)*-mutated cancers, EGFR inhibitors, including the third-generation tyrosine kinase inhibitor (TKI) osimertinib, significantly prolong progression-free survival and overall survival, and are the current standard of care. However, progression after EGFR inhibition invariably occurs, and further study has helped elucidate mechanisms of resistance. Abnormalities in the *mesenchymal-epithelial transition (MET)* oncogenic pathway have been implicated as common alterations after progression, with *MET* amplification as one of the most frequent mechanisms. Multiple drugs with inhibitory activity against MET, including TKIs, antibodies, and antibody–drug conjugates, have been developed and studied in advanced NSCLC. Combining MET and EGFR is a promising treatment strategy for patients found to have a *MET*-driven resistance mechanism. Combination TKI therapy and EGFR-MET bispecific antibodies have shown promising anti-tumor activity in early clinical trials. Future study including ongoing large-scale trials of combination EGFR-MET inhibition will help clarify if targeting this mechanism behind EGFR resistance will have meaningful clinical benefit for patients with advanced EGFR-mutated NSCLC.

Keywords: NSCLC; *EGFR*; *mesenchymal-epithelial transition*; *MET* amplification; tyrosine kinase inhibitor; antibody drug conjugate

Citation: Feldt, S.L.; Bestvina, C.M. The Role of MET in Resistance to EGFR Inhibition in NSCLC: A Review of Mechanisms and Treatment Implications. *Cancers* **2023**, *15*, 2998. https://doi.org/10.3390/cancers15112998

Academic Editors: Jan Trøst Jørgensen and Jens Mollerup

Received: 29 April 2023
Revised: 27 May 2023
Accepted: 29 May 2023
Published: 31 May 2023

1. Introduction

Identifying activating mutations and the development of targeted therapies for these mutations have made precision medicine a reality for the care of patients with non-small cell lung cancer (NSCLC). In patients with advanced or metastatic NSCLC, molecular testing is recommended by the National Comprehensive Cancer Network (NCCN) guidelines for established molecular biomarkers [1]. An *epidermal growth factor receptor (EGFR)* mutation is a commonly found oncogenic driver in advanced NSCLC [2]. Exon 19 deletions and exon 21 L858R substitutions are the most common *EGFR*-activating mutations, accounting for

80–90% of *EGFR*-positive tumors [3]. For patients with activating *EGFR*-mutated NSCLC, targeted therapy with an EGFR tyrosine kinase inhibitor (TKI) is standard of care [2].

First- and second-generation EGFR TKIs including erlotinib, afatinib, geftinib, and dacomitinib have inhibitory effects on EGFR. However, a common pattern seen when utilizing targeted therapy is the development of resistance to the targeted agent through additional mutations or alterations. This was seen after treatment with first- and second-generation EGFR inhibitors; resistance commonly developed through the T790M *EGFR* mutation [4]. The third generation EGFR TKI osimertinib was developed and shown to be able to overcome the T790M resistance mechanism and was initially approved as second-line therapy. In the FLAURA trial, osimertinib demonstrated superior efficacy to first and second generation EGFR TKIs, and is now recommended as first-line therapy in advanced *EGFR*-positive NSCLC [5,6].

However, despite the shifting treatment paradigm to first-line osimertinib, resistance inevitably develops. Next-line therapy for those who progress after osimertinib has typically consisted of platinum-based chemotherapy. The study of resistance mechanisms to osimertinib has revealed additional molecular targets for therapy, with *mesenchymal-epithelial transition (MET)* oncogene alterations as the most common mechanism. In analyses of patients in the FLAURA trial, in *EGFR*-positive NSCLC treated with first-line osimertinib who progress, *MET* amplification can be seen in 15% of patients [7]. In an analysis of circulating-tumor DNA from plasma samples at baseline and at disease progression after first-line treatment with osimertinib similarly found *MET* amplification as the most frequent genetic mechanism of resistance found, identified in 16% [8]. Identifying specific mechanisms of resistance in an individual tumor after progression following first-line osimertinib treatment and adapting treatment to target to the mechanism could prolong survival [9]. This is not exclusive to *EGFR* mutant tumors; *MET* amplification is also being identified in as a driver to resistance to ALK, RET, and ROS-1 fusion TKI treatment [10].

MET encodes for a receptor tyrosine kinase which is activated by hepatocyte growth factor (HGF), and is found primarily in epithelial cells [11]. Downstream *MET* signaling activates RAS-MAPK, PI3K, and STAT3 pathways, leading to cell migration, invasion, proliferation, and cell survival [10,12]. Increased *MET* signaling has been seen in many types of cancers, including NSCLC, including gastric cancers, colorectal cancer, and papillary renal carcinomas [13]. *MET* abnormalities occur through *MET* exon skipping and amplification [14,15]. With the further implication of the role of *MET* in resistance to EGFR TKIs and the development of MET targeted therapies, this has become a promising area of clinical development in the treatment of advanced NSCLC [16].

2. Molecular Mechanisms of Acquired Resistance

Further study of the genomic changes of patients who've progressed after treatment with osimertinib have elucidated a diverse range of mechanisms of resistance. This includes EGFR-dependent mechanisms, alterations in *MET, RET, BRAF, KRAS,* and *PI3K,* as well as histologic transformation. Patients may have more than one coexisting molecular mechanism contributing to resistance [17]. While the overall landscape of osimertinib resistance is diverse, *MET* alterations, particularly *MET* amplification, have been seen as the most frequent mechanism.

MET dysregulation most commonly occurs through amplification or exon 14 skipping mutations. Exon 14 skipping mutations are caused by a point mutation or deletion that leads to loss of exon 14. Without exon 14 appropriately transcribed, the MET protein is more stable and less prone to degradation, thereby increasing MET signaling [18]. In a treatment-naïve population with comprehensive genomic profiling, *MET exon 14* skipping mutations are found in about 3% of patients with NSCLC [19].

MET amplification results in an increase in the number of copies of *MET,* either through focal amplification or chromosome 7 polysomy. While focal *MET* amplification has been associated with oncogenicity, amplification through chromosome 7 polysomy is typically not. Focal *MET* amplification is found in 1–6% of treatment-naïve NSCLC [20], and has demonstrated sensitivity to MET inhibitors such as capmatinib [21]. *MET* amplification is seen more frequently after first-line treatment with osimertinib than in a treatment-naïve population and is the most common mechanism of resistance to osimertinib identified [7].

3. Testing of MET Alterations

Currently, there is not a uniform practice for how to perform *MET* testing at either time of diagnosis or at time of progression on targeted therapy. *MET* alterations can be tested for using a variety of methods, including both tumor tissue testing and liquid biopsy testing. *MET* amplification is traditionally diagnosed through fluorescence-in-situ-hybridization (FISH) of tissue obtained by biopsy. *MET* amplification by FISH is defined using two main strategies. The first determines gene copy number (GCN), with multiple cutoffs used, commonly a GCN of five or more, but also six or fifteen have been used as cutoffs. An alternative method is to control for chromosome 7 by using a ratio of *MET* per cell to chromosome 7 centromere (MET/CEP7). With this method, a *MET* to CEP7 ratio of greater than 2 is the typical definition of *MET* amplification [13].

NGS can also be used for the detection of *MET* amplification. As with FISH, there is not a consensus of copy number to define amplification. Unlike FISH, NGS does not control for chromosome 7 copy number [13]. There has been some discrepancy between *MET* amplification diagnosed with FISH compared to NGS, so a higher GCN cutoff of at least 10 is often used. With a higher cutoff, concordance with FISH is improved, but low to moderate *MET* amplification is less likely to be detected. At these levels, FISH has improved detection compared to NGS [22].

However, for detecting *MET exon 14* skipping mutations, NGS is overall the preferred strategy [23]. There is a suggestion that RNA-based NGS may be preferred to DNA-based NGS for in this setting, possibly due to heterogeneity in mutation variants resulting in exon 14 skipping [24]. Qt-PCR of *MET exon 14* skipping as a single gene testing could also be a reasonable approach [23].

IHC has also been studied for diagnosis of *MET* alterations. Different IHC cutoffs for determining high *MET* expression have been used in clinical trials and vary in both the minimum percent of tumor cells expressing *MET* and in level of IHC positivity (2+ vs. 3+). IHC appears to be less sensitive in detecting both exon 14 skipping mutations and *MET* amplification [25]. Thus, FISH and NGS are the preferred methods for detecting *MET* amplification.

The use of cell-free DNA (cfDNA) testing in non-small cell lung cancer has increased dramatically over the past five years. Liquid biopsy can provide either a complementary or an alternative method of genomic profiling in addition to or in place of tissue moleculars. cfDNA NGS can also be used to diagnose alterations of *MET*, though interpretation of *MET* amplification can be difficult using this platform [26]. An additional benefit of this method is that cfDNA does not have the same heterogeneity and potential sampling error of a solid tissue biopsy. However, tumors may not shed enough genetic material for alterations to be detected, and this may lead to missing alterations [13]. Cell-free DNA could have a role as a molecular marker of response to treatment. In the VISION trial of tepotinib in advanced NSCLC with *MET exon 14* skipping mutations, cfDNA was obtained baseline and on treatment. A total of 67% of patients had a molecular response, which had a high concordance with clinical response [27].

Without a consensus in method of detecting *MET* alterations, clinical trials use multiple of these methods to determine what is considered *MET*-altered NSCLC, and future study could benefit from a standardized approach. However, given the implication of *MET* in resistance mechanisms, rebiopsy or liquid biopsy for genomic profiling of resistance is warranted in patients who progress after EGFR inhibition and may reveal *MET* alterations for targeted treatment.

4. MET Inhibition

4.1. Non-Selective MET TKIs

TKIs with activity against MET are defined by how they bind MET as well as whether they are selective or non-selective for MET. Type 1 MET TKIs compete with ATP and are divided into Type 1a and Type 1b based on where they bind to MET. Type 1a binds the solvent front residue, which is not specific to MET, and are therefore non-selective MET inhibitors. Type 1b selectively binds MET. Type 2 MET TKIs differ in binding inactive MET and are also non-selective. Non-selective MET inhibitors include both crizotinib and cabozantinib. Crizotinib is a type 1a or non-selective MET TKI in that it inhibits MET in addition to ALK, ROS, and RON. It was initially approved for *ALK* or *ROS1* rearrangements in NSCLC [28,29]. In the phase I PROFILE 1001 trial, crizotinib was studied in advanced NSCLC with multiple other genetic alterations considering its non-specific inhibition. In 69 patients with *MET exon 14* alterations who received crizotinib, the objective response rate (ORR) was 32% (95% CI 21–45%) with a median progression free survival (PFS) of 7.3 months (95% CI 5.4–9.1 months). The most common adverse events seen with crizotinib included edema (51%), vision disorder (45%), diarrhea (39%) and vomiting (29%). Most common grade 3 or greater adverse events were transaminitis (4%) and dyspnea (4%) [30].

Cabozantinib is a type II non-selective MET TKI. In addition to MET activity, it also targets VEGFR, RET, TIE2, FLT-3, and KIT. Cabozantinib is currently approved for renal cell carcinoma, medullary thyroid carcinoma, and hepatocellular carcinoma. Studies of cabozantinib in NSCLC are more limited. In advanced NSCLC, cabozantinib has been studied in *EGFR* wildtype, but not mutant, patients. In a phase II trial, 42 patients with *EGFR* wildtype advanced non-squamous NSCLC who had received one or two previous treatments were randomized to receive erlotinib, cabozantinib, or combination erlotinib and cabozantinib. *MET* alterations were not specifically tested for. PFS was longer with cabozantinib alone (4.3 months, 95% CI 3.6–7.4) and with combination erlotinib and cabozantinib (4.7 months, 95% CI 2.4–7.4) than with erlotinib monotherapy (1.8 months; 95% CI 1.7–2.2 months). Adverse events with cabozantinib were most commonly fatigue (55%), diarrhea (50%), and nausea (45%). Of grade 3 or greater adverse events, hypertension (25%), fatigue (15%), and oral mucositis (10%) were most common [31].

4.2. Selective MET TKIs

Type 1b MET inhibitors are selective in their inhibition of MET. This class of drugs has had recent FDA approvals after demonstrating success in multiple recent clinical trials, based on an improvement in both efficacy as well as decreased toxicity given there is less off-target activity. This class includes tepotinib, capmatinib, and savolitinib.

Tepotinib is a selective MET inhibitor that is FDA approved in patients with NSCLC with *MET exon 14* skipping alterations. The phase II VISION trial studied tepotinib in cohorts of both *MET* amplified and *MET exon 14* skipping as a mix of first-, second-, and third-line therapy. *MET exon 14* skipping was determined by liquid biopsy or RNA NGS of tissue. In 152 patients treated with tepotinib, the ORR was 46% (95% CI 36–61%) and was similar when stratified by prior therapy, and a median PFS of 8.5 months (95% CI 6.7–11.0) was seen. Grade 3 or greater adverse events occurred in 28% of patients, most commonly peripheral edema (7%), increased lipase (3%) and amylase (2%) [27].

The VISION cohort of *MET* amplification consisted of 24 patients with amplification determined by liquid biopsy. Median PFS was 4.2 months (95% CI 1.4-NE) in this cohort. The overall ORR was 42% (95% CI 22–63), with a 71% (95% CI 29–96) response rate seen in treatment-naïve patients, 30% (95% CI 7–65) as second-line therapy, and 29% (95% CI 4–71) as third-line therapy. A total of 67% of patients had treatment-related adverse events, most commonly peripheral edema (38%), generalized edema (17%), and constipation (17%). Grade 3 or greater adverse events were experienced by 29% of patients, most commonly peripheral edema (8%) and generalized edema (8%) [32].

Capmatinib is a selective MET inhibitor that is FDA approved for NSCLC with *MET exon 14* skipping mutations. In the GEOMETRY mono-1 phase II clinical trial, patients with *MET* dysregulated advanced NSCLC were treated with capmatinib. In those with *MET exon 14* skipping mutations, the ORR was 41% (95% CI 29–53) with a median PFS of 5.4 months (95% CI 4.2–7.0). In patients who had not received prior treatment, the ORR was higher at 68% (95% CI 48–84) with a higher median PFS as well at 12.4 months (95% CI 8.2-NE). However, in patients with *MET* amplification, efficacy was limited to those with high gene copy numbers, with an ORR of 29% (95% CI 19–41) and a median PFS of 4.1 months (95% CI 2.9–4.8) in those with a gene copy number of greater than 10. Peripheral edema (51% of patients) and nausea (45%) were the most common adverse events. Grade 3 or higher adverse events were seen in 67% of patients; most commonly peripheral edema (9%) and dyspnea (7%) [21].

Savolitinib is a selective MET TKI that is approved in China for NSCLC with *MET exon 14* skipping mutations. Savolitinib was studied in a phase II trial in China in advanced or metastatic NSCLC with an exon 14 skipping mutation and that has progressed or had toxicity to a prior treatment. The ORR was 49.2% (95% CI 31.1–55.3) with a PFS of 6.9 months (95% CI 4.6–8.3). All patients had a treatment-related adverse event, most commonly peripheral edema (54%), nausea (46%), and transaminitis (37–39%). Grade 3 or greater treatment-related adverse events were seen in 46% of patients, most commonly transaminitis (10–13%) and peripheral edema (9%) [33].

4.3. MET Antibodies

MET antibody-directed therapy is another area of active development and clinical investigation. One of the first promising signals of efficacy was emibetuzumab, a humanized immunoglobulin G4 monoclonal bivalent MET specific monoclonal antibody. Emibetuzumab has both ligand dependent and independent effects, both blocking HGF from binding MET and leading to MET being internalized and degraded. In a phase II trial, patients with stage IV EGFR mutated NSCLC were randomized after 8 weeks of erlotinib to continuing erlotinib monotherapy or to erlotinib plus emibetuzumab 750 mg infusion every 2 weeks. There was no difference in progression free survival between the overall groups (9.3 months with combination therapy and 9.5 months with erlotinib monotherapy). A post-hoc analysis of those with high *MET* expression was performed, defined as IHC with *MET* expression level of 3+ in at least 90% of tumor cells, which 24 patients met. With this definition of high *MET* expression, progression free survival with emibetuzumab plus erlotinib was 20.7 months vs. 5.4 months with erlotinib alone (HR = 0.39, 90% CI 0.17–0.91). Toxicity was higher in the combination arm, with peripheral edema (11.3% vs. 0%) and mucositis (15.5% vs. 8.6%) occurring more frequently than in those who received erlotinib alone [34].

Onartuzumab is a recombinant, humanized monoclonal antibody against *MET*. Onartuzumab blocks interaction with HGF by binding to extracellular MET. Phase II and III trials have studied onartuzumab in combination with erlotinib and chemotherapy without demonstrating a benefit in PFS or overall survival (OS) [35–37]. The phase III *MET* Lung trial studied onartuzumab with erlotinib or placebo infusion plus erlotinib. 499 patients with locally advanced or metastatic NSCLC with *MET* positive status defined as IHC 2+ or greater in at least 50% of cells who had progressed after platinum-based chemotherapy were enrolled. Shorter OS was seen in in the onartuzumab arm at 6.8 months vs. 9.1 months with placebo (HR = 1.27, 95% CI 0.98–1.65). Peripheral edema (21.8% vs. 7.8%) and hypoalbuminemia (17.3% vs. 3.7%) were more common in the onartuzumab arm than erlotinib monotherapy [38]. Grade 4 (5.2% vs. 2.9%) and grade 5 (6.9% and 4.1%) adverse events were more common in the onartuzumab group.

4.4. MET Antibody-Drug Conjugates

Teliso-V is an antibody-drug conjugate of the telisotuzumab humanized monoclonal antibody which targets c-MET conjugated to the microtubule inhibitor monomethyl auristatin E (MMAE). In a phase I/Ib study, patients with NSCLC and a c-MET H-score of at least 150 or local lab reported *MET* amplification or exon 14 skipping mutation received

Teliso-V monotherapy. 40 c-MET+ patients were enrolled, and the ORR was 23% (95% CI 10.8–38.5) with median PFS of 5.2 months. A total of 65% of patients had grade 3 or greater adverse events, most commonly anemia (10%), fatigue (8%), and peripheral neuropathy (6%) [39]. Another MET-targeting antibody-drug conjugate (REGN5093-M114) is being studied in pre-clinical models and has shown promising anti-tumor activity in cells after progression following treatment with osimertinib and savolitinib [40].

5. Combination MET and EGFR Inhibitors

Considering the demonstrated benefit of EGFR inhibition in *EGFR*-mutated advanced NSCLC and the prevalence of *MET* dysregulation in resistance to third generation EGFR inhibitors, inhibition of EGFR and MET in combination has been a preferred strategy in clinical investigation in patients previously treated and progressed after initial EGFR inhibitor therapy. With uncertainty in how to treat patients with *MET*-amplification-mediated resistance to EGFR-TKI, real-world data does support the approach of combination EGFR and MET inhibition. 70 patients received either EGFR-TKI and crizotinib (n = 38), crizotinib monotherapy (n = 10), or chemotherapy (n = 22). PFS was longer in combination inhibition than with crizotinib monotherapy or chemotherapy (5.0 months vs. 2.3 and 2.9 months, respectively) [41]. More recently, combination EGFR/MET inhibition has been moved into studies in the first line.

5.1. EGFR TKI and MET TKI

With multiple EGFR and MET inhibitors available, trials have studied different combinations of these inhibitors to identify a dual-inhibition strategy with significant anti-tumor activity as well as an acceptable risk profile. Results of trials utilizing and EGFR TKI and a MET TKI are summarized in Table 1.

Table 1. Trials targeting both EGFR and MET using EGFR and MET TKIs.

EGFR TKI	MET TKI	Study (NCT ID, Name, Author, Year)	Population	Treatment	MET Alteration	N	Objective Response Rate (ORR)	Progression Free Survival (PFS)
Gefitinib	Capmatinib	NCT01610336, Wu et al., 2018 [41]	*EGFR+* NSCLC acquired resistance to first- or second-generation EGFR TKI	Gefitinib 250 mg once daily + Capmatinib 400 mg twice daily	IHC 3+, IHC 2+ plus *MET* GCN $\geq$ 5, or *MET* GCN $\geq$ 4	100	29%	5.5 months
Osimertinib	Savolitinib	NCT02143466, TATTON, Sequist et al., 2020 [42]	*EGFR+* NSCLC progressed after prior therapy	Osimertinib 80 mg once daily + Savolitinib 300 mg once daily	IHC 3+, *MET* GCN $\geq$ 5 or *MET*/CEP $\geq$ 2:1	138	33–67%	5.5–11.0 months
			EGFR+ T790M-NSCLC with no prior third-generation EGFR TKI			42	62%	9.0 months
Osimertinib	Savolitinib	NCT03944772, ORCHARD, Yu et al., 2021 [43]	*EGFR+* advanced NSCLC with progression on first-line osimertinib	Osimertinib 80 mg once daily + Savolitinib 300 or 600 mg once daily	*MET* amplification or exon 14 skipping by NGS	20	41%	Not reported
Gefitinib	Savolitinib	NCT02374645, Yang et al., 2021 [44]	*EGFR+* advanced NSCLC progressed on prior EGFR-TKI with *MET* amplification	Gefitinib 250 mg once daily + Savolitinib 600 mg once daily	*MET* amplification by FISH GCN $\geq$ 5 or *MET*/CEP $\geq$ 2:1)	51	31%	4.0 months
Erlotinib	Capmatinib	NCT01911507, McCoach et al., 2021 [45]	Advanced *MET*-positive NSCLC (Cohort A *EGFR+*)	Erlotinib 100–150 mg once daily + Capmatinib 100–600 mg twice daily	FISH GCN or *MET*/CEP outside of normal range, IHC 2-3+, +RT-PCR, or exon14 splice mutation	12	50%	Not reported

Table 1. *Cont.*

EGFR TKI	MET TKI	Study (NCT ID, Name, Author, Year)	Population	Treatment	MET Alteration	N	Objective Response Rate (ORR)	Progression Free Survival (PFS)
Gefitinib	Tepotinib	NCT01982955, INSIGHT 1, Liam et al., 2023 [46]	Advanced/metastatic *EGFR*+ NSCLC acquired resistance to first- or second-generation EGFR TKI T790M-, no prior MET therapy	Gefitinib 250 mg once daily + Tepotinib 500 mg once daily vs. Chemo-therapy	IHC 2+ or 3+, *MET* GCN $\geq$ 5, or *MET*/CEP $\geq$ 2:1	55	45% (vs. 33% with chemo-therapy)	4.9 months (vs. 4.4 months with chemotherapy)
					MET amplification by FISH GCN $\geq$5 or *MET*/CEP $\geq$2:1)	19	68% (vs. 43% with chemo-therapy)	16.6 months (vs. 4.2 months with chemo-therapy)
Osimertinib	Tepotinib	NCT03940703, INSIGHT 2, Mazieres et al., 2022 [47]	Advanced *EGFR*+ NSCLC with *MET* amplification after progression on first-line osimertinib	Osimertinib 80 mg once daily + Tepotinib 500 mg once daily	*MET* amplification by FISH GCN $\geq$ 5 or *MET*/CEP $\geq$ 2:1)	22	55%	Not reported

EGFR: Epidermal growth factor receptor; MET: Mesenchymal-epithelial transition; TKI: Tyrosine kinase inhibitor.

In a study of 104 treatment-naïve patients with de novo *EGFR* positive and *MET* over-expressed advanced NSCLC, EGFR TKI monotherapy (n = 48), EGFR TKI plus crizotinib (n = 9), EGFR TKI plus chemotherapy (n = 12), and chemotherapy alone (n = 35) were compared. *MET* overexpression was defined by IHC with above-median H-score and 2+ or 3+ staining in 50% or greater of tumor cells. Notably, EGFR TKIs varied and included geftinib, erlotinib, afatinib, and osimertinib. This study showed that those treated with an EGFR-TKI (monotherapy and combination with crizotinib) had a longer PFS than chemotherapy (8.0 months vs. 4.0 months, HR = 0.50, 95% CI 0.21–0.80). EGFR TKI monotherapy or with crizotinib had comparable PFS (8.0 months vs. 8.5 months, HR = 0.96, 95% CI 0.44–2.09). Grade 3 or greater rashes were more common in patients who received EGFR-TKI plus crizotinib (0% vs. 33.3%) [42].

In a phase 1b/II trial, patients with *EGFR*-positive, *MET*-amplified NSCLC who had progressed on prior EGFR TKI treatment received capmatinib in combination with gefitinib. *MET* amplification was defined as GCN of at least 5 and/or a *MET*/centromere ratio of 2 or greater or MET overexpression with at least 50% of tumor cells with moderate or strong IHC staining. The ORR across the phase Ib/II study was 27%. In patients with high *MET* amplification defined as gene copy number of at least 6 (36 of the total 100 patients), a higher ORR of 47% was seen. The most common adverse events were nausea (28%), peripheral edema (22%), decreased appetite (21%), and rash (20%). Grade 3 or greater adverse events were seen in 33% of patients, most commonly increased amylase or lipase (both in 6%) [48].

TATTON was a phase 1b study of locally advanced or metastatic *MET*-amplified, *EGFR*-mutated NSCLC who had progressed on EGFR TKIs. Osimertinib was studied in combination with multiple other targeted therapies: selumetinib (MEK1/2 inhibitor), durvalumab (anti-PD-L1 monoclonal antibody), and savolitinib. *MET* amplification was defined as FISH with GCN of 5 of greater or *MET*-CEP7 ratio 2 or greater, IHC with 3+ expression in 50% of cells, or NGS with 5 or greater copies in 20% of tumor cells. Cohorts were stratified based on exposure to prior third generation EGFR TKI, and in those without prior third generation TKI, whether *EGFR T790M* was present [49].

In patients previously treated with a third generation EGFR TKI, the ORR of savolitinib plus osimertinib was 33% (95% CI 22–46). However, in those who had not been previously treated with a third generation EGFR TKI, ORR was even higher, ranging from 62–67% and regardless of T790M status [43]. Nausea (67%), rash (56%), and vomiting (50%) were the most common adverse events [50].

The combination of osimertinib and savolitinib was also studied in *EGFR* mutant NSCLC previously treated with first-line osimertinib with *MET* alterations in the OR-CHARD study. *MET* amplification and exon 14 skipping mutations were included and identified by NGS of tumor biopsy. In 20 patients, an ORR of 41% was seen. A total of

30% had a grade 3 or greater adverse event, most commonly pneumonia and decreased neutrophil count (10% each) [51].

The phase II SAVANNAH trial is currently investigating savolitinib plus osimertinib as second-line therapy in patients with acquired resistance to osimertinib *MET* overexpression or amplification [44]. The phase II FLOWERS trial is ongoing and studying osimertinib with or without savolitinib as first-line therapy in patients with *MET*-amplified or overexpressed and *EGFR*-positive locally advanced or metastatic NSCLC [45].

In a phase 1b trial, combination savolitinib and gefitinib were studied in patients with *EGFR*-mutant *MET*-amplified advanced NSCLC in China after progression following prior EGFR TKI therapy. *MET* amplification was determined by FISH with a GCN or 5 or greater or *MET*-CEP7 of 2:1 or greater. ORR overall was 31%, with higher ORR seen in *EGFR T790M* negative (52%) and lower seen in *EGFR T790M* positive (9%). Most common adverse events were vomiting (46%), nausea (40%), and increased aspartate aminotransferase (39%). Most common grade 3 or greater adverse events included transaminitis (AST and ALT 7% each), and increased gamma-glutamyltransferase (5%) [46].

In a phase I/II dose escalation trial, combination capmatinib and erlotinib were studied in *MET*-positive NSCLC. This included a cohort of patients with *EGFR* mutations designated Cohort A with 12 patients. ORR in this cohort was 50%. Most common adverse events were rash (63%), fatigue (51%), and nausea (45.7%). Grade 3 or greater adverse events were seen in 34% of patients, most commonly decreased lymphocytes (9%), limb edema (6%), anorexia (6%), and increased lipase (6%) [52].

The INSIGHT-1 phase 1b/2 trial studied tepotinib plus gefitinib in patients with *EGFR*-mutated NSCLC that were T790M negative and *MET* overexpression (IHC 2+ or 3+) or *MET* amplified by FISH (GCN 5 or greater or *MET*/CEP 2 or greater). In the phase 2 trial, patients were randomly assigned to tepotinib plus gefitinib at the phase 1b determined dose of 500 mg or standard platinum doublet chemotherapy. Final analyses of phase 2 with 55 patients are now published, and in the group at large, median PFS was similar at 4.9 months with tepotinib and gefitinib vs. 4.4 months with chemotherapy. In the 19 patients with *MET* amplification, tepotinib plus geftinib had longer PFS (HR 0.13, 90% CI 0.04–0.43) and OS (HR 0.10, 90% CI 0.02–0.36) than chemotherapy. ORR was 66.7% with combination TKI therapy compared to 42.9% with chemotherapy [53]. Rates of treatment-related grade 3 or worse adverse events were similar between the groups (19% in the tepotinib/gefitinib group vs. 30% in chemotherapy), with increased amylase (16%) or lipase (13%) being the most common of grade 3 or greater adverse events with combination tepotinib and gefitinib [47].

INSIGHT 2 is an ongoing phase II trial of tepotinib plus osimertinib in advanced *EGFR*-mutant NSCLC with acquired resistance to first-line osimertinib and with *MET* amplification determined by FISH with GCN of 5 or greater or *MET*/CEP7 of 2 or greater [54]. Initial results from INSIGHT 2 were presented at ESMO 2022, and in 22 patients with at least 9 months follow up, an ORR of 54.5% was seen, as well as a similar safety profile. Primary analysis is planned for when all patients have had at least 9 months follow up [55].

5.2. EGFR TKI and MET Antibody

Results of combination EGFR- and MET-inhibition-utilizing antibodies and antibody drug conjugates are summarized in Table 2.

In a randomized phase II trial, patients with stage IV NSCLC with acquired resistance to erlotinib with increased *MET* expression were randomized 3:1 to either emibetuzumab with erlotinib or emibetuzumab monotherapy. Increased *MET* expression was defined IHC with as at least 10% of cells expressing MET at 2+. ORRs were low: 3% for emibetuzumab plus erlotinib (95% CI 0.4–10.5) and 4.3% for emibetuzumab monotherapy (95% CI 0.1–21.9). With combination therapy, fatigue (29%), diarrhea (25%), and nausea (23%) were the most common adverse events overall. Grade 3 or greater adverse events were seen in 24.1% of patients on combination therapy, most commonly dermatitis acneiform (6%) and hypoalbuminemia (3.6%) [56].

Table 2. Trials targeting both EGFR and MET using either antibodies or antibody drug conjugates.

Mechanism Studied	Study (NCT ID, Name, Author, Year)	Population	Treatment	MET Alteration	N	Objective Response Rate (ORR)	Progression Free Survival (PFS)
EGFR TKI + MET antibody	NCT01900652, Camidge et al., 2022 [54]	Metastatic stage IV NSCLC with acquired resistance to erlotinib	Erlotinib 150 mg once daily + Emibetuzumab 750 mg 1.5-h infusion once every 2 weeks	IHC 2+	83	3%	2.9 months
EGFR/MET bispecific antibody +/− EGFR TKI	NCT02609776 CHRYSALIS, Bauml et al. [56] and Leighl et al. [55]	Metastatic or unresectable *EGFR* mutant NSCLC progressed on osimetinib	Amivantamab 1050 mg or 1400 mg (if >80kg) once a week in cycle 1, every 2 weeks following monotherapy	N/A	121	19%	4.2 months
			+ lazertinib 240 mg once daily		45	36%	4.9 months
EGFR TKI + MET ADC	NCT02099058, Park et al., 2021 [56]	Advanced *EGFR*+ NSCLC progressed on prior EGFR TKI	Erlotinib 150 mg once daily + Telisotuzumab vedotin 2.7 mg/kg IV once every 3 weeks	IHC H-score ≥ 150	28	32.1%	5.9 months

EGFR: Epidermal growth factor receptor; MET: Mesenchymal-epithelial transition; TKI: Tyrosine kinase inhibitor; ADC: Antibody drug conjugate.

5.3. EGFR TKI and MET ADC

Teliso-V was studied in a phase 1b trial in combination with erlotinib in 42 patients with *EGFR* positive NSCLC who had progressed on a prior EGFR TKI. Patients were *c-MET* positive determined by histology H score of at least 150. Median PFS was 5.9 months (95% CI 2.8 to not reached). ORR was 30.6% (95% CI 16.3–48.1), and in those who were *c-MET* high defined as H score 225 or greater, ORR was 52%. The most common adverse events were neuropathies (57%) and dermatitis acneiform (38%), with grade 3 treatment-related AEs in 31% of patients, most frequently hypophosphatemia (7%) and peripheral sensory neuropathy (7%) [57].

5.4. EGFR-MET Bispecific Antibody

Amivantamab is a bispecific antibody for both EGFR and MET. Amivantamab was studied in patients with advanced NSCLC in cohorts based on *EGFR* and *MET* status [58]. In the CHRYSALIS phase 1 trial, amivantamab was studied alone or in combination lazertinib in patients with metastatic or unresectable NSCLC with *EGFR* mutations. Results have been presented at ASCO and ESMO. In patients who had progressed on osimertinib, an ORR of 36% was seen in those treated with combination amivantamab and lazertinib, with median PFS of 4.9 months. This is compared to ORR 19% and median PFS 4.2 months with amivantamab monotherapy [59,60]. CHRYSALIS 2 explored the combination of amivantamab and lazertinib in patients with *EGFR* mutant advanced NSCLC who had progressed after osimertinib and platinum-based chemotherapy. In this population, an ORR of 33% was seen [61].

In the CHRYSALIS trial, a cohort of patients with *EGFR exon 20* insertion mutations were also enrolled. In patients with locally advanced or metastatic NSCLC with *EGFR exon 20* insertion mutations who had progressed on or after platinum-based chemotherapy, an ORR of 40% (95% CI 29–51) was seen. Amivantamab received accelerated FDA approval in this setting based on these results [62].

In safety analyses of amivantamab, the most common adverse events were rash (86%), infusion-related reactions (66%), and paronychia (45%). Grade 3 or greater adverse events occurred in 35%, most commonly hypokalemia (5%), rash, pulmonary embolism, diarrhea, and neutropenia (all in 4% each) [58]. The MARIPOSA and MARIPOSA-2 phase III trials of amivantamab and lazertinib (third generation EGFR TKI) as first-line therapy for *EGFR*+ NSCLC are ongoing.

6. Future Directions

There are several areas of clinical study that can help to move the field forward for *MET* amplification as a resistance mechanism to EGFR TKIs. Establishing a universal definition of *MET* amplification or overexpression with a focus on FISH and NGS as more sensitive diagnostic methods will be important in standardizing reporting establishing clinical trial eligibility. While tissue and cfDNA NGS have had significant uptake in recent years, it is unclear to what degree *MET* FISH is being performed for NSCLC off protocol and as standard of care. The optimal method for targeting *MET* amplification or overexpression has yet to be determined. As detailed above, MET is being targeted by TKIs, antibodies, and antibody drug conjugates. It is unclear if one particular drug or modality will have a clear advantage as far as efficacy or toxicity, or if these decisions will need to be made on an individual patient basis. Additionally, as clinical trials are now being designed to move *MET*-directed therapy to frontline treatment for EGFR-mutated NSCLC patients, it is yet unknown if this will yield benefit for all patients, or if additional biomarker studies are needed to understand exactly which patients benefit from a combination approach.

7. Conclusions

Identifying driver mutations and targeted therapies has developed a new standard of care for many patients with NSCLC. By identifying the genetic drivers of EGFR inhibitor resistance in a similar way of identifying initial driver mutations, the same principle of targeting these genetic drivers could be a strategy to delay progression and potentially delay the need for chemotherapy. In studies of genetic alterations after progression on osimertinib, *MET* dysregulation, particularly *MET* amplification, has been identified as a common mechanism of resistance. However, identifying a standardized definition of *MET* amplification for trial eligibility is needed.

Multiple drugs that inhibit MET have been studied in advanced NSCLC, including tyrosine kinase inhibitors, antibodies, and antibody-drug conjugates. Combining EGFR inhibition and MET-directed therapy has been a promising area of clinical study. In early clinical trials, both combination TKI therapy, and combination TKI and antibody, antibody-drug-conjuate, and bi-specific EGFR-MET antibody have shown anti-tumor activity. Large-scale clinical trials of combination EGFR-MET inhibition are needed to understand the clinical benefit of this treatment strategy, as well as to consider this strategy as potential first-line therapy in advanced *EGFR* positive NSCLC.

Author Contributions: Conceptualization, S.L.F. and C.M.B.; methodology, S.L.F. and C.M.B.; software, S.L.F. and C.M.B.; validation, S.L.F. and C.M.B.; formal analysis, S.L.F. and C.M.B.; investigation, S.L.F. and C.M.B.; resources, S.L.F. and C.M.B.; data curation, S.L.F. and C.M.B.; writing—original draft preparation, S.L.F. and C.M.B.; writing—review and editing, S.L.F. and C.M.B.; visualization, S.L.F. and C.M.B.; supervision, S.L.F. and C.M.B.; project administration, S.L.F. and C.M.B.; funding acquisition, S.L.F. and C.M.B. All authors have read and agreed to the published version of the manuscript.

Funding: This research received no external funding.

Institutional Review Board Statement: Not applicable.

Informed Consent Statement: Not applicable.

Data Availability Statement: No new data were created or analyzed in this study. Data sharing is not applicable to this article.

Conflicts of Interest: C.M.B. reports research funding to the institution from AstraZeneca and BMS; personal consulting/advisory board fees from AstraZeneca, BMS, CVS, Daiichi Sankyo, Genentech, Jazz, JNJ, Novartis, Pfizer, Regeneron, Sanofi, Seattle Genetics, Takeda, TEMPUS; and speakers bureau from Merck. The funders had no role in the design of the study; in the collection, analyses, or interpretation of data; in the writing of the manuscript; or in the decision to publish the results. S.L.F. declares no conflict of interest.

References

1. NCCN Guidelines. Non-Small Cell Lung Cancer. Version 3. 2022. Available online: http://www.nccn.org (accessed on 8 January 2023).
2. Kris, M.G.; Johnson, B.E.; Berry, L.D.; Kwiatkowski, D.J.; Iafrate, A.J.; Wistuba, I.I.; Varella-Garcia, M.; Franklin, W.A.; Aronson, S.L.; Su, P.F.; et al. Using Multiplexed Assays of Oncogenic Drivers in Lung Cancers to Select Targeted Drugs. *JAMA* **2014**, *311*, 1998–2006. [CrossRef] [PubMed]

3. Sequist, L.V.; Yang, J.C.-H.; Yamamoto, N.; O'Byrne, K.; Hirsh, V.; Mok, T.; Geater, S.L.; Orlov, S.; Tsai, C.-M.; Boyer, M.; et al. Phase III Study of Afatinib or Cisplatin Plus Pemetrexed in Patients With Metastatic Lung Adenocarcinoma With EGFR Mutations. *J. Clin. Oncol.* **2013**, *31*, 3327–3334. [CrossRef] [PubMed]

4. Lim, S.M.; Syn, N.L.; Cho, B.C.; Soo, R.A. Acquired resistance to EGFR targeted therapy in non-small cell lung cancer: Mechanisms and therapeutic strategies. *Cancer Treat. Rev.* **2018**, *65*, 1–10. [CrossRef] [PubMed]

5. Soria, J.-C.; Ohe, Y.; Vansteenkiste, J.; Reungwetwattana, T.; Chewaskulyong, B.; Lee, K.H.; Dechaphunkul, A.; Imamura, F.; Nogami, N.; Kurata, T.; et al. Osimertinib in Untreated EGFR-Mutated Advanced Non–Small-Cell Lung Cancer. *N. Engl. J. Med.* **2018**, *378*, 113–125. [CrossRef]

6. Ramalingam, S.S.; Yang, J.C.-H.; Lee, C.K.; Kurata, T.; Kim, D.-W.; John, T.; Nogami, N.; Ohe, Y.; Mann, H.; Rukazenkov, Y.; et al. Osimertinib As First-Line Treatment of *EGFR* Mutation–Positive Advanced Non–Small-Cell Lung Cancer. *J. Clin. Oncol.* **2018**, *36*, 841–849. [CrossRef]

7. Leonetti, A.; Sharma, S.; Minari, R.; Perego, P.; Giovannetti, E.; Tiseo, M. Resistance mechanisms to osimertinib in EGFR-mutated non-small cell lung cancer. *Br. J. Cancer* **2019**, *121*, 725–737. [CrossRef]

8. Chmielecki, J.; Gray, J.E.; Cheng, Y.; Ohe, Y.; Imamura, F.; Cho, B.C.; Lin, M.-C.; Majem, M.; Shah, R.; Rukazenkov, Y.; et al. Candidate mechanisms of acquired resistance to first-line osimertinib in EGFR-mutated advanced non-small cell lung cancer. *Nat. Commun.* **2023**, *14*, 1070. [CrossRef]

9. Choudhury, N.J.; Marra, A.; Sui, J.S.; Flynn, J.; Yang, S.-R.; Falcon, C.J.; Selenica, P.; Schoenfeld, A.J.; Rekhtman, N.; Gomez, D.; et al. Molecular Biomarkers of Disease Outcomes and Mechanisms of Acquired Resistance to First-Line Osimertinib in Advanced EGFR-Mutant Lung Cancers. *J. Thorac. Oncol.* **2022**, *18*, 463–475. [CrossRef]

10. Coleman, N.; Hong, L.; Zhang, J.; Heymach, J.; Hong, D.; Le, X. Beyond epidermal growth factor receptor: MET amplification as a general resistance driver to targeted therapy in oncogene-driven non-small-cell lung cancer. *ESMO Open* **2021**, *6*, 100319. [CrossRef]

11. Trusolino, L.; Bertotti, A.; Comoglio, P.M. MET signalling: Principles and functions in development, organ regeneration and cancer. *Nat. Rev. Mol. Cell Biol.* **2010**, *11*, 834–848. [CrossRef]

12. Comoglio, P.M.; Trusolino, L.; Boccaccio, C. Known and novel roles of the MET oncogene in cancer: A coherent approach to targeted therapy. *Nat. Rev. Cancer* **2018**, *18*, 341–358. [CrossRef]

13. Guo, R.; Luo, J.; Chang, J.; Rekhtman, N.; Arcila, M.; Drilon, A. MET-dependent solid tumours—Molecular diagnosis and targeted therapy. *Nat. Rev. Clin. Oncol.* **2020**, *17*, 569–587. [CrossRef]

14. Cortot, A.B.; Kherrouche, Z.; Descarpentries, C.; Wislez, M.; Baldacci, S.; Furlan, A.; Tulasne, D. Exon 14 Deleted MET Receptor as a New Biomarker and Target in Cancers. *J. Natl. Cancer Inst.* **2017**, *109*, djw262. [CrossRef]

15. Vuong, H.G.; Ho, A.T.N.; Altibi, A.M.; Nakazawa, T.; Katoh, R.; Kondo, T. Clinicopathological implications of MET exon 14 mutations in non-small cell lung cancer—A systematic review and meta-analysis. *Lung Cancer* **2018**, *123*, 76–82. [CrossRef]

16. Lee, M.; Jain, P.; Wang, F.; Ma, P.C.; Borczuk, A.; Halmos, B. MET alterations and their impact on the future of non-small cell lung cancer (NSCLC) targeted therapies. *Expert Opin. Ther. Targets* **2021**, *25*, 249–268. [CrossRef]

17. Ríos-Hoyo, A.; Moliner, L.; Arriola, E. Acquired Mechanisms of Resistance to Osimertinib—The Next Challenge. *Cancers* **2022**, *14*, 1931. [CrossRef]

18. Reungwetwattana, T.; Liang, Y.; Zhu, V.; Ou, S.-H.I. The race to target MET exon 14 skipping alterations in non-small cell lung cancer: The Why, the How, the Who, the Unknown, and the Inevitable. *Lung Cancer* **2016**, *103*, 27–37. [CrossRef]

19. Schrock, A.B.; Frampton, G.M.; Suh, J.; Chalmers, Z.R.; Rosenzweig, M.; Erlich, R.L.; Halmos, B.; Goldman, J.; Forde, P.; Leuenberger, K.; et al. Characterization of 298 Patients with Lung Cancer Harboring MET Exon 14 Skipping Alterations. *J. Thorac. Oncol.* **2016**, *11*, 1493–1502. [CrossRef]

20. Schildhaus, H.-U.; Schultheis, A.M.; Rüschoff, J.; Binot, E.; Merkelbach-Bruse, S.; Fassunke, J.; Schulte, W.; Ko, Y.-D.; Schlesinger, A.; Bos, M.; et al. *MET* Amplification Status in Therapy-Naïve Adeno- and Squamous Cell Carcinomas of the Lung. *Clin. Cancer Res.* **2015**, *21*, 907–915. [CrossRef]

21. Wolf, J.; Seto, T.; Han, J.-Y.; Reguart, N.; Garon, E.B.; Groen, H.J.; Tan, D.S.; Hida, T.; de Jonge, M.; Orlov, S.V.; et al. Capmatinib in *MET* Exon 14–Mutated or *MET*-Amplified Non–Small-Cell Lung Cancer. *N. Engl. J. Med.* **2020**, *383*, 944–957. [CrossRef]

22. Peng, L.-X.; Jie, G.-L.; Li, A.-N.; Liu, S.-Y.; Sun, H.; Zheng, M.-M.; Zhou, J.-Y.; Zhang, J.-T.; Zhang, X.-C.; Zhou, Q.; et al. MET amplification identified by next-generation sequencing and its clinical relevance for MET inhibitors. *Exp. Hematol. Oncol.* **2021**, *10*, 52. [CrossRef] [PubMed]

23. Kim, E.K.; Kim, K.A.; Lee, C.Y.; Kim, S.; Chang, S.; Cho, B.C.; Shim, H.S. Molecular Diagnostic Assays and Clinicopathologic Implications of MET Exon 14 Skipping Mutation in Non–small-cell Lung Cancer. *Clin. Lung Cancer* **2018**, *20*, e123–e132. [CrossRef] [PubMed]

24. Davies, K.D.; Lomboy, A.; Lawrence, C.A.; Yourshaw, M.; Bocsi, G.T.; Camidge, D.R.; Aisner, D.L. DNA-Based versus RNA-Based Detection of MET Exon 14 Skipping Events in Lung Cancer. *J. Thorac. Oncol.* **2019**, *14*, 737–741. [CrossRef]

25. Guo, R.; Berry, L.D.; Aisner, D.L.; Sheren, J.; Boyle, T.; Bunn, P.A.; Johnson, B.E.; Kwiatkowski, D.J.; Drilon, A.; Sholl, L.M.; et al. MET IHC Is a Poor Screen for MET Amplification or MET Exon 14 Mutations in Lung Adenocarcinomas: Data from a Tri-Institutional Cohort of the Lung Cancer Mutation Consortium. *J. Thorac. Oncol.* **2019**, *14*, 1666–1671. [CrossRef] [PubMed]
26. Mondelo-Macía, P.; Rodríguez-López, C.; Valiña, L.; Aguín, S.; León-Mateos, L.; García-González, J.; Abalo, A.; Rapado-González, O.; Suárez-Cunqueiro, M.; Díaz-Lagares, A.; et al. Detection of MET Alterations Using Cell Free DNA and Circulating Tumor Cells from Cancer Patients. *Cells* **2020**, *9*, 522. [CrossRef]
27. Paik, P.K.; Felip, E.; Veillon, R.; Sakai, H.; Cortot, A.B.; Garassino, M.C.; Mazieres, J.; Viteri, S.; Senellart, H.; Van Meerbeeck, J.; et al. Tepotinib in Non–Small-Cell Lung Cancer with MET Exon 14 Skipping Mutations. *N. Engl. J. Med.* **2020**, *383*, 931–943. [CrossRef]
28. Shaw, A.T.; Kim, D.-W.; Nakagawa, K.; Seto, T.; Crinó, L.; Ahn, M.-J.; De Pas, T.; Besse, B.; Solomon, B.J.; Blackhall, F.; et al. Crizotinib versus Chemotherapy in Advanced *ALK*-Positive Lung Cancer. *N. Engl. J. Med.* **2013**, *368*, 2385–2394. [CrossRef]
29. Shaw, A.T.; Ou, S.-H.I.; Bang, Y.-J.; Camidge, D.R.; Solomon, B.J.; Salgia, R.; Riely, G.J.; Varella-Garcia, M.; Shapiro, G.I.; Costa, D.B.; et al. Crizotinib in ROS1-Rearranged Non–Small-Cell Lung Cancer. *N. Engl. J. Med.* **2014**, *371*, 1963–1971. [CrossRef]
30. Drilon, A.; Clark, J.W.; Weiss, J.; Ou, S.-H.I.; Camidge, D.R.; Solomon, B.J.; Otterson, G.A.; Villaruz, L.C.; Riely, G.J.; Heist, R.S.; et al. Antitumor activity of crizotinib in lung cancers harboring a MET exon 14 alteration. *Nat. Med.* **2020**, *26*, 47–51. [CrossRef]
31. Neal, J.W.; Dahlberg, S.E.; Wakelee, H.A.; Aisner, S.C.; Bowden, M.; Huang, Y.; Carbone, D.P.; Gerstner, G.J.; Lerner, R.E.; Rubin, J.L.; et al. Erlotinib, cabozantinib, or erlotinib plus cabozantinib as second-line or third-line treatment of patients with EGFR wild-type advanced non-small-cell lung cancer (ECOG-ACRIN 1512): A randomised, controlled, open-label, multicentre, phase 2 trial. *Lancet Oncol.* **2016**, *17*, 1661–1671. [CrossRef]
32. Le, X.; Paz-Ares, L.G.; Van Meerbeeck, J.; Viteri, S.; Galvez, C.C.; Baz, D.V.; Kim, Y.-C.; Kang, J.-H.; Schumacher, K.-M.; Karachaliou, N.; et al. Tepotinib in patients (pts) with advanced non-small cell lung cancer (NSCLC) with *MET* amplification (*MET*amp). *J. Clin. Oncol.* **2021**, *39*, 9021. [CrossRef]
33. Lu, S.; Fang, J.; Li, X.; Cao, L.; Zhou, J.; Guo, Q.; Liang, Z.; Cheng, Y.; Jiang, L.; Yang, N.; et al. Once-daily savolitinib in Chinese patients with pulmonary sarcomatoid carcinomas and other non-small-cell lung cancers harbouring MET exon 14 skipping alterations: A multicentre, single-arm, open-label, phase 2 study. *Lancet Respir. Med.* **2021**, *9*, 1154–1164. [CrossRef]
34. Scagliotti, G.; Moro-Sibilot, D.; Kollmeier, J.; Favaretto, A.; Cho, E.K.; Grosch, H.; Kimmich, M.; Girard, N.; Tsai, C.-M.; Hsia, T.-C.; et al. A Randomized-Controlled Phase 2 Study of the MET Antibody Emibetuzumab in Combination with Erlotinib as First-Line Treatment for EGFR Mutation–Positive NSCLC Patients. *J. Thorac. Oncol.* **2019**, *15*, 80–90. [CrossRef]
35. Han, K.; Chanu, P.; Jonsson, F.; Winter, H.; Bruno, R.; Jin, J.; Stroh, M. Exposure–Response and Tumor Growth Inhibition Analyses of the Monovalent Anti-c-MET Antibody Onartuzumab (MetMAb) in the Second- and Third-Line Non-Small Cell Lung Cancer. *AAPS J.* **2017**, *19*, 527–533. [CrossRef]
36. Hirsch, F.R.; Govindan, R.; Zvirbule, Z.; Braiteh, F.; Rittmeyer, A.; Belda-Iniesta, C.; Isla, D.; Cosgriff, T.; Boyer, M.; Ueda, M.; et al. Efficacy and Safety Results From a Phase II, Placebo-Controlled Study of Onartuzumab Plus First-Line Platinum-Doublet Chemotherapy for Advanced Squamous Cell Non–Small-Cell Lung Cancer. *Clin. Lung Cancer* **2017**, *18*, 43–49. [CrossRef]
37. Wakelee, H.; Zvirbule, Z.; de Braud, F.G.M.; Kingsley, C.D.; Mekhail, T.; Lowe, T.; Schütte, W.; Lena, H.; Lawler, W.; Braiteh, F.; et al. Efficacy and Safety of Onartuzumab in Combination With First-Line Bevacizumab- or Pemetrexed-Based Chemotherapy Regimens in Advanced Non-Squamous Non–Small-Cell Lung Cancer. *Clin. Lung Cancer* **2017**, *18*, 50–59. [CrossRef]
38. Spigel, D.R.; Edelman, M.J.; O'byrne, K.; Paz-Ares, L.; Mocci, S.; Phan, S.; Shames, D.S.; Smith, D.; Yu, W.; Paton, V.E.; et al. Results From the Phase III Randomized Trial of Onartuzumab Plus Erlotinib Versus Erlotinib in Previously Treated Stage IIIB or IV Non–Small-Cell Lung Cancer: METLung. *J. Clin. Oncol.* **2017**, *35*, 412–420. [CrossRef]
39. Camidge, D.R.; Morgensztern, D.; Heist, R.S.; Barve, M.; Vokes, E.E.; Goldman, J.W.; Hong, D.S.; Bauer, T.M.; Strickler, J.H.; Angevin, E.; et al. Phase I Study of 2- or 3-Week Dosing of Telisotuzumab Vedotin, an Antibody–Drug Conjugate Targeting c-Met, Monotherapy in Patients with Advanced Non–Small Cell Lung Carcinoma. *Clin. Cancer Res.* **2021**, *27*, 5781–5792. [CrossRef]
40. Oh, S.Y.; Lee, Y.W.; Lee, E.J.; Kim, J.H.; Park, Y.J.; Heo, S.G.; Yu, M.R.; Hong, M.H.; DaSilva, J.; Daly, C.; et al. Preclinical Study of a Biparatopic METxMET Antibody–Drug Conjugate, REGN5093-M114, Overcomes MET-driven Acquired Resistance to EGFR TKIs in EGFR-mutant NSCLC. *Clin. Cancer Res.* **2022**, *29*, 221–232. [CrossRef]
41. Liu, L.; Qu, J.; Heng, J.; Zhou, C.; Xiong, Y.; Yang, H.; Jiang, W.; Zeng, L.; Zhu, S.; Zhang, Y.; et al. A Large Real-World Study on the Effectiveness of the Combined Inhibition of EGFR and MET in EGFR-Mutant Non-Small-Cell Lung Cancer After Development of EGFR-TKI Resistance. *Front. Oncol.* **2021**, *11*, 722039. [CrossRef]
42. Mi, J.; Huang, Z.; Zhang, R.; Zeng, L.; Xu, Q.; Yang, H.; Lizaso, A.; Tong, F.; Dong, X.; Yang, N.; et al. Molecular characterization and clinical outcomes in EGFR-mutant de novo MET-overexpressed advanced non-small-cell lung cancer. *ESMO Open* **2022**, *7*, 100347. [CrossRef] [PubMed]
43. Hartmaier, R.J.; Markovets, A.A.; Ahn, M.J.; Sequist, L.V.; Han, J.-Y.; Cho, B.C.; Yu, H.A.; Kim, S.-W.; Yang, J.C.-H.; Lee, J.-S.; et al. Osimertinib + Savolitinib to Overcome Acquired MET-Mediated Resistance in Epidermal Growth Factor Receptor–Mutated, *MET*-Amplified Non–Small Cell Lung Cancer: TATTON. *Cancer Discov.* **2022**, *13*, 98–113. [CrossRef]

44. Oxnard, G.R.; Cantarini, M.; Frewer, P.; Hawkins, G.; Peters, J.; Howarth, P.; Ahmed, G.F.; Sahota, T.; Hartmaier, R.; Li-Sucholeiki, X.; et al. SAVANNAH: A Phase II trial of osimertinib plus savolitinib for patients (pts) with EGFR-mutant, MET-driven (MET+), locally advanced or metastatic non-small cell lung cancer (NSCLC), following disease progression on osimertinib. *J. Clin. Oncol.* **2019**, *37* (Suppl. 15), TPS9119. [CrossRef]
45. Li, A.; Chen, H.-J.; Yang, J.-J. Design and Rationale for a Phase II, Randomized, Open-Label, Two-Cohort Multicenter Interventional Study of Osimertinib with or Without Savolitinib in De Novo MET Aberrant, EGFR-Mutant Patients with Advanced Non-Small-Cell Lung Cancer: The FLOWERS Trial. *Clin. Lung Cancer* **2022**, *24*, 82–88. [CrossRef] [PubMed]
46. Yang, J.-J.; Fang, J.; Shu, Y.-Q.; Chang, J.-H.; Chen, G.-Y.; He, J.X.; Li, W.; Liu, X.-Q.; Yang, N.; Zhou, C.; et al. A phase Ib study of the highly selective MET-TKI savolitinib plus gefitinib in patients with EGFR-mutated, MET-amplified advanced non-small-cell lung cancer. *Investig. New Drugs* **2020**, *39*, 477–487. [CrossRef]
47. Wu, Y.-L.; Cheng, Y.; Zhou, J.; Lu, S.; Zhang, Y.; Zhao, J.; Kim, D.-W.; Soo, R.A.; Kim, S.-W.; Pan, H.; et al. Tepotinib plus gefitinib in patients with EGFR-mutant non-small-cell lung cancer with MET overexpression or MET amplification and acquired resistance to previous EGFR inhibitor (INSIGHT study): An open-label, phase 1b/2, multicentre, randomised trial. *Lancet Respir. Med.* **2020**, *8*, 1132–1143. [CrossRef]
48. Wu, Y.-L.; Zhang, L.; Kim, D.-W.; Liu, X.; Lee, D.H.; Yang, J.C.-H.; Ahn, M.-J.; Vansteenkiste, J.F.; Su, W.-C.; Felip, E.; et al. Phase Ib/II Study of Capmatinib (INC280) Plus Gefitinib After Failure of Epidermal Growth Factor Receptor (EGFR) Inhibitor Therapy in Patients With *EGFR*-Mutated, MET Factor–Dysregulated Non–Small-Cell Lung Cancer. *J. Clin. Oncol.* **2018**, *36*, 3101–3109. [CrossRef]
49. Sequist, L.V.; Han, J.-Y.; Ahn, M.-J.; Cho, B.C.; Yu, H.; Kim, S.-W.; Yang, J.C.-H.; Lee, J.S.; Su, W.-C.; Kowalski, D.; et al. Osimertinib plus savolitinib in patients with EGFR mutation-positive, MET-amplified, non-small-cell lung cancer after progression on EGFR tyrosine kinase inhibitors: Interim results from a multicentre, open-label, phase 1b study. *Lancet Oncol.* **2020**, *21*, 373–386. [CrossRef]
50. Oxnard, G.R.; Yang, J.C.-H.; Yu, H.; Kim, S.-W.; Saka, H.; Horn, L.; Goto, K.; Ohe, Y.; Mann, H.; Thress, K.S.; et al. TATTON: A multi-arm, phase Ib trial of osimertinib combined with selumetinib, savolitinib, or durvalumab in EGFR-mutant lung cancer. *Ann. Oncol.* **2020**, *31*, 507–516. [CrossRef]
51. Yu, H.A.; Ambrose, H.; Baik, C.; Cho, B.C.; Cocco, E.; Goldberg, S.B.; Goldman, J.W.; Kraljevic, S.; de Langen, A.J.; Okamoto, I.; et al. Le ORCHARD Osimertinib + Savolitinib Interim Analysis: A Biomarker-Directed Phase II Platform Study in Patients (Pts) with Advanced Non-Small Cell Lung Cancer (NSCLC) Whose Disease Has Progressed on First-Line (1L) Osimertinib. *Ann. Onc.* **2021**, *32*, S949–S1039. [CrossRef]
52. McCoach, C.E.; Yu, A.; Gandara, D.R.; Riess, J.W.; Vang, D.P.; Li, T.; Lara, P.N.; Gubens, M.; Lara, F.; Mack, P.C.; et al. Phase I/II Study of Capmatinib Plus Erlotinib in Patients With MET-Positive Non–Small-Cell Lung Cancer. *JCO Precis. Oncol.* **2021**, *1*, 177–190. [CrossRef]
53. Liam, C.K.; Ahmad, A.R.; Hsia, T.-C.; Zhou, J.; Kim, D.-W.; Soo, R.A.; Cheng, Y.; Lu, S.; Shin, S.W.; Yang, J.C.-H.; et al. Randomized Trial of Tepotinib Plus Gefitinib versus Chemotherapy in *EGFR*-Mutant NSCLC with EGFR Inhibitor Resistance Due to *MET* Amplification: INSIGHT Final Analysis. *Clin. Cancer Res.* **2023**, *29*, 1879–1886. [CrossRef]
54. Smit, E.F.; Dooms, C.; Raskin, J.; Nadal, E.; Tho, L.M.; Le, X.; Mazieres, J.; Hin, H.S.; Morise, M.; Zhu, V.W.; et al. INSIGHT 2: A phase II study of tepotinib plus osimertinib in *MET*-amplified NSCLC and first-line osimertinib resistance. *Futur. Oncol.* **2022**, *18*, 1039–1054. [CrossRef]
55. Mazieres, J.; Kim, T.; Lim, B.; Wislez, M.; Dooms, C.; Finocchiaro, G.; Hayashi, H.; Liam, C.; Raskin, J.; Tho, L.; et al. LBA52 Tepotinib + osimertinib for EGFRm NSCLC with MET amplification (METamp) after progression on first-line (1L) osimertinib: Initial results from the INSIGHT 2 study. *Ann. Oncol.* **2022**, *33*, S1419–S1420. [CrossRef]
56. Camidge, D.R.; Moran, T.; Demedts, I.; Grosch, H.; Mileham, K.; Molina, J.; Juan-Vidal, O.; Bepler, G.; Goldman, J.W.; Park, K.; et al. A Randomized, Open-Label Phase II Study Evaluating Emibetuzumab Plus Erlotinib and Emibetuzumab Monotherapy in MET Immunohistochemistry Positive NSCLC Patients with Acquired Resistance to Erlotinib. *Clin. Lung Cancer* **2022**, *23*, 300–310. [CrossRef]
57. Camidge, D.R.; Barlesi, F.; Goldman, J.W.; Morgensztern, D.; Heist, R.; Vokes, E.; Spira, A.; Angevin, E.; Su, W.-C.; Hong, D.S.; et al. Phase Ib Study of Telisotuzumab Vedotin in Combination With Erlotinib in Patients With c-Met Protein–Expressing Non–Small-Cell Lung Cancer. *J. Clin. Oncol.* **2023**, *41*, 1105–1115. [CrossRef]
58. Park, K.; Haura, E.B.; Leighl, N.B.; Mitchell, P.; Shu, C.A.; Girard, N.; Viteri, S.; Han, J.-Y.; Kim, S.-W.; Lee, C.K.; et al. Amivantamab in EGFR Exon 20 Insertion–Mutated Non–Small-Cell Lung Cancer Progressing on Platinum Chemotherapy: Initial Results From the CHRYSALIS Phase I Study. *J. Clin. Oncol.* **2021**, *39*, 3391–3402. [CrossRef]
59. Bauml, J.; Chul Cho, B.; Park, K.; Hyeong Lee, K.; Kyung Cho, E.; Kim, D.-W.; Kim, S.-W.; Haura, E.B.; Sabari, J.F.; Sanborn, R.E.; et al. Amivantamab In Combination With Lazertinib For The Treatment OF Osimertinib-Relapsed, Chemotherapy-Naive EGFR Mutant (EGFRm) Non-Small Cell Lung Cancer (NSCLC) and Potential Biomarkers For Response. *J. Clin. Oncol.* **2021**, *39*, 9006. [CrossRef]
60. Leighl, N.; Shu, C.; Minchom, A.; Felip, E.; Cousin, S.; Cho, B.; Park, K.; Han, J.-Y.; Boyer, M.; Lee, C.; et al. 1192MO Amivantamab monotherapy and in combination with lazertinib in post-osimertinib EGFR-mutant NSCLC: Analysis from the CHRYSALIS study. *Ann. Oncol.* **2021**, *32*, S951–S952. [CrossRef]

61. Shu, C.A.; Goto, K.; Ohe, Y.; Besse, B.; Lee, S.-H.; Wang, Y.; Griesinger, F.; Yang, J.C.-H.; Felip, E.; Sanborn, R.E.; et al. Amivantamab and Lazertinib in Patients with EGFR-Mutant Non-Small Cell Lung Cancer (NSCLC) after Progression on Osimertinib and Platinum-Based Chemotherapy: Updated Results from CHRYSALIS-2. *J. Clin. Oncol.* **2022**, *40*, 9006. [CrossRef]
62. Chon, K.; Larkins, E.; Chatterjee, S.; Mishra-Kalyani, P.S.; Aungst, S.; Wearne, E.; Subramaniam, S.; Li, Y.; Liu, J.; Sun, J.; et al. FDA Approval Summary: Amivantamab for the Treatment of Patients with Non–Small Cell Lung Cancer with *EGFR* Exon 20 Insertion Mutations. *Clin. Cancer Res.* **2023**, OF1–OF5. [CrossRef] [PubMed]

Article

Molecular Characteristics of Radon Associated Lung Cancer Highlights MET Alterations

Gabriele Gamerith [1,†], Marcel Kloppenburg [2,†], Finn Mildner [1], Arno Amann [1], Sabine Merkelbach-Bruse [3], Carina Heydt [3], Janna Siemanowski [3], Reinhard Buettner [3], Michael Fiegl [1,4], Claudia Manzl [5,*,‡] and Georg Pall [1,*,‡]

[1] Department of Haematology and Oncology, Clinic of Internal Medicine V, Medical University of Innsbruck, 6020 Innsbruck, Austria
[2] Clinic of Otorhinolaryngology—Head & Neck Surgery, Medical University of Innsbruck, 6020 Innsbruck, Austria
[3] Institute of Pathology, University Hospital Cologne, 50937 Cologne, Germany
[4] Clinic Hochrum, 6063 Rum, Austria
[5] Institute of Pathology, Neuropathology and Molecularpathology, Medical University of Innsbruck, 6020 Innsbruck, Austria
* Correspondence: claudia.manzl@i-med.ac.at (C.M.); georg.pall@tirol-kliniken.at (G.P.)
† These authors contributed equally to this work.
‡ These authors contributed equally to this work.

Simple Summary: Lung cancer (LC) is the leading cause of cancer death worldwide. After smoking, one of the most prominent risk factors for LC development is radon (Rn) exposure. In our study we analysed and compared the genetic landscape of LC patients from a Rn exposed village with local matched non-exposed patients. Within the concordant genetic landscape, an increase in genetic MET proto-oncogene, receptor tyrosine kinase (MET) alteration in the Rn-exposed cohort was monitored, underlining the importance of routine MET testing and potential to enable a more effective treatment for this specific subgroup.

Abstract: Effective targeted treatment strategies resulted from molecular profiling of lung cancer with distinct prevalent mutation profiles in smokers and non-smokers. Although Rn is the second most important risk factor, data for Rn-dependent driver events are limited. Therefore, a Rn-exposed cohort of lung cancer patients was screened for oncogenic drivers and their survival and genetic profiles were compared with data of the average regional population. Genetic alterations were analysed in 20 Rn-exposed and 22 histologically matched non-Rn exposed LC patients using targeted Next generation sequencing (NGS) and Fluorescence In Situ Hybridization (FISH). Sufficient material and sample quality could be obtained in 14/27 non-exposed versus 17/22 Rn-exposed LC samples. Survival was analysed in comparison to a histologically and stage-matched regional non-exposed lung cancer cohort ($n = 51$) for hypothesis generating. Median overall survivals were 83.02 months in the Rn-exposed and 38.7 months in the non-exposed lung cancer cohort ($p = 0.22$). Genetic alterations of both patient cohorts were in high concordance, except for an increase in MET alterations and a decrease in TP53 mutations in the Rn-exposed patients in this small hypothesis generating study.

Keywords: radon exposure; lung cancer; genetic profile

Citation: Gamerith, G.; Kloppenburg, M.; Mildner, F.; Amann, A.; Merkelbach-Bruse, S.; Heydt, C.; Siemanowski, J.; Buettner, R.; Fiegl, M.; Manzl, C.; et al. Molecular Characteristics of Radon Associated Lung Cancer Highlights MET Alterations. *Cancers* **2022**, *14*, 5113. https://doi.org/10.3390/cancers14205113

Academic Editors: Jan Trøst Jørgensen and Jens Mollerup

Received: 15 September 2022
Accepted: 11 October 2022
Published: 19 October 2022

Publisher's Note: MDPI stays neutral with regard to jurisdictional claims in published maps and institutional affiliations.

1. Introduction

Besides the main risk factor, smoking, which is responsible for approximately 80% of LCs, another 10–25% of LC are diagnosed in never-smokers (LCINS) [1–4]. Rn exposure is the most important risk factor in never-smokers [5,6], and the second most important in LC of smokers, and is therefore associated with 9% of all deaths caused by LC and 2% of all cancer deaths in Europe [7]. Rn 222, an environmental radioactive pollutant-gas, is mainly released from the decay of uranium 238 in rock and soil causing alpha and beta emissions [7,8]. As it is electrically charged, it can attach to natural aerosol or dust and tends to be deposited on the bronchial epithelial cells. This explains its organ specific risk for lung cancer due to the cells' exposure to this local radiation [7,9], which is well known to create molecular changes, such as DNA double-strand breaks, mutations, translocation or gene deletions [10].

Indoor Rn accumulates by structural defects in basements [11]. The correlation between high indoor Rn-exposure and LC was extensively studied in the 1990s and 2000s and a significant excess of LC due to this residential exposure was confirmed, for example, in a review of 7148 LC patients compared to 14,208 controls by Darby et al. [7]. The World Health Organization (WHO) recommends levels below 100 Bq/m^3 [5], even though some studies observed an increase in LC risk already for 50 Bq/m^3 [12,13]. Most find a linear risk increase of 11–16% per 100 Bq/m^3 [7,14,15], but also a non-linear dose-response relationship has been reported [16]. Even though Rn is the most important risk factor for LCINS and the second important in smokers, few works describe molecular profiles and no detailed outcome data are available. Most available data represent uranium miners, who suffer a higher risk of LC associated with the Rn exposure in mines [17–19]. Nevertheless, these cohorts are biased—most miners are male smokers and many other radioactive chemicals and carcinogens, such as arsenic, silica or diesel, exist in these mines. Within those cohorts, TP53 and EGFR are the most abundantly studied genes. The largest investigation found that epidermal growth factor receptor (EGFR), tumor protein p53 (TP53), NK2 homeobox1 (NKX 2.1), phosphatase and tensin homolog (PTEN), chromodomain helicase DNA binding protein 7 (CHD7), discoidin domain receptor tyrosine kinase 2 (DDR2), lysine methyltransferase 2C (MLL3, approved abbr. is KMT2C), chromodomain helicase DNA binding protein 5 (CHD5), FAT atypical cadherin 1 (FAT1) and serine/threonine/tyrosine interacting like 2 (DUSP27, now: STYXL2), LAK receptor tyrosine kinase (ALK), ret proto-oncogene (RET), AKT serine/threonine kinase 1 (AKT1), B-Raf proto-oncogene, serine/threonine kinase (BRAF), catenin beta 1(CTNNB1), erb-b2 receptor tyrosine kinase 2 (ERBB2), KRAS proto-oncogene, GTPase (KRAS), mitogen-activated protein kinase kinase 1 (MAP2K1), MET proto-oncogene, receptor tyrosine kinase (MET), NRAS proto-oncogene, GTPase (NRAS), phosphatidylinositol-4,5-bisphosphate 3-kinase catalytic subunit alpha (PIK3CA) out of 37 cancer susceptive genes related to LCINS were associated with Rn exposure [20]. The results of those molecular investigations are summarized in Table 1 [7,19–25].

Table 1. Overview of previous studies.

Author (Ref.)	Year	Region	Subjects	Gene
S. Darby et al. [7]	2005	Europa	7142 patients + 14,208 controls	-
K. Vahakangas et al. [17]	1992	Europa	19 uranium miners	TP53
J.A. Taylor et al. [19]	1994	America	52 uranium miners	TP53
Q. Yang et al. [21]	2000	Germany	79 uranium miners	TP53
A. Ruano-Ravina et al. [22]	2009	Worldwide	578 individuals	TP53
A. Ruano-Ravina et al. [23]	2016	Spain	323 Patients	EGFR/ALK
M. Taga et al. [24]	2012	USA	70 women	EGFR
Bonner et al. [25]	2006	USA	270 individuals	GSTM1
J.R. Choi et al. [20]	2017	Asia	19 patients	EGFR/TP53/NKX 2.1/PTEN/CHD7/DDR2/MLL3/CHD5/FAT1/DUSP27

Therefore, we investigated a local cohort with high Rn-exposure for survival and genetic alterations of oncogenic drivers in comparison to a regional non-exposed LC cohort.

2. Materials and Methods

2.1. Patients and Patient Material

This study was approved by the regional ethical board (AN 1018/2018). Histologically confirmed, adult LC patients from 1995 to 2009 with resident addresses in Umhausen, the Rn-exposed region, were included and their Rn-exposure was assessed based on results from a previous study on longitudinal Rn-exposure in this region. In total 20 patients were identified and outcomes, characteristics and treatments, as well as risk factors of the patients were evaluated based on the documented patient's histories. For 2/20, a second lung tumour sample was available and analysed, accounting for analyses of 22 samples from 20 patients. To identify the non-exposed LC cohort the TYROL registry of LC patients resident in Tyrol and the FUZZY matching tool of IBM SPSS Statistics were used. We matched the patients based on their age at diagnosis and gender. In addition, for molecular testing, histology and year of diagnosis was most important, whereas for survival we added the Union for International Cancer control stage based on clinical tumour assessment (cUICC). Smoking history could not be adequately matched based on missing information in this historical cohort.

2.2. Molecular Analyses

2.2.1. Fluorescence In Situ Hybridization (FISH)

In order to detect gene fusions or amplifications of specific oncogenic drivers, FISH analyses on formalin-fixed paraffin-embedded tumour tissue (FFPE) were performed. In brief, 1–3 μm tumour sections were used, and after tissue pre-treatment nuclear DNA and the respective probes were denaturated for 5 min at 80 °C followed by a hybridization step at 37 °C for 18 h. The following probes were used: ALK (ALK Dual Color Break apart Probe, Zytovision, Bremerhaven, Germany), RET (RET Dual Colour Break apart Probe, Zytovision), ROS1 (ROS1 Dual Colour Break apart Probe, Zytovision) and MET (MET/CEN17 probe, Vysis, Downers Grove, IL, USA). Tissue was mounted in a DAPI containing mounting media (Zytovision) and interphase nuclei were monitored using a ZEISS axioplan2 microscope equipped with a 63× oil objective (Zeiss, Oberkochen, Germany) and a Progress GRYPHAX SUBRA camera/software (Jenoptik, Jena, Germany). Evaluation was performed as previously described [26–29].

2.2.2. Parallel Sequencing (Next Generation Sequencing, NGS)

For DNA extraction three to six 10 μm-thick sections were cut from FFPE tissue. Sections were deparaffinized and the tumour areas were macro-dissected from unstained slides using a marked haematoxylin-eosin (H&E) stained slide as a reference. After proteinase K digestion, the DNA was isolated with the Maxwell® 16 FFPE Plus Tissue LEV DNA Purification Kit (Promega, Mannheim, Germany) on the Maxwell® 16 (Promega) following manufacturer's instructions. For Next Generation Sequencing, the DNA concentration and integrity was measured using a quantitative real-time PCR (qPCR). Multiplex PCR-based parallel sequencing was performed on all FFPE samples (Table S1). Isolated DNA was amplified with an Ion AmpliSeq Custom DNA Panel (Thermo Fisher Scientific, Waltham, MA, USA) and the Ion AmpliSeq Library Kit 2.0 (Thermo Fisher Scientific) following manufacturer's instructions. The gene panel comprised relevant exons of the following genes: AKT1, ALK, BRAF, CTNNB1, DDR2, EGFR, ERBB2, KRAS, MAP2K1, MET, NRAS, PIK3CA, PTEN and TP53 (Table S2: detailed investigated exons and HGNC full names of genes). After end-repair and adenylation, NEXTflex DNA Barcodes were ligated (Bio Scientific, Austin, TX, USA). Barcoded libraries were amplified, final library products were quantified, diluted and pooled in equal amounts. Finally, 12 pM of the constructed libraries were sequenced on the MiSeq (Illumina, San Diego, CA, USA) with a MiSeq reagent kit V2 (300-cycles) (Illumina) following manufacturer's recommendations.

2.2.3. Measurement of Indoor Radon Levels

All Rn measurements were performed within the study "Radon und Lungenkrebs im Bezirk Imst/Tirol" by W. Oberaigner et al. [30] in 2002, where long-term measurements in basements and ground floors indoors were used. Ennemoser et al. measured concentrations with medians ranging from 361–3750 Bq/m^3 (basements) and 210–1.160 Bq/m^3 (ground floors) in summer and winter, respectively [31]. The maximum Rn concentration which was measured was 274.000 Bq/m^3 [31].

2.3. Data/Statistical Analysis

Data were analysed with IBM SPSS Statistics 24 (IBM Corp., Armonk, NY, USA). For survival analyses Kaplan-Meier plots and log-rank tests were performed.

NGS data were exported as FASTQ files. Alignment and annotation were done using a modified version of a previously described method [32]. BAM files were visualized in the Integrative Genomics Viewer (http://www.broadinstitute.org/igv/ (accessed on 1 June 2021), Cambridge, MA, USA). A 5% cut-off for variant calls was used and results were only interpreted if the coverage was >200×.

3. Results

3.1. Patients Characteristics

Patient characteristics are given in detail in Table 2. For the Rn-exposed patients we were able to match 20 patients for molecular testing based on histology and year of diagnosis besides age and gender. In the overall non-exposed LC cohort used for the survival analyses, 52 patients based on cUICC were matched including those for molecular testing.

Table 2. Patients' characteristics.

	Rn-Exposed LC Cohort $n = 20$	Non-Exposed LC Cohort $n = 52$
Mean age at diagnosis	63.0 years	63.4 years
Gender (female/male)	9/11	21/31
Alive/dead	12/8	20/31 *
UICC		
I	3 (20.0%)	12 (23.1%)
II	3 (20.0%)	9 (17.3%)
III	4 (26.7%)	17 (32.7%)
IV	5 (33.3%) *	14 (26.9%)
ECOG		
0–1	10 (76.9%)	39 (86.7%)
2	1 (7.7%)	5 (11.1%)
>2	2 (15.4%) *	1 (2.2%) *
Histology		
Adenocarcinoma	8 (40.0%)	27 (52.0%)
Squamous cell carcinoma	10 (50.0%)	22 (42.3%)
Adenosquamous cell carcinoma	1 (5.0%)	0 (0.0%)
Large cell carcinoma	0 (0.0%)	2 (3.8%)
Large cell neuroendocrine carcinoma	1 (5.0%)	0 (0.0%)
NOS	0 (0.0%)	1 (1.9%)
Smoking		
Smoker or former smoker	6 (30.0%)	34 (65.5%)
Never smoker	4 (20.0%)	10 (19.2%)
Unknown	10 (50.0%)	8 (15.3%)

* information for some patients is missing. ECOG = Eastern Cooperative Oncology Group.

In total, the mean age of patients at time of diagnosis was 63.0 in the Rn-exposed and 63.4 years in in the non-exposed LC cohort with a gender distribution of 9f:11m and 21f:31m, respectively. According to cUICC stages (at diagnosis) the Rn-exposed cohort consisted of three patients in stage I (20%), three in stage II (20%), four in stage III (26.7%) and five in stage IV (33.3%). The stage of the remaining patients (n = 5) was not available. The non-exposed LC cohort comprised 12 patients in stage I (23.1%), 9 in stage II (17.3%), 17 in stage III (32.7%) and 14 in stage IV (26.9%). ECOG data were not available for all patients; however, the distribution was similar between the two groups, with 76.9 and 86.7% with ECOG 0–1, 7.7 and 11.1% with ECOG 2. Single patients had an ECOG > 2, two in the Rn-exposed cohort and one in the non-exposed LC cohort, respectively. 40% of Rn-exposed tumours were classified as adeno-, 50% as squamous cell carcinomas. In the overall non-exposed LC cohort, we had a slight bias towards adenocarcinomas in comparison to the Rn-exposed patients 52.0 vs. 40.0%, respectively, and 42.3 percent were squamous carcinomas, besides singular rare histological subtypes (Table 2). Smoking history of analysed patients was not comparable between the two cohorts, as information for half of the patients in the Rn-exposed group was missing. However, the percentage of patients with proven/evidenced smoking history in the non-exposed LC cohort (77.2%) was numerically higher than in the Rn-exposed group (60%).

3.2. Molecular Profile

In order to monitor genetic alterations in the two patient cohorts, an NGS based testing for AKT1, ALK, BRAF, CTNNB1, DDR2, EGFR, ERBB2, KRAS, MAP2K1, MET, NRAS, PIK3CA, PTEN and TP53 genes was performed and translocations of ALK, ROS and RET and amplification of MET were analysed by FISH (Figure 1; Table 3).

Table 3. Basic overview of genetic alterations.

	Cohort	Frequency (*n*)	Percent (%)
Non-exposed	no alterations	1	7
	KRAS_mut	1	7
	TP53_mut	8	57
	TP53/ERBB2mut	1	7
	PTEN mut	1	7
	ALK-rearr	2	14
Rn-exposed	no alterations	6	35
	KRAS_mut	2	12
	TP53_mut	3	18
	ROS-rearr	1	6
	MET-amplification	1	6
	MET_mutation	2	12
	MET T1010I, ALK rearr	1	6
	MET T1010I, TP53_mut	1	6

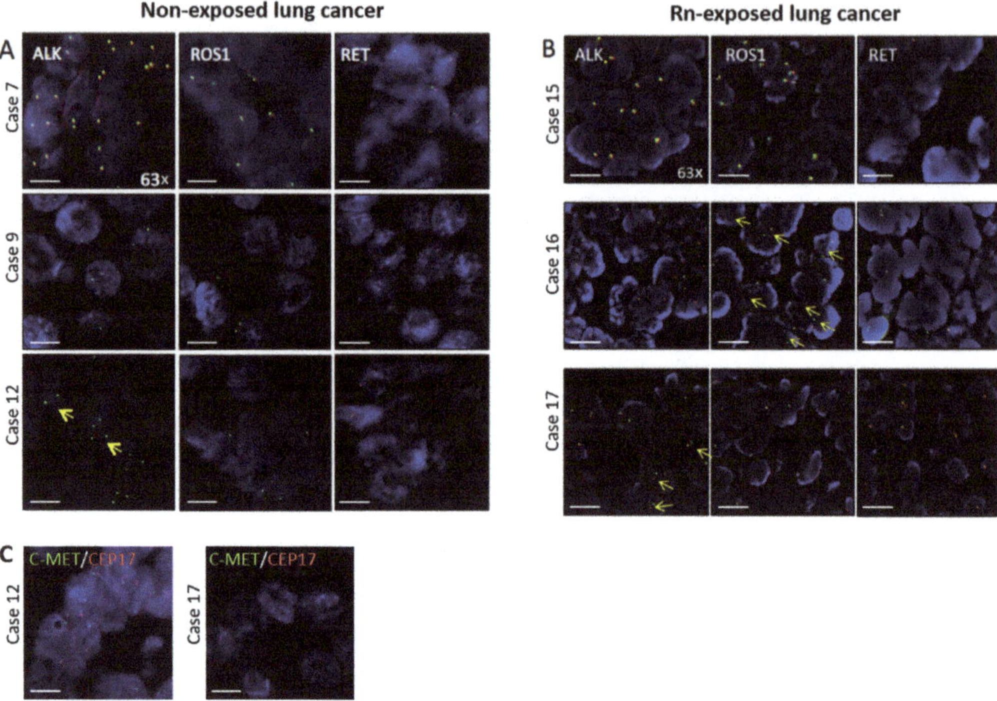

Figure 1. Fluorescence in-situ hybridization images showing ALK, ROS1 and RET signals of representative cases of (**A**) non-exposed LC group (*n* = 3) and (**B**) Rn-exposed group (*n* = 3). Yellow arrows highlight split signals (1F 1O 1G; insertion) or single red (in the case of ALK and ROS break apart probes) and single green signals (RET probe), both indicating insertion with deletion. (**C**) FISH images showing c-MET signal (green) and CEP17 signals (red) in two exemplary non-amplified cases. Images were taken using an Axioplan 2 (Zeis) microscope equipped with a 63× oil objective and a progress GRYPHAX SUBRA camera/software (Jenoptik). Scale bar = 50 μm.

Sufficient material and sample quality could be obtained in 14/27 non-exposed versus 17/22 Rn-exposed LC samples for NGS read out and FISH analyses, respectively (see Table 3). Additionally, some samples were feasible for one method, but not for the other technique. This might be due to low DNA quality and/or quantity or an under- or over-fixation with formalin of the tissue. In two Rn-exposed patients two samples from different time points within the course of their disease were analysed—the first case was a squamous cell carcinoma without alteration at both time points. A second lung cancer primary was detected in one patient 6.5 years after the first diagnosis including a switch in histology from squamous cell carcinoma without any alterations to an adenocarcinoma with a MET mutation.

In general, a higher number of patients with genetic alterations of these known driver genes were detected in the non-exposed LC cohort compared to the Rn-exposed cohort (Figure 2, Table 3), but the latter had more different alterations. Genetic alterations in PTEN and ERBB2 genes were each found in one patient of the non-exposed LC cohort. The patient showing an activating exon 20 ERBB2 alteration (c.2332_2340dup; p.G778_P780dup) also showed a deletion in TP53 (c.276–1_391del), with an unknown consequence for protein function. The case with PTEN mutation showed two different alterations in this gene, i.e., c.274G>C (p.D92H), resulting in a loss of function and c.860C>G (p.S287), with a potential loss of function of PTEN.

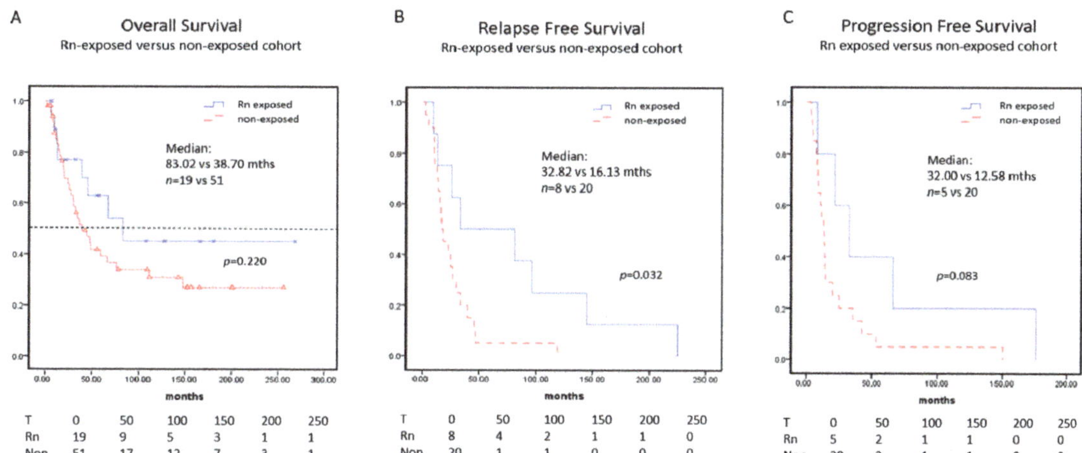

Figure 2. Kaplan–Meier survival analysis-non-exposed (red dotted line) versus Rn-exposed subgroup (blue line). (**A**) overall survival (n = 51 vs. 19; median survival of 38.7 vs. 83.0 mths), (**B**) relapse free survival (n = 20 vs. 8; median survival of 16.13 vs. 32.82 mths) and (**C**) progression-free survival (n = 20 vs. 5; with median survival of 12.85 vs. 32 mths) of non-exposed and Rn-exposed groups.

Mutations in the oncogene KRAS were determined in 1/14 (7%) and 2/17 (12%) in the non-exposed and Rn-exposed cohort, respectively. All detected mutations were in codon 12 (exon 2), representing a base exchange leading to G12C, V or D. Exclusive genetic alterations in the tumour suppressor gene TP53 were monitored in 8/14 (57%) and 3/17 (18%) patients of the non-exposed and Rn-exposed cohorts, respectively. In the Rn-exposed cohort less TP53 mutations were depicted while a higher number of KRAS mutations and MET alterations were detected compared to the non-exposed LC cohort. One case of the non-exposed LC cohort showed a duplication in ERBB2 gene in parallel and in one patient of the Rn-exposed cohort a concomitant variation of MET, c.3029C>T; p.T1010I was found. Interestingly, another case in the Rn-exposed cohort expressed the same T1010I mutation in the MET gene and also in this case it was a concomitant alteration. In this tumour, with the histological subtype of adenocarcinoma, an additional translocation in the anaplastic lymphoma kinase (ALK) could be monitored by FISH analysis, the only one in the Rn-exposed cohort (1/17; 6%). In comparison in the non-exposed LC cohort, 2/14 (14%) harboured an ALK translocation. Mutations in the MET gene were only detected in theRn-exposed group, 2/17 (6%) as concomitant alterations (p.T1010I) and 2/17 (6%) as sole mutation/insertion-deletion (c.3082 + 3A>T; c.2942–15_2942–4delinsACACA) of the juxtamembrane domain, but all four mutations were located in exon 14 of MET. Additionally, in one tumour sample of the Rn-exposed cohort an amplification in MET was determined with FISH analysis (6%). Testing for further translocations (fusion proteins) in ROS1 and RET one ROS1 positive case in the Rn-exposed group could be monitored, while no positive result for RET translocation was found.

Summarizing, investigation of this hotspot panel of genetic alterations resulted in (i) a higher number of non-mutated patients in the Rn-exposed cohort compared to the non-exposed group (32 vs. 7%), with (ii) an increased number of TP53 alterations in the non-exposed LC cohort and (iii) an unexpectedly high amount of MET alterations solely in the Rn-exposed patient cohort (Table 3).

3.3. Number of Mutations in Correlation with Level of Rn-Exposure

The Rn exposure-levels of 12/20 patients of our Rn-exposed cohort varied between 209 Bq/m^3 and 29,970 Bq/m^3. Due to the limited available data concerning the Rn exposure

(63%), i.e., doses and time, no correlation of number and kind of genetic alteration and radon exposure could be assessed.

3.4. Survival

As shown in Figure 2 the Rn-exposed group had a longer median overall survival (mOS) with 83 months compared to the non-exposed group with a mOS of 39 months, but this difference did not reach statistical significance due to the low patient numbers ($p = 0.22$). Similarly, relapse free survival was longer in the Rn-exposed cohort (median 81 vs. 33 months, $p = 0.84$). Stage dependent survivals are given in Supplementary Table S3.

Stratification of survival data in different genetic sub-groups reflects well-known characteristics with short overall survivals in historical cohorts and worse survivals for a the mainly TP53 and KRAS mutated subgroup or a sub-cohort with no specific drivers tested, whereas MET mutated or amplified patients displayed longer survivals (see Figure 3).

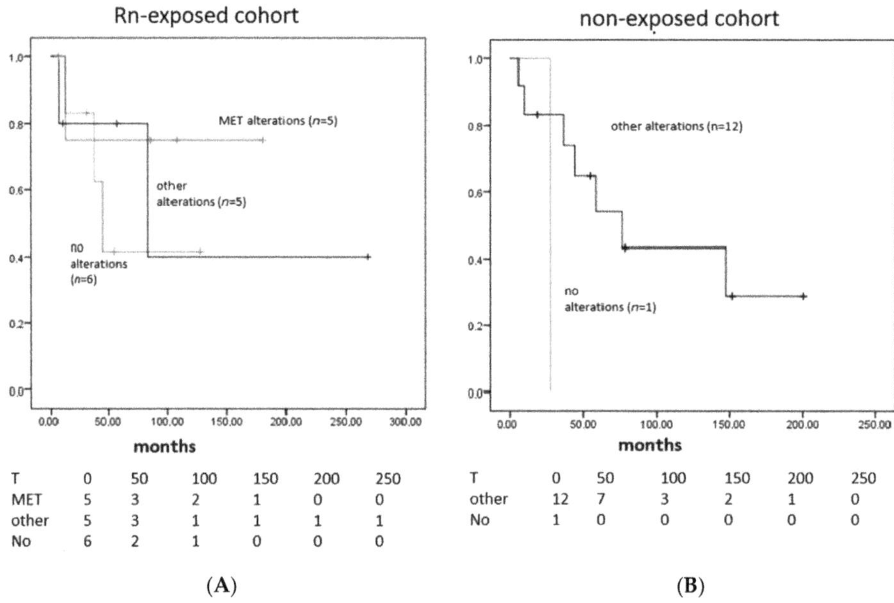

T	0	50	100	150	200	250
MET	5	3	2	1	0	0
other	5	3	1	1	1	1
No	6	2	1	0	0	0

T	0	50	100	150	200	250
other	12	7	3	2	1	0
No	1	0	0	0	0	0

(A)　　　　　　　　　　　(B)

Figure 3. Overall survival analysis stratified by summarized genetic alterations in patients of (**A**) Rn-exposed LC cohort and (**B**) non-exposed LC cohort. The red line represents MET alterations, and the black line, all other alterations (mainly TP53 and KRAS); the blue line depicts those without alterations.

Statistical analyses were not performed due to low patient numbers and the results need to be interpreted with caution based on the limitations in patient numbers. Nevertheless, these findings warrant further analyses of survivals in Rn-exposed patients and especially in MET mutated cohorts, even if no conclusions can be drawn from our study.

4. Discussion

Rn is the most important risk factor for LC in never smokers and the second in smokers [5,6], but the knowledge about specific mutations or mutation-frequencies in potentially Rn-related NSCLC remains low and no outcome data on Rn-related lung cancer patients are available so far. The best-investigated gene is TP53, but none of the Rn-specific mutations that were detected in previous studies could be validated by others [17–19,21,22]. In this study we screened for alterations in ALK, AKT1, BRAF, CTNNB1, DDR2, EGFR, ERBB2, KRAS, MAP2K1, MET, NRAS, PIK3CA, RET, ROS1, PTEN and TP53 genes in

Rn-exposed NSCLC patients from a village in Tyrol, where high Rn exposure due to the soil texture occurred.

Rn exposures measured for our cohort with >300 Bq/m^3 in 81.8% of the available cases were higher than in other studies [20,23,33] and far exceed the WHO recommended limit of 100 Bq/m^3 [5]. This emphasizes Umhausen as an area with extensive environmental Rn exposure.

Our Rn-exposed cohort presented with a median survival of 83 months compared to our local matched non-exposed group with 39 months. This result is limited by the low numbers and retrospective character of this analysis with lack of statistical significance. Some clinical data in this retrospective series were missing, including ECOG status and smoking habits in several cases. A higher awareness of cancer risk might contribute to early detections and better survivals, but clinical data and stages for the survival analyses were matched between the cohorts and patient characteristics were well balanced. Hence, a genetic background seems reasonable. On the one hand, in the Rn-exposed LC patients less TP53 mutations were detected, which are known negative prognostic factors and might translate into better survival [34]. In addition, TP53 mutations are linked to platin resistance, which was one of the only available standard drugs at the time of patient treatment courses [35]. In addition, radiation causing a higher rate of DNA damage might overwhelm functioning repair mechanisms including TP53, which might also explain tumorgenesis even in TP53 wt individuals and supports a different genetic background. On the other hand, MET mutations were increased, but their specific clinical prognostic relevance remains to be determined. Of note, at time of diagnosis and initial treatment no targeted drugs were available, and hence were not applied or were used very late in the course of the disease. Furthermore, several other metabolic, genomic and clinical patterns were not assessed. Nevertheless, to our knowledge, this is the first study to show data concerning survivals and genetic mutations in Rn-exposed LC subgroups, which hint towards a distinct genetic background and therapeutic responsiveness. This should encourage future studies of this specific patient population.

Concerning the molecular alterations, TP53 mutations were less common in our Rn-exposed cohort (20%) in comparison to NSCLC patients of our non-exposed group (64%) and in the AACR project (50.3%) [36]. Our results are in line with the findings of a large review of Ruano-Ravina et al. including 578 individuals reporting a frequency of TP53 mutations in 26% in uranium miners and 24% in environmental Rn-exposed patients [22], and with Choi et al. in a Rn-exposed Asian cohort (21%) [20]. In summary, Rn-exposed people showed less TP53 mutations and neither we nor others found specific Rn-induced mutations.

KRAS mutations occurred in 10% of the Rn-exposed cohort and in 7% in the non-exposed cohort; both percentages are below pre-described frequencies of 29.7% of NSCLC-patients in the AACR project [36]. McDonald et al. described KRAS mutations as more frequent in Rn-exposed uranium miners [37]. Choi et al. described KRAS mutations in 5.0% of their Rn-exposed study population [20]. The limited numbers hamper our results, but differences might well be due to differences in the investigated cohorts. Uranium miners are exposed to several carcinogens and most of them were smokers and had adenocarcinomas, which harbour more KRAS mutations than found in squamous cell carcinoma in lung [36]. This is even more interesting, as both our cases occurred in squamous cell carcinomas. The G12V and G12C mutations in our patients are well known and frequent mutations in LC and colorectal adenocarcinoma [36,38]. The mutations are associated with a worse prognosis [39] but are unlikely to have been caused exclusively by Rn-exposure. Recent developments offer a targeted drug for this cohort, which encourages standard testing [40]. For re-arrangements in ALK, ROS1 and RET neither higher incidence, nor Rn-specificity was found. Only one, the ROS1 + case, received a documented targeted treatment with crizotinib in 3rd line, due to diagnoses prior to any targeted treatment options.

The analyses of MET gene alterations showed the most interesting result. While in the non-exposed lung cancer cohort no tumour had an alteration in the MET gene, five patients

(25%) of the Rn-exposed cohort showed alterations in this gene, whereof four tumours showed a substitution and missense mutation (for detail see Table S1) and in one tumour there was an amplification of MET, although classified as low level. In addition, the results of the AACR-Project, identifying 5.8% of patients with MET mutations and 0.1% with MET amplifications [36], underlines the low frequency of MET alterations in lung cancers. However, the T1010I variation that we found in two cases is very rare. Of note, this T1010I MET mutation is a reported SNP (rs56391007) with 1.2% in a European sub-population [41]; however, the clinical significance is still under discussion and reported data are conflicting. A correlation to Rn-associated tumours remains to be clarified in a bigger study. This specific mutation was additionally described as relatively frequent in differentiated thyroid carcinoma [42], and an increase of growth factor independent proliferation and motility in in-vitro tumour cell lines with a T1010I mutation [43] were described. In contrast, Voortman et al. described the T1010I mutation as not resulting in enhanced Met phosphorylation [44], which questions its role in oncogenesis [44]. Both patients in the Rn-exposed cohort with T1010I had an additional genetic alteration in another investigated gene: one tumour had an ALK translocation and one tumour a mutation in TP53. This could diminish the relevance, at least of the T1010I mutation, as an oncogenic driver and therefore the importance remains elusive. So far, only one study conducted by Choi et al. investigated MET mutations in Rn-exposed people. They described an alteration frequency of 5% in MET in this cohort of 19 LC patients. All four MET alterations occurred in Exon-14, and Exon-14 skipping mutations are a well-known driver of alteration in LC. Therefore, our findings might be clinically relevant, as capmatinib [45], tepotinib [46] and savolitinib [47], are specific MET-inhibitors. Quite recently, the FDA and/or EMA approved therapeutics for MET-Exon-14 skipping alterations in NSCLC, and crizotinib [48,49], a tyrosine kinase inhibitor (TKI) targeting ALK, MET and ROS is under clinical development with inclusion criteria of MET-mutations.

Of note, for the T1010I mutation bearing and MET amplified patients the smoking status was unknown, but both other MET mutations occurred in former smokers, who quit smoking at least two years prior to diagnosis.

Several limitations apply to our analysis. The limited number of patients precludes definitive conclusions and renders the results hypothesis generating. Furthermore, the retrospective data collection, with, at least in part, missing information on important prognostic factors (i.e., performance-status, smoking-history), might have confounded the survival-outcome.

Hence, our results encourage further studies in a larger cohort of Rn-exposed LC patients to highlight the potential correlation between Rn exposure and MET mutations as well as KRAS mutations and inclusion of Rn-exposed LC patients within the screening for trials investigating MET and KRAS inhibitors.

5. Conclusions

Within a Tyrolean NSCLC cohort exposed to high environmental Rn levels (81.8% >300 Bq/m^3) a long median survival of 83 months was observed. This might be due to less TP53 mutations compared to reported numbers and a local control group, as well as an increased number of genetic alterations in the MET-gene, even though results are limited by the small patient numbers and the retrospective character of this study. Nevertheless, alterations in the MET gene in general and the T1010I mutation specifically are very rare in LC patients and hence an association with Rn-exposure seems reasonable.

Supplementary Materials: The following supporting information can be downloaded at: https://www.mdpi.com/article/10.3390/cancers14205113/s1, Table S1: Primer for targeted NGS; Table S2: Specific genes and exons for targeted NGS panel: NGS_LUN3_#3; Table S3: Median overall survivals in months of the radon exposed, matched non-exposed cohort.

Author Contributions: Conceptualization, G.P., G.G. and C.M.; Validation, G.G., C.M. and G.P.; Formal Analysis, G.G., M.K., M.F., A.A., S.M.-B. and C.M.; Investigation, G.G., M.K., S.M.-B., C.H., J.S. and C.M.; Resources, G.P., S.M.-B., R.B. and C.M.; Data Curation, G.G., M.K., S.M.-B. and C.M.; Writing—Original Draft Preparation, G.G., M.K. and C.M.; Writing—Review and Editing, G.G., M.K., F.M., S.M.-B., R.B., C.M. and G.P.; Visualization, G.G., M.K. and C.M.; Supervision, R.B. and G.P.; Project Administration, G.G. and C.M.; Funding Acquisition, C.M. and G.P. All authors have read and agreed to the published version of the manuscript.

Funding: This research was partly funded by the "Verein für Tumorforschung", grant number F012019.

Institutional Review Board Statement: The study was conducted according to the guidelines of the Declaration of Helsinki, and was approved by the Ethics Committee of the Medical University of Innsbruck on 03 may 2018, AN 1018/2018.

Informed Consent Statement: Informed consent was obtained from all subjects, as needed according to the local ethical board.

Data Availability Statement: The data presented in this study are available on request from the corresponding author. The data are not publicly available due to ethical restrictions.

Acknowledgments: The authors thank Julia Heppke and Iris Oberauer for their great technical support.

Conflicts of Interest: The authors declare no conflict of interest.

References

1. Parkin, D.M.; Boyd, L.; Walker, L.C. The fraction of cancer attributable to lifestyle and environmental factors in the UK in 2010. *Br. J. Cancer* **2011**, *105* (Suppl. S2), S77–S81. [CrossRef] [PubMed]
2. Parkin, D.M.; Bray, F.; Ferlay, J.; Pisani, P. Global cancer statistics 2002. *CA Cancer J. Clin.* **2005**, *55*, 74–108. [CrossRef] [PubMed]
3. Couraud, S.; Zalcman, G.; Milleron, B.; Morin, F.; Souquet, P.J. Lung cancer in never smokers—A review. *Eur. J. Cancer* **2012**, *48*, 1299–1311. [CrossRef] [PubMed]
4. Ferlay, J.; Shin, H.R.; Bray, F.; Forman, D.; Mathers, C.; Parkin, D.M. Estimates of worldwide burden of cancer in 2008: GLOBOCAN 2008. *Int. J. Cancer* **2010**, *127*, 2893–2917. [CrossRef]
5. World Health Organization. WHO Handbook on Indoor Radon: A Public Health Perspective. 2009. Available online: https://www.ncbi.nlm.nih.gov/books/NBK143216/ (accessed on 6 July 2022).
6. Ruano-Ravina, A.; Pereyra, M.F.; Castro, M.T.; Perez-Rios, M.; Abal-Arca, J.; Barros-Dios, J.M. Genetic susceptibility, residential radon, and lung cancer in a radon prone area. *J. Thorac. Oncol.* **2014**, *9*, 1073–1080. [CrossRef] [PubMed]
7. Darby, S.; Hill, D.; Auvinen, A.; Barros-Dios, J.M.; Baysson, H.; Bochicchio, F.; Deo, H.; Falk, R.; Forastiere, F.; Hakama, M.; et al. Radon in homes and risk of lung cancer: Collaborative analysis of individual data from 13 European case-control studies. *BMJ* **2005**, *330*, 223. [CrossRef]
8. United Nations Scientific Committee on the Effects of Atomic Radiation. *Sources and Effects of Ionizing Radiation*; United Nations Pubns.: New York, NY, USA, 2000; Volume 1. Available online: https://www.unscear.org/docs/publications/2000/UNSCEAR_2000_Report_Vol.I.pdf (accessed on 6 July 2022).
9. Al-Zoughool, M.; Krewski, D. Health effects of radon: A review of the literature. *Int. J. Radiat. Biol.* **2009**, *85*, 57–69. [CrossRef]
10. Robertson, A.; Allen, J.; Laney, R.; Curnow, A. The cellular and molecular carcinogenic effects of radon exposure: A review. *Int. J. Mol. Sci.* **2013**, *14*, 14024–14063. [CrossRef]
11. Appleton, J.D. Radon: Sources, health risks, and hazard mapping. *Ambio* **2007**, *36*, 85–89. [CrossRef]
12. Barros-Dios, J.M.; Ruano-Ravina, A.; Perez-Rios, M.; Castro-Bernardez, M.; Abal-Arca, J.; Tojo-Castro, M. Residential radon exposure, histologic types, and lung cancer risk. A case-control study in Galicia, Spain. *Cancer Epidemiol. Biomarkers Prev.* **2012**, *21*, 951–958. [CrossRef]
13. Barros-Dios, J.M.; Barreiro, M.A.; Ruano-Ravina, A.; Figueiras, A. Exposure to residential radon and lung cancer in Spain: A population-based case-control study. *Am. J. Epidemiol.* **2002**, *156*, 548–555. [CrossRef] [PubMed]
14. Lubin, J.H.; Wang, Z.Y.; Boice, J.D., Jr.; Xu, Z.Y.; Blot, W.J.; De Wang, L.; Kleinerman, R.A. Risk of lung cancer and residential radon in China: Pooled results of two studies. *Int. J. Cancer* **2004**, *109*, 132–137. [CrossRef] [PubMed]
15. Krewski, D.; Lubin, J.H.; Zielinski, J.M.; Alavanja, M.; Catalan, V.S.; Field, R.W.; Klotz, J.B.; Létourneau, E.G.; Lynch, C.F.; Lyon, J.L. A combined analysis of North American case-control studies of residential radon and lung cancer. *J. Toxicol. Environ. Health A* **2006**, *69*, 533–597. [CrossRef] [PubMed]
16. Duan, P.; Quan, C.; Hu, C.; Zhang, J.; Xie, F.; Hu, X.; Yu, Z.; Gao, B.; Liu, Z.; Zheng, H. Nonlinear dose-response relationship between radon exposure and the risk of lung cancer: Evidence from a meta-analysis of published observational studies. *Eur. J. Cancer Prev.* **2015**, *24*, 267–277. [CrossRef]
17. Vahakangas, K.H.; Metcalf, R.A.; Welsh, J.A.; Bennett, W.P.; Harris, C.C.; Samet, J.M.; Lane, D.P. Mutations of p53 and ras genes in radon-associated lung cancer from uranium miners. *Lancet* **1992**, *339*, 576–580. [CrossRef]

18. Vahakangas, K.H. TP53 mutations in workers exposed to occupational carcinogens. *Hum. Mutat.* **2003**, *21*, 240–251. [CrossRef]
19. Taylor, J.A.; Watson, M.A.; Devereux, T.R.; Michels, R.Y.; Saccomanno, G.; Anderson, M. p53 mutation hotspot in radon-associated lung cancer. *Lancet* **1994**, *343*, 86–87. [CrossRef]
20. Choi, J.R.; Koh, S.B.; Park, S.Y.; Kim, H.R.; Lee, H.; Kang, D.R. Novel Genetic Associations between Lung Cancer and Indoor Radon Exposure. *J. Cancer Prev.* **2017**, *22*, 234–240. [CrossRef]
21. Yang, Q.; Wesch, H.; Mueller, K.M.; Bartsch, H.; Wegener, K.; Hollstein, M. Analysis of radon-associated squamous cell carcinomas of the lung for a p53 gene hotspot mutation. *Br. J. Cancer* **2000**, *82*, 763–766. [CrossRef]
22. Ruano-Ravina, A.; Faraldo-Valles, M.J.; Barros-Dios, J.M. Is there a specific mutation of p53 gene due to radon exposure? A systematic review. *Int. J. Radiat. Biol.* **2009**, *85*, 614–621. [CrossRef]
23. Ruano-Ravina, A.; Torres-Duran, M.; Kelsey, K.T.; Parente-Lamelas, I.; Leiro-Fernandez, V.; Abdulkader, I.; Abal-Arca, J.; Montero-Martínez, C.; Vidal-García, I.; Amendo, M. Residential radon, EGFR mutations and ALK alterations in neversmoking lung cancer cases. *Eur. Respir. J.* **2016**, *48*, 1462–1470. [CrossRef] [PubMed]
24. Taga, M.; Mechanic, L.E.; Hagiwara, N.; Vahakangas, K.H.; Bennett, W.P.; Alavanja, M.C.; Welsh, J.A.; Khan, M.A.; Lee, A.; Diasio, R. EGFR somatic mutations in lung tumors: Radon exposure and passive smoking in former- and never-smoking U.S. women. *Cancer Epidemiol. Biomarkers Prev.* **2012**, *21*, 988–992.21. [CrossRef] [PubMed]
25. Bonner, M.R.; Bennett, W.P.; Xiong, W.; Lan, Q.; Brownson, R.C.; Harris, C.C.; Field, R.W.; Lubin, J.H.; Alavanja, M.C. Radon, secondhand smoke, glutathione-S-transferase M1 and lung cancer among women. *Int. J. Cancer* **2006**, *119*, 1462–1467. [CrossRef] [PubMed]
26. Kim, H.; Shim, H.S.; Kim, L.; Kim, T.J.; Kwon, K.Y.; Lee, G.K.; Chung, J.H. Guideline Recommendations for Testing of ALK Gene Rearrangement in Lung Cancer: A Proposal of the Korean Cardiopulmonary Pathology Study Group. *Korean J. Pathol.* **2014**, *48*, 1–9. [CrossRef]
27. Warth, A.; Muley, T.; Dienemann, H.; Goeppert, B.; Stenzinger, A.; Schnabel, P.A.; Schirmacher, P.; Penzel, R.; Weichert, W. ROS1 expression and translocations in non-small-cell lung cancer: Clinicopathological analysis of 1478 cases. *Histopathology* **2014**, *65*, 187–194. [CrossRef]
28. Lee, S.E.; Lee, B.; Hong, M.; Song, J.Y.; Jung, K.; Lira, M.E.; Mao, M.; Han, J.; Kim, J.; Choi, Y.L. Comprehensive analysis of RET and ROS1 rearrangement in lung adenocarcinoma. *Mod. Pathol.* **2015**, *28*, 468–479. [CrossRef]
29. Ach, T.; Zeitler, K.; Schwarz-Furlan, S.; Baader, K.; Agaimy, A.; Rohrmeier, C.; Zenk, J.; Gosau, M.; Reichert, T.E.; Brockhoff, G.; et al. Aberrations of MET are associated with copy number gain of EGFR and loss of PTEN and predict poor outcome in patients with salivary gland cancer. *Virchows Arch.* **2013**, *462*, 65–72. [CrossRef]
30. Oberaigner, W.; Kreienbrock, L.; Schaffrath-Rosario, A.; Kreuzer, M.; Wellmann, J.; Keller, G.; Gerken, M.; Langer, B.; Wichmann, H.E. *Radon und Lungenkrebs im Bezirk Imst/Österreich, Reihe Fortschritte in der Umweltmedizin*; Ecomed: Landsberg am Lech, Germany, 2002.
31. Ennemoser, O.; Ambach, W.; Brunner, P.; Schneider, P.; Oberaigner, W.; Purtscheller, F.; Stingl, V.; Keller, G. Unusually high indoor radon concentrations from a giant rock slide. *Sci. Total Environ.* **1994**, *151*, 235–240. [CrossRef]
32. Peifer, M.; Fernandez-Cuesta, L.; Sos, M.L.; George, J.; Seidel, D.; Kasper, L.H.; Plenker, D.; Leenders, F.; Sun, R.; Zander, T.; et al. Integrative genome analyses identify key somatic driver mutations of small-cell lung cancer. *Nat. Genet.* **2012**, *44*, 1104–1110. [CrossRef]
33. Lorenzo-Gonzalez, M.; Ruano-Ravina, A.; Peon, J.; Pineiro, M.; Barros-Dios, J.M. Residential radon in Galicia: A cross-sectional study in a radon-prone area. *J. Radiol. Prot.* **2017**, *37*, 728–741. [CrossRef]
34. Jiao, X.D.; Qin, B.D.; You, P.; Cai, J.; Zang, Y.S. The prognostic value of TP53 and its correlation with EGFR mutation in advanced non-small cell lung cancer, an analysis based on cBioPortal data base. *Lung Cancer* **2018**, *123*, 70–75. [CrossRef]
35. Zhang, X.; Qi, Z.; Yin, H.; Yang, G. Interaction between p53 and Ras signaling controls cisplatin resistance via HDAC4- and HIF-1α-mediated regulation of apoptosis and autophagy. *Theranostics* **2019**, *9*, 1096–1114. [CrossRef] [PubMed]
36. AACR Project GENIE Consortium. AACR Project GENIE: Powering Precision Medicine through an International Consortium. *Cancer Discov.* **2017**, *7*, 818–831. [CrossRef] [PubMed]
37. McDonald, J.W.; Taylor, J.A.; Watson, M.A.; Saccomanno, G.; Devereux, T.R. p53 and K-ras in radon-associated lung adenocarcinoma. *Cancer Epidemiol. Biomarkers Prev.* **1995**, *4*, 791–793. [PubMed]
38. Jeanson, A.; Tomasini, P.; Souquet-Bressand, M.; Brandone, N.; Boucekine, M.; Grangeon, M.; Chaleat, S.; Khobta, N.; Milia, J.; Mhanna, L.; et al. Efficacy of Immune Checkpoint Inhibitors in KRAS-Mutant Non-Small Cell Lung Cancer (NSCLC). *J. Thorac. Oncol.* **2019**, *14*, 1095–1101. [CrossRef]
39. Hayama, T.; Hashiguchi, Y.; Okamoto, K.; Okada, Y.; Ono, K.; Shimada, R.; Ozawa, T.; Toyoda, T.; Tsuchiya, T.; Iinuma, H.; et al. G12V and G12C mutations in the gene KRAS are associated with a poorer prognosis in primary colorectal cancer. *Int. J. Color. Dis.* **2019**, *34*, 1491–1496. [CrossRef]
40. Hong, D.S.; Fakih, M.G.; Strickler, J.H.; Desai, J.; Durm, G.A.; Shapiro, G.I.; Falchook, G.S.; Price, T.J.; Sacher, A.; Denlinger, C.S.; et al. KRAS Inhibition with Sotorasib in Advanced Solid Tumors. *N. Engl. J. Med.* **2020**, *383*, 1207–1217. [CrossRef] [PubMed]
41. Reference SNP Cluster Report: rs56391007. 2022. Available online: https://www.ncbi.nlm.nih.gov/snp/rs56391007?horizontal_tab=true (accessed on 6 July 2022).
42. Wasenius, V.M.; Hemmer, S.; Karjalainen-Lindsberg, M.L.; Nupponen, N.N.; Franssila, K.; Joensuu, H. MET receptor tyrosine kinase sequence alterations in differentiated thyroid carcinoma. *Am. J. Surg. Pathol.* **2005**, *29*, 544–549. [CrossRef]

43. Ma, P.C.; Kijima, T.; Maulik, G.; Fox, E.A.; Sattler, M.; Griffin, J.D.; Johnsen, B.E.; Salgia, R. c-MET mutational analysis in small cell lung cancer: Novel juxtamembrane domain mutations regulating cytoskeletal functions. *Cancer Res.* **2003**, *63*, 6272–6281.

44. Voortman, J.; Harada, T.; Chang, R.P.; Killian, J.K.; Suuriniemi, M.; Smith, W.I.; Meltzer, P.S.; Lucchi, M.; Wang, Y.; Giaconne, G. Detection and therapeutic implications of c-Met mutations in small cell lung cancer and neuroendocrine tumors. *Curr. Pharm. Des.* **2013**, *19*, 833–840. [CrossRef]

45. Vansteenkiste, J.F.; Van De Kerkhove, C.; Wauters, E.; Van Mol, P. Capmatinib for the treatment of non-small cell lung cancer. *Expert Rev. Anticancer Ther.* **2019**, *19*, 659–671. [CrossRef] [PubMed]

46. Wu, Z.X.; Li, J.; Dong, S.; Lin, L.; Zou, C.; Chen, Z.S. Tepotinib hydrochloride for the treatment of non-small cell lung cancer. *Drugs. Today* **2021**, *57*, 265–275. [CrossRef] [PubMed]

47. Huang, C.; Zou, Q.; Liu, H.; Qiu, B.; Li, Q.; Lin, Y.; Liang, Y. Management of Non-small Cell Lung Cancer Patients with MET Exon 14 Skipping Mutations. *Curr. Treat. Options Oncol.* **2020**, *21*, 33. [CrossRef] [PubMed]

48. Drilon, A.E.; Camidge, D.R.; Ou, S.-H.I.; Clark, J.W.; Socinski, M.A.; Weiss, J.; Riely, G.J.; Winter, M.; Wang, S.C.; Monti, K.; et al. Efficacy and safety of crizotinib in patients (pts) with advanced MET exon 14-altered non-small cell lung cancer (NSCLC). *J. Clin. Oncol.* **2016**, *34* (Suppl. 15), 108. [CrossRef]

49. Landi, L.; Chiari, R.; Tiseo, M.; D'Inca, F.; Dazzi, C.; Chella, A.; Delmonte, A.; Bonanno, L.; Giannarelli, D.; Cortinovis, D.L.; et al. Crizotinib in MET deregulated or ROS1 rearranged pretreated non-small-cell lung cancer (METROS): A phase II, prospective, multicentre, two-arms trial. *Clin. Cancer Res.* **2019**, *25*, 7312–7319. [CrossRef]

Systematic Review

A Systematic Review of Mesenchymal Epithelial Transition Factor (*MET*) and Its Impact in the Development and Treatment of Non-Small-Cell Lung Cancer

Embla Bodén [1,2,3], Fanny Sveréus [1,2,3], Franziska Olm [1,2,3,4] and Sandra Lindstedt [1,2,3,4,*]

[1] Department of Clinical Sciences, Lund University, 22184 Lund, Sweden; embla.boden_janson@med.lu.se (E.B.); fa4743sv-s@student.lu.se (F.S.); franziska.olm@med.lu.se (F.O.)
[2] Wallenberg Center for Molecular Medicine, Lund University, 22184 Lund, Sweden
[3] Lund Stem Cell Center, Lund University, 22184 Lund, Sweden
[4] Department of Cardiothoracic Surgery and Transplantation, Skåne University Hospital, 22242 Lund, Sweden
* Correspondence: sandra.lindstedt_ingemansson@med.lu.se

Simple Summary: Lung cancer is the type of cancer that kills the most people in the world each year. It is difficult to diagnose lung cancer in the early stages and there are only few treatment options available once the cancer has spread. The mesenchymal epithelial transition factor (*MET*) gene is of importance in lung cancer development, and mutations in this gene are related to poor prognosis. Consequently, it is important to develop new treatment options that specifically target the *MET* protein. In this systematic review, we aimed to summarize the existing knowledge on the impact of *MET* on lung cancer development and the effect of currently available medications. Our hope is that the findings of this systematic review will deepen the understanding of other researchers, possibly providing a guiding hand as to what may be most interesting to focus on in future research projects on this subject.

Citation: Bodén, E.; Sveréus, F.; Olm, F.; Lindstedt, S. A Systematic Review of Mesenchymal Epithelial Transition Factor (*MET*) and Its Impact in the Development and Treatment of Non-Small-Cell Lung Cancer. *Cancers* **2023**, *15*, 3827. https://doi.org/10.3390/cancers15153827

Academic Editors: Jan Trøst Jørgensen and Jens Mollerup

Received: 14 June 2023
Revised: 20 July 2023
Accepted: 25 July 2023
Published: 27 July 2023

Abstract: Lung cancer represents the leading cause of annual cancer-related deaths worldwide, accounting for 12.9%. The available treatment options for patients who experience disease progression remain limited. Targeted therapeutic approaches are promising but further understanding of the role of genetic alterations in tumorigenesis is imperative. The *MET* gene has garnered great interest in this regard. The aim of this systematic review was to analyze the findings from multiple studies to provide a comprehensive and unbiased summary of the evidence. A systematic search was conducted in the reputable scientific databases Embase and PubMed, leading to the inclusion of twenty-two articles, following the PRISMA guidelines, elucidating the biological role of *MET* in lung cancer and targeted therapies. The systematic review was registered in PROSPERO with registration ID: CRD42023437714. *MET* mutations were detected in 7.6–11.0% of cases while *MET* gene amplification was observed in 3.9–22.0%. Six studies showed favorable treatment outcomes utilizing *MET* inhibitors compared to standard treatment or placebo, with increases in PFS and OS ranging from 0.9 to 12.4 and 7.2 to 24.2 months, respectively, and one study reporting an increase in ORR by 17.3%. Furthermore, patients with a higher mutational burden may derive greater benefit from treatment with *MET* tyrosine kinase inhibitors (TKIs) than those with a lower mutational burden. Conversely, two studies reported no beneficial effect from adjunctive treatment with a *MET* targeted therapy. Given these findings, there is an urgent need to identify effective therapeutic strategies specifically targeting the *MET* gene in lung cancer patients.

Keywords: non-small cell lung cancer; mesenchymal epithelial transition factor; *MET*; targeted therapies; genetic alterations; biomarker; systematic review

1. Introduction

Lung cancer is one of the most common malignancies, causing 12.9% of cancer-related deaths worldwide, resulting in 1.3 million deaths annually [1–4]. Between 80.0 and 85.0% of lung cancer cases are non-small cell lung cancer (NSCLC) [3,5,6]. Late diagnosis is a major problem, contributing to the short median survival of approximately 18 months and the overall 5-year survival rate for lung cancer of 15–21%, depending on gender [7,8]. Up to 75.0% of patients with newly diagnosed NSCLC have locally advanced or metastasized disease at diagnosis, with a 5-year survival rate below 5.0% [1]. To this day, locally advanced or metastasized NSCLC is commonly treated with platinum-based chemotherapy, offering modest efficacy, with response rates of 20.0–30.0% and a plethora of side effects, or immunotherapy [3,5,9]. Targeted therapies for several types of cancer, including NSCLC, have emerged as a beneficial option for subsets of patients. Current treatment guidelines for advanced NSCLC call for broad molecular profiling to identify and guide the choice of potential targeted therapy options [10]. The proportion of patients with NSCLC receiving next generation sequencing (NGS) is low, which is in part due to insufficient availability of tumor tissue at the time of diagnosis. Approximately 60.0–65.0% of patients undergo testing for mutations in the epidermal growth factor receptor gene (*EGFR*) and less than 25.0% of patients are tested for alterations in the mesenchymal epithelial transition factor gene (*MET*) [11].

EGFR is a transmembrane receptor that is involved in several signaling pathways; it promotes cell proliferation and is anti-apoptotic. Overexpression of the *EGFR* gene is a well-known pathological mechanism in NSCLC, present in 43.0–89.0% of NSCLC cases, which can lead to poorer outcomes [12]. Several *EGFR* TKIs exist; however acquired resistance to these therapies is common and alterations in the *MET* gene have been proven to be a contributing factor [13]. The protein *MET* is a transmembrane receptor tyrosine kinase (RTK) with a central role in cell motility, morphogenesis, proliferation, survival, and invasion (Figure 1A) [5,14–16]. Alterations in the *MET* gene, such as gene copy number (GCN) gain, mutation, or overexpression of the protein, have been reported in NSCLC [17]. The only known ligand to *MET* is hepatocyte growth factor (HGF) [6]. HGF is found in healthy lung tissue but is often overexpressed in NSCLC. Aberrant signaling through the HGF/*MET* pathway has clinically been linked to oncogenic potential and poor outcomes in NSCLC, with shortened overall survival (OS) and progression-free survival (PFS). However, studies showing the opposite also exist, introducing a contradiction in this area of research. One proven cause of unfavorable outcomes in *MET* altered lung cancer is acquired resistance to *EGFR* TKIs, underlining the need for more efficient targeted therapies [1,5,6,14,15,17–20]. The frequency of any form of dysregulation of *MET* in NSCLC ranges from 3.0 to 7.0%. Sporadic GCN gain of *MET* is detected in 1.0–4.0% of wild type *EGFR* NSCLC cases. *MET* exon 14 skipping mutations occur in approximately 3.0% of NSCLC cases [2,3,18,20]. Amplification of the *MET* gene is the most common type of dysregulated signaling in NSCLC with acquired resistance to *EGFR* TKIs, with reported frequencies between 5.0 and 26.0% [20,21]. Alterations in *MET* have been shown to upregulate the expression of *EGFR* ligands, which in turn increases *EGFR* signaling, promoting cell proliferation, angiogenesis, and apoptosis [15]. Increased expression of HGF can also promote resistance to *EGFR* TKIs by supporting clonal selection of a subpopulation of cells with *MET* amplification [1]. Despite the fact that acquired resistance to *EGFR* TKIs is very common, they remain the preferred first-line treatment for locally advanced or metastatic *EGFR* mutation-positive NSCLC [15,22,23]. For patients with acquired resistance to *EGFR* TKIs caused by upregulation or amplification of *MET*, it may be beneficial to treat with a combination of inhibitors of both *MET* and *EGFR*, as this has been shown to have a synergistic inhibitory effect on the proliferation of cancer cells [2,21,22]. A combination of *EGFR* and *MET* TKIs has been shown to possibly delay the occurrence of resistance to *EGFR* TKIs [24]. For an overview of the *EGFR* and *MET* TKIs discussed in this article, see Table 1 (Figure 1B).

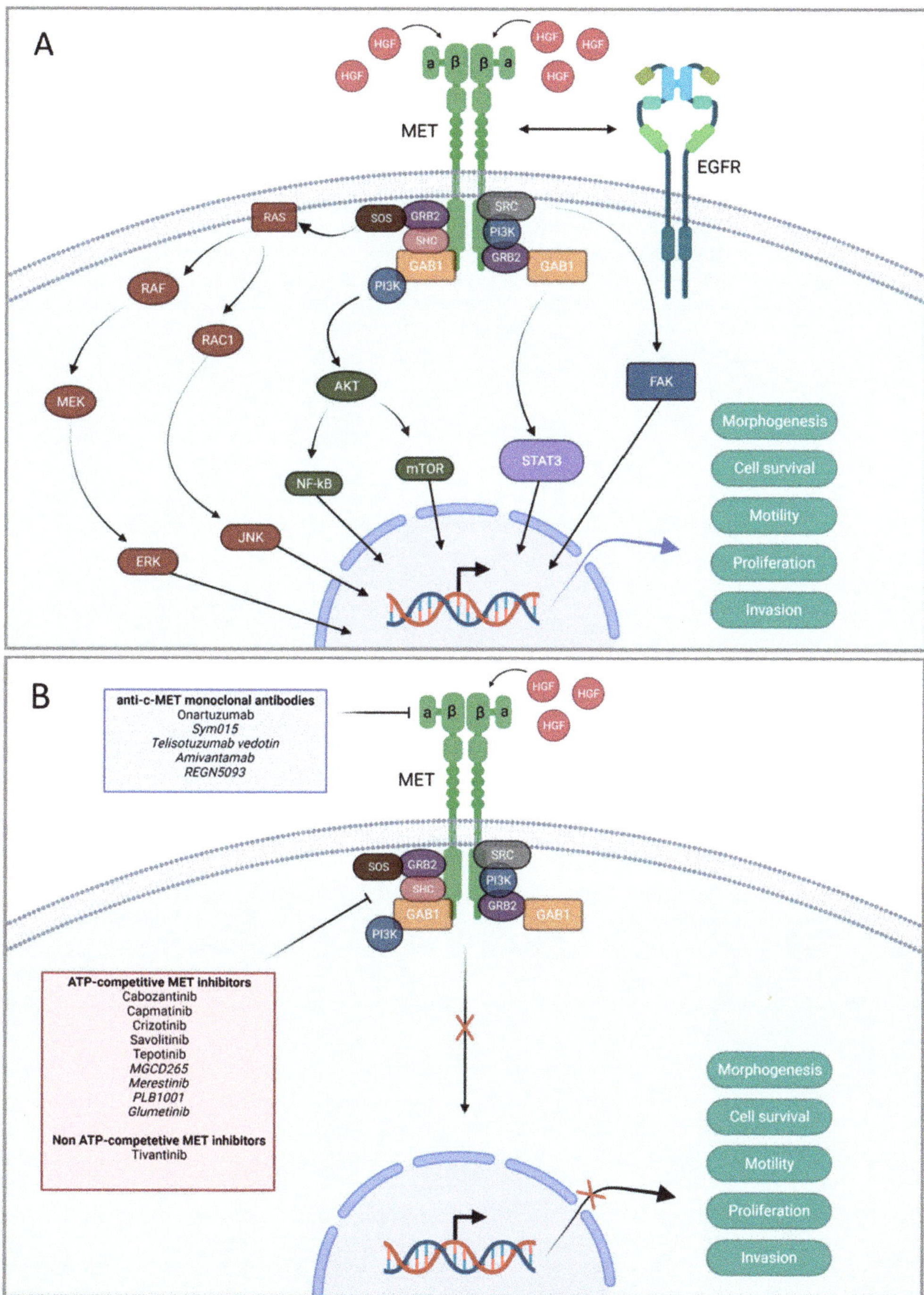

Figure 1. Schematic of the *MET* signaling pathway and the sites of action for *MET* targeted therapies. (**A**) The *MET* signaling pathway with downstream intracellular signaling and transcription of genes

leading to enhanced morphogenesis, cell survival, motility, proliferation, and invasion. *MET* is activated by its ligand, HGF, and the *MET* receptor can interact in various ways with the *EGFR* receptor. (**B**) The *MET* targeted therapies presented according to specific targets, either extracellular or intracellular. The pipe-line drugs are highlighted in cursive. MET: mesenchymal epithelial transition factor, HGF: hepatocyte growth factor, EGFR: epidermal growth factor receptor, SRC: v-src sarcoma (Schmidt-Ruppin A-2) viral oncogene homolog, PI3K: phosphatidylinositol 3-kinase, GRB2: growth factor receptor-bound protein 2, GAB1: GRB2-associated binding protein 1, SHC: src homology 2 domain-containing, SOS: son of sevenless, RAS: rat sarcoma, RAF: rapidly accelerated fibrosarcoma, MEK: MAPK effector kinase, ERK: extracellular signal-regulated kinase, RAC1: ras-related C3 botulinum toxin substrate 1, JNK: janus kinase 1, AKT: ak strain transforming, NF-kB: nuclear factor kappa B, mTOR: mammalian target of rapamycin, FAK: focal adhesion kinase, STAT3: signal transducer and activator of transcription 3, ATP: adenosine triphosphate. Created in biorender.com.

Table 1. Overview of included *EGFR* and *MET* TKIs. *EGFR* and *MET* targeted therapies included in this systematic review. *EGFR*: epidermal growth factor receptor, *MET*: mesenchymal epithelial transition factor, TKI: tyrosine kinase receptor, *VEGFR2*: vascular endothelial growth factor receptor 2, *RET*: ret proto-oncogene, *ROS1*: ROS proto-oncogene 1, *KIT*: CD117, *TIE-2*: tyrosine kinase with immunoglobin and EGF homology domains 2, *AXL*: AXL receptor tyrosine kinase, RTK: receptor tyrosine kinase, NSCLC: non-small cell lung cancer, HGF: hepatocyte growth factor, ATP: adenosine triphosphate, CYP2C19: cytochrome P450 2C19.

Drug Name	Effect	References
Afatinib	Binds covalently and irreversibly to the kinase domain of *EGFR*.	Arrieta et al. [1]
Cabozantinib	A small molecule TKI that targets *MET*, *VEGFR2*, *RET*, *ROS1*, *KIT*, *TIE-2*, and *AXL*. Binds intracellularly to *MET*.	Neal et al. [2] Landi et al. [18]
Capmatinib	A highly selective intracellular *MET* inhibitor.	Schuler et al. [20] Sequist et al. [22]
Crizotinib	An intracellular *MET*/*ALK*/*ROS1* RTK inhibitor with high specificity for *MET*.	Landi et al. [18]
Dacomitinib	A small irreversible pan-human *EGFR* inhibitor.	Jänne et al. [21]
Erlotinib	A reversible, small-molecule *EGFR* TKI.	Spigel et al. [6]
Gefitinib	A reversible *EGFR* TKI.	Arrieta et al. [1] Wu et al. [23]
Onartuzumab	A recombinant, fully humanized, one-armed anti-*MET* monovalent monoclonal antibody. Binds to the extracellular domain of *MET* without activating it and without dimerizing.	Hirsch et al. [17] Kishi et al. [15]
Osimertinib	A CNS-active, irreversible *EGFR* TKI.	Sequist et al. [22]
Rociletinib	An irreversible *EGFR* TKI targeting mutated form of the *EGFR* gene.	Arrieta et al. [1]
Savolitinib	A small molecule, ATP competitive, selective *MET* TKI.	Sequist et al. [22]
Tepotinib	A highly selective, ATP competitive *MET* inhibitor.	Wu et al. [23]
Tivantinib	A selective, non-ATP-competitive *MET* inhibitor metabolized by CYP2C19.	Yoshioka et al. [5]

An issue with these types of targeted therapies remains the inherent genetic heterogeneity of NSCLC. Several mutations may co-exist in the same patient and complicate the interpretation of specific drug effects [3,25]. An additional aspect to take into consideration is the variability in methodologies employed to quantify aberrant *MET* expression. Some studies use immunohistochemistry (IHC) with the definition of *MET* positivity ranging from 1+ to 3+ and *MET* negativity varying from 0 to 1+ [2,17,18,20]. Other studies apply gene copy number (GCN) $\geq$ 5 or *MET*/centromere 7 ratio $\geq$ 2.0 as the lower limit for defined *MET* positivity [20].

Genetic profiling carried out by NGS is a rapidly progressing area of research within oncology. The methods used are becoming increasingly cost and time efficient, allowing for improved individual genetic testing and tailored therapy options [26]. The aim of this systematic review was to provide a comprehensive and unbiased summary of the evidence investigating the role of *MET* and aberrant expression of the protein in lung cancer development and treatment thereof.

2. Methods

2.1. Search Strategy

The Preferred Reporting Items for Systematic reviews and Meta-Analyses (PRISMA) guidelines were used in conducting this systematic review; for the full PRISMA 2020 Checklist, see Appendix B [27] (Tables A1 and A2). A protocol was not established. The systematic review was registered in PROSPERO with registration ID CRD42023437714. Medically relevant databases containing existing literature were systematically searched by the authors for articles related to *MET*, lung cancer, and targeted *MET* therapies. The systematic search was conducted in Embase and PubMed in February of 2023 with the help of two search queries (Figure 2, Appendix A) based on the population, intervention, comparison, and outcome (PICO) model. Only clinical trials and randomized clinical trials were included.

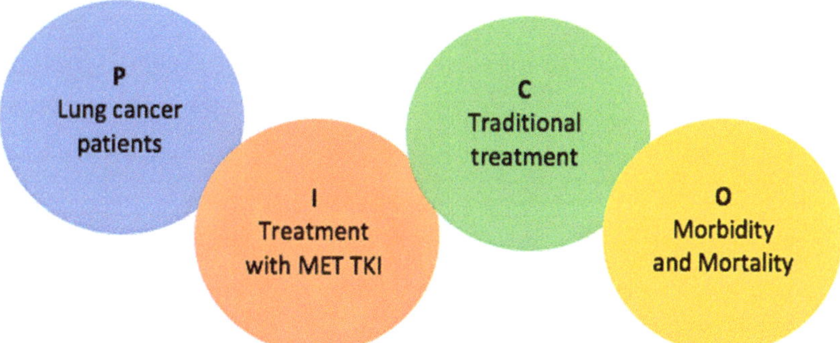

Figure 2. Illustration of the PICO model used to create the search queries used in this systematic review. P: population, I: intervention, C: comparison, O: outcome, *MET*: mesenchymal epithelial transition factor, TKI: tyrosine kinase inhibitor.

2.2. Exclusion and Inclusion Criteria

After the initial search in Embase and PubMed, all duplicates were removed. All remaining articles were screened independently by two of the authors, E.B. and F.S., for relevance based on the title, abstract, and full article. Articles were included if one or both parties deemed it relevant to the subject. Articles that were published in 2012 or earlier were excluded as well as supplemental materials and conference abstracts. A full text screening of all eligible articles was performed independently by the two parties, excluding irrelevant and ongoing trials. A flow diagram of exclusion steps is presented in Figure 3.

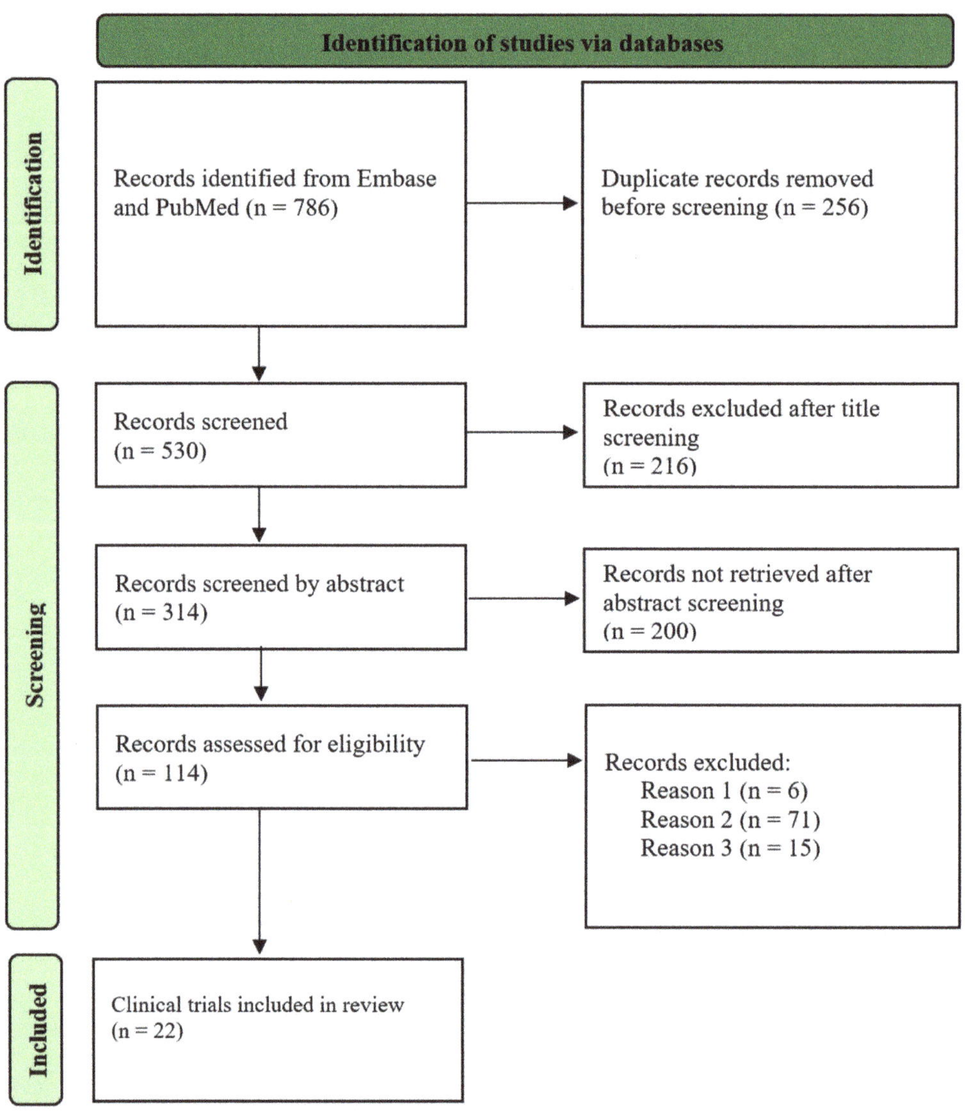

Reason 1 = published 2013 and later

Reason 2 = supplements or conference abstracts

Reason 3 = excluded due to irrelevance after reading full article

Figure 3. Flow diagram of inclusion and exclusion steps according to the PRISMA guidelines.

2.3. Data Extraction and Processing

The included papers were grouped by investigated drug; the main characteristics are described in Tables 2 and 3. The main outcome measures presented and analyzed in this article are PFS and OS. Data on prevalence of *MET* mutations and amplifications in lung cancer were also analyzed and included in this article.

Table 2. Characteristics of included articles reporting the prevalence and impact of *MET* and *HGF* expression. Summary of characteristics of the six included articles reporting numbers regarding the prevalence of aberrant *MET* expression and results regarding the impact of the protein HGF on outcome. NSCLC: non-small cell lung cancer, EGFR: epidermal growth factor receptor, WT: wild type, NS: mutational status not specified, TKI: tyrosine kinase inhibitor, HGF: hepatocyte growth factor, ORR: objective response rate, PFS: progression-free survival, OS: overall survival, MET: mesenchymal epithelial transition factor, GCN: gene copy number, METex14: MET exon 14 skipping mutations.

	Total Participants	Cancer Type	Mutational Status	Active Drug	Results
Arrieta et al., 2016 [1]	$n = 66$	NSCLC	*EGFR* WT *EGFR*+	Afatinib (*EGFR* TKI)	Reduced levels of HGF led to improved ORR, PFS, and OS.
Helman et al., 2018 [25]	$n = 77$	NSCLC	*EGFR*+	Rociletinib (*EGFR* TKI)	Prevalence of *MET* alteration was 15.0%. *MET* amplification was seen in 7.6% of these, 4.5% had focal amplification, and 3.1% had aneuploidy.
Matsumoto et al., 2014 [19]	$n = 47$	NSCLC	*EGFR* WT	Erlotinib (*EGFR* TKI)	Expression of HGF resulted in poor response to erlotinib with shorter PFS. *MET* mutational status did not correlate to response to erlotinib or PFS.
Okamoto et al., 2014 [3]	$n = 295$	NSCLC	NS	Chemotherapy	Totally, 21 patients (7.6%) had *MET* mutations. *MET* amplifications were present in 9 (3.9%) cases. Median GCN was 8.8 among *MET* amplified patients.
Palmero et al., 2021 [11]	$n = 186$	NSCLC	NS	None	Totally, 22.0% of patients had *MET* amplifications and 11.0% had *MET*ex14 mutations.
Sacher et al., 2016 [28]	$n = 22$	NSCLC	*EGFR* WT *EGFR*+	Erlotinib	Totally, 45.0% of subjects harbored a *MET* alteration. *MET* amplification was present in 9.0% of the patients.

Table 3. Characteristics of included articles reporting effect of targeted therapies. Summary of characteristics and results of the 16 included articles that reported the effect of specific targeted therapies. NSCLC: non-small cell lung cancer, EGFR: epidermal growth factor receptor, WT: wild type, MET: mesenchymal epithelial transition factor, NS: mutational status not specified, IHC: immunohistochemistry, GCN: gene copy number, CEP7: centromere 7, METex14: MET exon 14 skipping mutations, NGS: next generation sequencing, TKI: tyrosine kinase inhibitor, VEGF-A: vascular endothelial growth factor A, PFS: progression-free survival, OS: overall survival, ns: not significant, ORR: objective response rate, 2/3L: second- or third-line.

	Study Design	Total Partici-pants	Cancer Type	Mutational Status	Definition of *MET+*	Active Drug	Results
Han et al., 2017 [14]	Phase III	$n = 636$	NSCLC	*EGFR+/WT, MET+/−*	Not defined	Onartuzumab (*MET* TKI) + erlotinib (*EGFR* TKI) vs. erlotinib	High dose onartuzumab resulted in longer PFS (4.4 months) compared to low dose (2.5 months) and erlotinib (2.5 months). No significant difference in OS.
Hirsch et al., 2017 [17]	Phase II	$n = 106$	NSCLC	*MET+/−*	*MET* IHC2+ or IHC3+	Onartuzumab (*MET* TKI) + chemotherapy vs. chemotherapy + placebo	Median PFS was 5 months for onartuzumab and 5.2 months for placebo (ns). Median OS was 10.8 months for onartuzumab and 7.9 months for placebo (ns).
Jänne et al., 2016 [21]	Phase I	$n = 67$	NSCLC	*EGFR+/WT*	*MET* IHC2+ or IHC3+, *MET* GCN $\geq$ 2.1	Crizotinib (*MET* TKI) + dacomitinib (*EGFR* TKI)	Median PFS was 3 months with 61.0% stable disease in the escalation phase. Median PFS was 2.1 months with 32.0% stable disease in the expansion phase.
Kishi et al., 2019 [15]	Phase II	$n = 61$	NSCLC	*EGFR+ MET+*	*MET* IHC2+ or IHC3+, the total number of *MET* genes in 20 cancer cells $\geq$ 90	Onartuzumab (*MET* TKI) + erlotinib (*EGFR* TKI)	Median PFS was 8.5 months, median OS 15.6 months, and ORR 68.9%.
Landi et al., 2019 [18]	Phase II	$n = 26$	NSCLC	*EGFR* WT *MET+*	*MET*-CEP7/ratio $\geq$ 2.2, *MET*ex14 mutation	Crizotinib (*MET* TKI)	ORR of 27.0%, median PFS 4.4 months, and median OS 5.4 months.
Moro-Sibilot et al., 2019 [29]	Phase II	$n = 53$	NSCLC	*EGFR* WT *EGFR+ MET+/−*	*MET* IHC2+ or IHC3+, *MET* GCN $\geq$ 6, *MET* exon skipping mutations in exon 14, 16–19 determined by NGS	Crizotinib (*MET* TKI)	ORR of 16.0%, median PFS of 3.2 months, and median OS of 7.7 months in the *MET* GCN $\geq$ 6 cohort. ORR of 10.7%, median PFS of 2.4 months, and median OS of 8.1 months in the *MET* mutated cohort.

Table 3. *Cont.*

	Study Design	Total Participants	Cancer Type	Mutational Status	Definition of *MET*+	Active Drug	Results
Neal et al., 2016 [2]	Phase II	$n = 111$	NSCLC	*EGFR* WT	Tested through IHC, positive if *MET* was expressed in either membrane or cytoplasm	Cabozantinib (*MET* TKI) + erlotinib (*EGFR* TKI) vs. cabozantinib vs. erlotinib	Median PFS of 1.8 months and OS of 5.1 months for erlotinib. PFS was 4.7 months and OS was 12.3 months for erlotinib + cabozantinib. PFS was 4.3 months and OS was 9.2 months (ns) for cabozantinib. No association between *MET* IHC+ and PFS in the cabozantinib group.
Ou et al., 2017 [24]	Phase I	$n = 26$	NSCLC	NS	Mutational status not mentioned	Crizotinib (*MET* TKI) + erlotinib (*EGFR* TKI)	Two patients had partial response, 8 had stable disease, and 10 had progressive disease.
Scagliotti et al., 2018 [30]	Phase III	$n = 109$	NSCLC	*EGFR*+ *MET*+/−	*MET* IHC2+ or IHC3+, *MET* GCN $\geq$ 4	Tivantinib (*MET* TKI) + erlotinib (*EGFR* TKI) vs. erlotinib + placebo	Greater overall response rate (60.7%) and median PFS (13.0 months) for tivantinib + erlotinib compared to erlotinib + placebo (43.4%, 7.5 months). Similar median OS between groups (25.5 months for tivantinib and 20.3 months for placebo).
Schuler et al., 2020 [20]	Phase I	$n = 55$	NSCLC	*EGFR* WT *MET*+	*MET* H-score $\geq$ 150, *MET*/CEP7 $\geq$ 2.0, *MET* GCN $\geq$ 5, *MET* IHC 2+ or IHC3+	Capmatinib (*MET* TKI)	Median PFS was 3.7 months. In patients with *MET* GCN $\geq$ 6 median PFS was 9.3 months.
Sequist et al., 2020 [22]	Phase Ib	$n = 180$	NSCLC	*EGFR*+ *MET*+	*MET* GCN $\geq$ 5, *MET*/CEP7 ratio $\geq$ 2, *MET* IHC3+ or $\geq$20.0% tumor cells in NGS	Savolitinib (*MET* TKI) + osimertinib (*EGFR* TKI)	Partial response in 89 patients treated with savolitinib and osimertinib. PFS was 7.6 months and 9.1 months in two different subgroups.
Seto et al., 2021 [31]	Phase II	$n = 45$	NSCLC	*EGFR* WT *MET*+	*MET*ex14 mutation, *MET* amplification	Capmatinib (*MET* TKI)	In a subcohort with *MET*ex14 mutations receiving second or third-line (2/3L) treatment with capmatinib, overall response rate was 36.4%. In a second cohort with *MET* GCN $\geq$ 10, overall response rate to 2/3L capmatinib was 45.5%.

Table 3. *Cont.*

	Study Design	Total Participants	Cancer Type	Mutational Status	Definition of *MET*+	Active Drug	Results
Spigel et al., 2013 [6]	Phase II	n = 136	NSCLC	*EGFR*+/WT *MET*+/−	*MET* IHC2+ or IHC3+	Onartuzumab (*MET* TKI) + erlotinib (*EGFR* TKI) vs. erlotinib	No difference in PFS and OS between groups. In a *MET*+ subgroup, PFS was significantly longer for onartuzumab compared to erlotinib + placebo (2.9 vs. 1.5 months). Longer OS in the *MET*+ subgroup receiving onartuzumab (12.6 vs. 3.8 months).
Wakelee et al., 2017 [32]	Phase II	n = 259	NSCLC	*EGFR*+/WT *MET*+/−	*MET* IHC2+ or ICH3+	Cohort 1: onartuzumab (*MET* TKI) + bevacizumab (*VEGF-A* TKI) + chemotherapy vs. placebo + bevacizumab + chemotherapy. Cohort 2: onartuzumab + chemotherapy vs. placebo + chemotherapy	Cohort 1: longer median PFS on placebo (6.8 months) compared to onartuzumab (5.0 months). A *MET*+ subgroup had a median PFS of 4.8 months and OS of 9.9 months on onartuzumab vs. 6.9 months and 16.5 months on placebo. Cohort 2: median PFS of 4.9 months and median OS of 8.5 months on onartuzumab vs. 5.1 months and 13.7 months on placebo. In a *MET*+ subgroup, median PFS was 5.0 months and OS was 8.0 months on onartuzumab vs. 5.0 months and 7.6 months on placebo.
Wu et al., 2020 [23]	Phase Ib/II	n = 55	NSCLC	*EGFR*+ *MET*+	*MET* IHC2+ or IHC3+, *MET* GCN ≥ 5	Tepotinib (*MET* TKI) + gefitinib (*EGFR* TKI) vs. chemotherapy	Significantly longer OS (37.3 months vs. 13.1 months) and PFS (16.6 months vs. 4.2 months) for tepotinib + gefitinib in patients with *MET* IHC3+ or *MET* GCN ≥ 5.
Yoshioka et al., 2015 [5]	Phase III	n = 303	NSCLC	*EGFR* WT *MET*+/−	IHC with moderate/strong intensity ≥ 50.0% of tumor cells, *MET* GCN ≥ 4	Tivantinib (*MET* TKI) + erlotinib (*EGFR* TKI) vs. erlotinib + placebo	Significantly longer PFS for tivantinib + erlotinib (2.9 months) compared to erlotinib + placebo (2 months). No effect on OS.

3. Bias

Publication bias is a factor inevitably impacting systematic reviews due to positive selection bias in the publication of research articles. This may lead to overestimation of the therapeutic effect of the investigated drugs. In this systematic review, several articles reported no effect of the evaluated drugs on PFS and OS among patients with lung cancer. By simultaneous screening of all articles by separate parties, the risk of bias in the process of selecting articles to be included in this systematic review was minimized. Furthermore, the Cochrane Risk of Bias Tool "robvis" was adapted to create risk of bias plots for all RCT's and clinical trials included in this systematic review, see Appendix C (Figures A1 and A2) [33]. None of the authors of this systematic review declare any conflicts of interest related to the article.

4. Results

This systematic review identified a total of 786 eligible articles by the two search queries (Appendix A) employed following the PRISMA guidelines. Exclusion of 256 duplicates resulted in 530 remaining articles. After the assessment of article titles, 314 articles remained followed by further exclusion based on the abstract, resulting in 114 remaining articles. Six articles were excluded based on publication year and 71 studies were excluded due to being conference abstracts or supplements. Consequently, 37 full articles remained for the final full text screening. A total of 22 articles were judged to be relevant to this systematic review (Figure 3). Six of the included articles present data regarding the prevalence and significance of aberrant signaling through the HGF/*MET* pathway in lung cancer. The remaining 16 articles present data on the effect of specific targeted therapies on mortality and morbidity of lung cancer patients.

4.1. Epidemiology

The reported patient characteristics varied greatly between the included studies, see Appendix C (Table A3). Landi et al. reported the lowest mean age in all the studies (56.0 years) while Moro-Sibilot et al. reported the highest in their subgroup of patients with *MET* mutations (72.0 years) [18,29]. In regard to the included subjects' gender, Matsumoto et al. included a vast overweight of women compared to men in their clinical trial (80.3% women), while Okamoto et al. reported the highest percentage of men included (77.0%) [3,19]. Similar to age and gender, smoking history also varied between the studies. The highest reported prevalence of never smokers was seen in the clinical trial conducted by Matsumoto et al., with 89.1% of the patients being never smokers in an *EGFR* WT population [19]. Interestingly, both of the *MET*-negative cohorts in a RCT by Spigel et al. held smaller fractions of patients with no smoking history (7.0% in the treated subgroup and 3.0% in the placebo subgroup) compared to the *MET+* cohorts in the same study (20.0% in the treated subgroup and 19.0% in the placebo subgroup) [6]. In the study by Moro-Sibilot et al., the authors reported double the amount of patients with a no smoking history in the *MET* mutated cohort compared to the *MET* amplified cohort (*MET* mutated: 48.0%, *MET* amplified: 24.0%) (Table A3) [29].

4.2. Prevalence of Aberrant MET Expression

The reported prevalence of the different *MET* gene alterations varies greatly between the articles included in this systematic review. In a prospective clinical trial by Palmero et al., genetic testing of treatment naïve NSCLC patients was carried out to identify genetic alterations. Next generation sequencing (NGS) of circulating cell free tumor DNA in blood was compared to standard-of-care tissue-based biopsy testing in order to determine the patients' mutational burden. The results showed no difference between the methods' ability to identify mutations in NSCLC. Out of 186 tested patients, 11.0% harbored *MET* exon 14 skipping mutations (*MET*ex14) and 22.0% harbored amplification of the *MET* gene [11].

In a retrospective clinical study by Okamoto et al., 295 patients were tested for *MET* mutations and 229 patients were tested for *MET* amplification. The results showed a prevalence of 7.6% for mutations in the *MET* gene and 3.9% for *MET* amplification, with a median GCN of 8.8. The median OS was found to be non-significantly longer among patients without *MET* amplification [3].

Sacher et al. carried out a single arm, single center clinical trial, testing the *EGFR* TKI erlotinib on 22 NSCLC patients with varying *EGFR* mutational statuses. All included patients were naïve to any systemic anti-cancer treatment. *MET* amplification was determined by fluorescence in situ hybridization (FISH) and *MET* positivity was defined as GCN $\geq$ 4. Protein expression was assessed through IHC and *MET* positivity was defined as an H-score $\geq$ 100. The prevalence of any aberrant *MET* expression in this study was 45.0%, while *MET* amplification specifically was seen in 9.0% of the patients [28].

Helman et al. conducted a retrospective study that performed NGS on plasma samples from patients suffering from NSCLC harboring *EGFR* mutations. NGS was carried out at study enrolment before receiving treatment and after disease progression to evaluate the effect of the *EGFR* tyrosine kinase inhibitor (TKI) rociletinib. Prior to treatment with rociletinib, 15.0% of the NGS screened patients were positive for genetic alterations in the *MET* gene. After disease progression on treatment with rociletinib, another 7.6% of the patients with acquired resistance to the drug had acquired amplification of the *MET* gene. This amplification was found to be caused by focal amplification in 4.5% of the patients and by aneuploidy in the remaining 3.1% [25].

In a phase II study by Arrieta et al., the levels of the HGF protein were analyzed in 66 patients with either *EGFR* mutated or *EGFR* wild type lung adenocarcinoma. All patients received treatment with the *EGFR* TKI afatinib after disease progression on first-line platinum-based chemotherapy. Patients with reduction of HGF levels after treatment with afatinib had a significantly longer PFS, OS, and objective response rate (ORR) compared to patients with higher levels of HGF after treatment with afatinib. These differences were most prominent among patients with *EGFR* mutated disease. The authors of the study suggest that HGF has a direct role in acquired resistance to *EGFR* TKIs, making HGF an interesting target in this field of research [1].

In a different phase II study by Matsumoto et al., HGF levels were measured and *MET* mutational status was determined in 47 patients with *EGFR* wild type NSCLC treated with erlotinib. Tumors expressing HGF had a poor response to erlotinib, and the patients had a shorter median PFS compared to the HGF negative study population. *MET* mutational status did not impact PFS or the response to erlotinib in this clinical trial [19].

4.3. Targeted Therapies

Sixteen of the papers included in this systematic review are studies investigating targeted therapies for the *MET* gene. The different *MET* TKIs evaluated in the clinical trials are presented in Table 3. The most common *MET* TKIs were onartuzumab and crizotinib, occurring in five and four of the included articles each.

In a phase II study conducted by Hirsch et al., treatment with the combination of onartuzumab, paclitaxel, and carboplatin/cisplatin was compared to treatment with only paclitaxel and carboplatin/cisplatin in 106 NSCLC patients. The included patients were all *EGFR* wild type, with or without mutations in the *MET* gene. The results showed no significant benefit of added treatment with onartuzumab, regardless of *MET* mutational status, with similar PFS (5 months in the onartuzumab group and 5.2 in the placebo group) and OS (10.8 months vs. 7.9 months) in the two groups. In this study, *MET* mutational status was evaluated by immunohistochemistry, with IHC3+ and IHC2+ considered *MET* positive, and IHC1+ and IHC0 considered *MET* negative [17].

Similar results were found in a phase II study conducted by Wakelee et al., including 259 NSCLC patients in two cohorts. Cohort 1 tested the addition of onartuzumab vs. placebo to treatment with bevacizumab, carboplatin/cisplatin, and paclitaxel. In cohort 2, patients received either onartuzumab or placebo in addition to carboplatin/cisplatin and pemetrexed. The subjects had varying mutational statuses regarding both *EGFR* and *MET*. *MET* positivity was defined as IHC3+ or IHC2+. In cohort 1, the overall median PFS was 5.0 months in the onartuzumab group compared to 6.8 months in the placebo group.

In a *MET*+ subgroup, median PFS was 4.8 months and median OS was 9.9 months in the onartuzumab arm compared to 6.9 months and 16.5 months in the placebo arm. Cohort 2 revealed similar results, with a median PFS of 5.1 months in the placebo group compared to 4.9 months in the onartuzumab group. Median OS was 13.7 months in the placebo treated group and 8.5 months in the onartuzumab treated group. In the *MET*+ subgroup of cohort 2, median PFS was 5.0 in for both onartuzumab and placebo while median OS was marginally longer in the onartuzumab arm (8.0 months) compared to the placebo arm (7.6 months) [32].

In another phase III study evaluating onartuzumab, 636 NSCLC patients with varying *EGFR* and *MET* mutational statuses were included and treated with either erlotinib and onartuzumab or erlotinib and placebo. In this study, high doses of onartuzumab were associated with a longer median PFS compared to lower doses of erlotinib or placebo (high dose PFS = 4.37; low dose PFS = 2.5 months; placebo PFS = 2.5 months). No significant differences were found in OS regardless of *MET* mutational status [14].

In contrast to this, a phase II study by Spigel et al. performed on NSCLC patients found that both median PFS and OS were significantly longer in a *MET* positive subgroup (defined as *MET* IHC3+ or IHC2+) treated with onartuzumab and erlotinib compared to a subgroup treated with erlotinib and placebo (dual treatment PFS = 2.9 months, OS = 12.6 months; single treatment PFS = 1.5 months, OS = 3.8 months). All 136 included participants were *EGFR* wild type but had varying *MET* mutational statuses. Interestingly, the patients without any mutations in the *MET* gene had earlier progression when treated with the combination of onartuzumab and erlotinib compared to erlotinib and placebo [6].

A third phase II study, titled the global METLung study (OAM4971g), presented results on treatment with onartuzumab and erlotinib without comparing the effects to a control group. This study included 61 patients with *MET* and *EGFR* mutation positive NSCLC but was terminated early due to lack of efficacy of onartuzumab. Patients with *MET* IHC3+ or IHC2+ as well as a total number of *MET* genes in 20 cancer cells $\geq$ 90 as determined by a gene amplification assay were considered *MET* positive. The preliminary results showed a median PFS of 8.5 months, a median OS of 15.6 months, and an overall response rate of 68.9% in the patients treated with onartuzumab and erlotinib [15].

In a phase II clinical trial, Landi et al. evaluated treatment with the targeted *MET* inhibitor crizotinib in an *EGFR* wild type NSCLC population of 26 patients with either *MET* amplification or *METex14* mutation. The median PFS was 4.4 months, median OS was 5.4 months, and the ORR was 27.0%. *MET* amplification was defined as a *MET*/centromere 7 (CEP7) ratio > 2.2 [18]. Jänne et al. treated 67 NSCLC patients with mixed *EGFR* mutational statuses with crizotinib and dacomitinib in a phase I clinical trial. The included patients had experienced progression on first-line treatment with either chemotherapy or another targeted therapy and were included in either an escalation phase cohort or in an expansion phase cohort. The median PFS was 3.0 months in the escalation cohort, with 61.0% of these patients having stable disease during the treatment. In the expansion cohort, the median PFS was 2.1 months and 32.0% of the patients had stable disease during treatment. No association was seen between overexpression of *MET* and PFS. *MET* positivity was defined as IHC3+, IHC2+, or *MET* GCN $\geq$ 2.1 [21].

In another phase II clinical trial conducted by Moro-Sibilot et al., 53 NSCLC patients with varying *EGFR* mutational statuses were treated with crizotinib. The study included two cohorts with aberrant *MET* expression, one with *MET* GCN $\geq$ 6 (n = 25) and one with positivity for *MET* exon skipping mutations in exon 14 or 16–19 (n = 28). In the *MET* amplified cohort (GCN $\geq$ 6), the ORR was 16.0%, the median PFS was 3.2 months, and the median OS was 7.7 months. In the *MET* mutated cohort, the OS was longer while the PFS and ORR were inferior (ORR = 10.7%, median PFS = 2.4 months, median OS = 8.1 months) [29].

In a phase I study by Ou et al., crizotinib was evaluated in combination with erlotinib on 26 NSCLC patients with previous progression of disease on one or two prior treatments with chemotherapy. The patients' *EGFR* and *MET* mutational statuses were not reported in this article. Only 20 of the 26 included patients were evaluated for response to treatment. Of these, two had partial response, eight had stable disease, and ten had progressive disease as defined by the RECIST version 1.1 guidelines [24].

One study evaluated the *MET* targeted drug savolitinib in combination with osimertinib in a phase Ib clinical trial. A total of 180 patients with *EGFR* mutation positive and *MET* amplified NSCLC received treatment with savolitinib and osimertinib after previous treatment with one, two, or three different *EGFR* TKIs. In this study, *MET* amplification was defined as *MET* GCN $\geq$ 5, *MET*/CEP7 ratio $\geq$ 2, *MET* IHC3+, or *MET* expression in $\geq$20.0% of tumor cells as determined by NGS. Patients were stratified into one of two cohorts according to type of *EGFR* mutation and number of prior *EGFR* TKI treatments. Only 161 patients were eligible for final evaluation of treatment effect. Of these, 89 patients had a partial response according to the RECIST version 1.1 guidelines. One of the two cohorts included 138 patients and had a median PFS of 7.6 months, while the other cohort, including 23 patients, had a median PFS of 9.1 months [22].

In a phase Ib/II trial by Wu et al., standard platinum doublet chemotherapy was compared to treatment with the targeted drugs tepotinib and gefitinib. In this clinical trial, 55 patients were included, all with NSCLC, positive for *EGFR* mutation, and *MET* overexpression or amplification. *MET* overexpression was defined as IHC2+ or IHC3+, and amplification was defined as GCN $\geq$ 5. All patients included in the phase II part of the trial had acquired resistance to other first or second-generation *EGFR* TKIs. There were no significant differences in median PFS or OS when comparing chemotherapy to targeted therapy. Subgroup analyses were carried out on groups of patients harboring either *MET* amplification or high *MET* overexpression defined as IHC3+. In the *MET* amplified group, both PFS and OS were significantly longer in the targeted therapy treated group compared to the chemotherapy treated group (targeted therapy median PFS = 16.6, chemotherapy median PFS = 4.2; targeted therapy OS = 37.7, chemotherapy OS = 13.1). In the *MET* IHC3+ group, the median PFS was 8.3 months, and the OS was 37.3 months compared to a PFS of 4.4 months and an OS of 17.9 in the chemotherapy group [23].

A phase I study evaluating capmatinib in NSCLC patients with *MET* amplification or *MET* overexpression showed that patients with a high *MET* GCN or *MET*ex14 mutations may benefit from treatment with *MET* inhibitors. Of 55 enrolled patients treated with capmatinib, 26 were included in a dose expansion group with varying *EGFR* mutational statuses and either *MET* overexpression or amplification. The remaining 29 patients were all *EGFR* wild type and had high overexpression of *MET*, defined as IHC3+. Complete response was observed in one of the 55 patients and partial response was seen in ten patients. The median PFS was 3.7 months for the entire cohort, whereas the *MET* IHC3+ group had a median PFS of 5.1 months. An even greater PFS was seen among patients with *MET* GCN $\geq$ 6 (median PFS of 9.3 months). Furthermore, in four patients with *MET*ex14 mutations, a reduction in the tumor burden between 14.0 and 83.0% could be seen. This trial was terminated early due to disease progression and frequent adverse events [20].

A phase II trial by Seto et al. tested the effect of capmatinib on 45 *EGFR* wild type NSCLC patients divided into several cohorts depending on *MET* mutational burden and prior systemic anti-cancer treatment. Aberrant *MET* expression was defined as the presence of the *MET*ex14 mutation or amplification of the *MET* gene. The results showed that treatment with capmatinib as a second or third-line option in *MET*ex14 positive subjects (n = 11) yielded an overall response rate of 36.4%. The overall response rate in a cohort of *MET* amplified patients with GCN $\geq$ 10 (n = 11) was found to be 45.5%. In contrast to this, in a cohort with *MET* GCN $\geq$ 4 but <6 (n = 10), the overall response rate was only 10.0%. The remainder of the patients were further subdivided into considerably smaller cohorts with inconclusive results [31].

Yoshioka et al. conducted a phase III clinical trial comparing treatment with tivantinib and erlotinib to treatment with erlotinib and placebo in 303 *EGFR* wild type NSCLC patients. All included patients had received one or two prior treatments, one of them being platinum-based chemotherapy. The patients had varying *MET* mutational statuses, including high and low expression of *MET* as well as elevated and normal *MET* GCN. The authors found no significant difference in the median OS when comparing the dual treatment group to the erlotinib plus placebo group. In the patients treated with both tivantinib and erlotinib, a significantly longer median PFS could be seen compared to the placebo group (dual treatment PFS = 2.9 months, placebo PFS = 2.0 months). Furthermore, high expression of HGF, defined as an H-score $\geq$ 200 as measured by IHC, was associated with a significant benefit in OS in the tivantinib plus erlotinib group compared to the placebo group. This clinical trial was terminated early due to increased incidence of interstitial lung disease, with 14 cases in the dual treatment group and 6 cases in the placebo group [5].

This is coherent with the results of a phase III study conducted by Scagliotti et al., which also compared erlotinib plus tivantinib to erlotinib plus placebo in 109 *EGFR* mutated NSCLC patients with varying *MET* mutational statuses. Here the median PFS was 13.0 months and the median OS was 25.5 months in the dual treatment group compared to a median PFS of 7.5 months and a median OS of 20.3 months in the erlotinib plus placebo group [30].

In a phase II clinical trial, Neal et al. assessed treatment with cabozantinib in *EGFR* wild type NSCLC patients with varying *MET* mutational statuses. This study comprised three treatment arms, one with both erlotinib and cabozantinib, one with only cabozantinib, and one with only erlotinib. All patients had previously been treated with one or two other therapeutic agents prior to enrolment in this trial. The median PFS was significantly longer in the erlotinib plus cabozantinib arm (median PFS = 4.7 months) and in the cabozantinib arm (median PFS = 4.3 months) compared to the erlotinib arm (median PFS = 1.8 months). The median OS was also longer in the erlotinib plus cabozantinib arm (median OS = 13.3 months) compared to the erlotinib arm (median OS = 5.1 months). The median OS in the cabozantinib arm was 9.2 months and was not significantly longer compared to the erlotinib arm. In this clinical trial, no association was found between *MET* IHC status and median PFS in the patient groups receiving treatment with cabozantinib or erlotinib plus cabozantinib [2].

In summary, many of the targeted drugs mentioned in this review show promising results that encourage further research. Onartuzumab, as an added treatment to erlotinib, has been shown in several trials to lead to a longer PFS and OS in NSCLC patients, compared to treatment with placebo and erlotinib. One clinical trial showed a reversed effect of onartuzumab as an added treatment to chemotherapy, as compared to placebo, with a shortened PFS and OS among the patients treated with onartuzumab. Crizotinib alone has been proven to greater increase OS among NSCLC patients harboring *MET* mutations compared to patients with *MET* amplified cancer genotypes. Crizotinib, as an addition to the *EGFR* TKIs dacomitinib or erlotinib, has resulted in a greater fraction of patients with stable disease. Treatment with savolitinib in combination with osimertinib has led to a fraction of patients with partial response of more than 55% in cohorts of NSCLC patients with previous progression of disease on *EGFR* TKIs. Tepotinib and gefitinib, as a treatment for NSCLC patients harboring high overexpression or amplification of *MET*, has been shown to prolong both the PFS and OS compared to treatment with chemotherapy. Similarly, in NSCLC patients with a high *MET* GCN or high levels of overexpression of *MET*, treatment with capmatinib has prolonged PFS and led to a reduced tumor burden and higher overall response rates compared to cohorts with lower grades of overexpression or amplification. The addition of tivatinib to treatment with erlotinib has been proven to be effective and has led to increases in PFS and OS compared to placebo. Lastly, use of cabozantinib alone or in combination with erlotinib has led to significantly longer PFS compared to treatment with erlotinib alone. For an overview of the results, see Figure A3 (Appendix C).

5. Discussion

Lung cancer continues to be the leading cause of global annual cancer-related mortality, and the need for more efficient therapies is evident. Targeted therapies are a promising area of research, offering new possibilities for inhibiting genetic alterations that are involved in driving tumorigenesis, but to this day, there are no known clinically useful biomarkers [34–37]. Lung cancer patients harboring alterations in the *MET* gene detected in circulating DNA in blood or tumor tissue have been shown to have poorer outcomes.

In a study by Andreasson et al., the authors were able to show that the *MET* protein could be found in exhaled breath as well as in blood plasma, and that the expression of the protein diminished after surgical removal of the lung cancer, underlining the proteins role in lung cancer tumorigenesis [34].

There is a pressing need for novel and more efficient therapies that can target the *MET* signaling pathway, but for this to be possible, the understanding of how abundant certain genetic alterations are, and how they affect and drive lung cancer development, needs to be further deepened. Therefore, this systematic review aimed to summarize and enlighten the current status of genetic mapping and prevalence of different alterations in the *MET* gene in NSCLC patients, as well as existing and pipe-line targeted therapies towards this protein.

The articles reviewed in this paper reported that *MET* mutations were found in 7.6–11.0% of lung cancer patients and that amplification was found in 3.9–22.0% of cases [3,11]. One study reported that genetic alterations in the *MET* gene were present in 15.0% of the lung cancer cases analyzed [25]. These numbers are in line with previously published data [38–40]. One study presented results on prevalence, which differed significantly from the rest of the included papers, with a reported frequency of aberrant *MET* expression of 45.0% [28]. The reported prevalence of aberrant *MET* expression varies in different studies. This may, in part, be due to the fact that different studies analyze samples from different genotypes of lung cancer, with differing characteristics, using different quantification methods and definitions. Some of the articles included in this systematic review analyzed the prevalence of aberrant *MET* signaling in *EGFR* mutated patients, while other studies do not specify *EGFR* mutational status [3,11,25]. It has previously been shown that alterations in the *MET* gene are a known mechanism of resistance to *EGFR* TKIs; therefore, considering the impact of previous treatments on genetic analyses results may offer valuable insights for included patients [22].

The varying patient characteristics in the included studies make it difficult to predict any specific groups of the population that are at increased risk of developing NSCLC with *MET* mutated genotypes. Some trials do not specify *MET* mutational status or present their data on patient characteristics in cohorts of patients with varying *MET* mutational statuses [6,21,24]. However, some trends can be seen among the studies presenting results on patient characteristics in cohorts with specified *MET* mutations [6,15,18,20,22,23,29,31]. The median age among patients harboring *MET* mutations ranges from 56.0 to 72.0 years. *MET* mutations appear equally among both genders, with the greatest difference being reported by a single study being 32% men and 68% women [29]. Cohorts with *MET* positive patients reported smoking history as "never smokers" ranging from 19 to 72%, compared to cohorts consisting of *MET* negative patients, with the lowest reported percentage of never smokers being 3.0% in the study by Spigel et al. [6]. This could be indicative of a trend with higher fractions of *MET* positive NSCLC patients being never smokers. However, more research that is more directly aimed at this research question needs to be conducted to verify these results.

Five of the included studies evaluated the *MET* TKI onartuzumab. Spigel et al. found a significantly longer OS and PFS in *MET* positive patients receiving treatment with both onartuzumab and erlotinib compared to patients receiving erlotinib alone. Interestingly, they also found that the *MET* negative patients suffered earlier progression of disease and increased mortality when treatment with onartuzumab was added compared to treatment with placebo alone [6]. This is in line with the findings of Han et al., who showed a significantly longer PFS among patients treated with high doses of onartuzumab together with erlotinib compared to single treatment with erlotinib or erlotinib plus lower doses of onartuzumab [14]. Kishi et al. presented similar results regarding OS to what Spigel et al. presented; however, there was a remarkable difference in PFS between the two studies (Kishi et al. PFS = 8.5 months, Spigel et al. PFS = 2.9 months) [6,15]. Hirsch et al. and Wakelee et al. were not able to show a longer OS or PFS with the addition of onartuzumab to standard chemotherapy in lung cancer patients with varying *MET* mutational statuses [17]. *MET* has previously been suggested to act as both a suppressor and an oncogene, which is in line with the presented results [6]. These findings underline the importance of genetic testing upon lung cancer diagnosis in order to customize targeted therapies and postpone or prevent progression caused by treatment with ill-suited therapies.

Two separate trials investigated the effect of dual treatment with tivantinib and erlotinib compared to single treatment with erlotinib. They both found a prolongation in the median PFS in the dual treatment group compared to treatment with erlotinib alone, although to very differing extents. No difference was seen in the median OS [5,30]. One study included only *EGFR* wild type subjects while the other had only *EGFR* mutated subjects. The variation in *EGFR* mutational status is likely a contributing factor to the varying results of the studies included in this review. Some clinical trials included only *EGFR* wild type lung cancers, others included only *EGFR* positive cases or a combination of wild type and positive cases. Two studies evaluated the effect of additional treatment with onartuzumab to erlotinib in patients with varying *EGFR* mutational statuses [6,14]. As erlotinib is an *EGFR* TKI, it seems likely that the patients with an *EGFR* positive cancer genotype would benefit more from treatment with erlotinib than the patients with an *EGFR* wild type genotype, making it difficult to evaluate the actual effect of the added onartuzumab or capmatinib treatment. Therefore, in these studies, the effect of added treatment with onartuzumab or capmatinib should be interpreted carefully and further research in larger clinical trials is needed to determine the effect.

Multiple studies performed on different *MET* TKIs have presented varying results on PFS, ranging from 2.1 to 9.1 months, without comparisons to control groups [18,20–22,29]. Interestingly, two of these articles showed that treatment with the targeted drug crizotinib had a greater effect in a *ROS1* translocated NSCLC cohort compared to *MET* altered cohorts [18,29]. Other studies evaluating crizotinib did not report *ROS1* mutational status, rendering the results more difficult to interpret [21,24]. Neal et al. reported that treatment with cabozantinib and erlotinib and single treatment with cabozantinib was superior to treatment with erlotinib alone in lung cancer patients with *EGFR* wild type lung cancer. These results support the statement that patients with *EGFR* wild type lung cancer may not respond to treatment with *EGFR* TKIs alone. In the clinical trial conducted by Neal et al., the authors did not find any association between *MET* IHC+ status and prolonged PFS in the treatment groups receiving cabozantinib. It is not possible to draw any conclusions regarding the association between prolonged time to progression and death, and inhibition of the *MET* receptor specifically, as cabozantinib has multiple gene targets [2,22]. Several clinical trials present similar results, indicating that patients with a high mutational burden in the *MET* gene (defined as high overexpression, high GCN gain, and high IHC+ status) benefit more from treatment with *MET* TKIs compared to patients with a lower mutational burden [5,20,23,31]. This further emphasizes the importance of thorough genetic mapping of cancer patients prior to treatment.

An important aspect in trials testing novel drugs is the fact that the included cohorts differ greatly in cancer genotype [6,14]. This puts the reliability and comparability of the results into question. Another aspect is the fact that the drugs presented in these trials typically show a gain in PFS and OS months at best [5,6,14,17]. This needs to be put in relation with the potential side effects and adverse events that the patients experience during treatment [5,20]. *MET* TKIs are still considered experimental, and treatment with targeted *MET* inhibitors is currently a second, third, or even fourth-line treatment option. It is known that cancers generally harbor fewer driver mutations in the beginning of tumorigenesis compared to the later stages of disease and so it is possible that earlier implementation of *MET* inhibitors could provide a greater effect on PFS and OS [41,42]. However, the timing of treatment will need to be investigated in further clinical trials.

The different cancer genotypes also contribute to potential bias in reporting more or less promising pipe-line drugs. However, there are several interesting ongoing clinical trials evaluating different pipe-line drugs on more targeted patient groups harboring different *MET* alterations (Table 4) [43]. Two of these ongoing trials have already presented results from their studies. The NCT02544633 trial tested the *MET* inhibitor MGCD265 on patients with NSCLC and *MET* activating mutations or *MET* amplifications. The patients were divided into four study arms, one with *MET* activating mutations in tumor tissue, one with *MET* activating mutations in circulating tumor DNA in the bloodstream, one with *MET* amplification in tumor tissue, and one with *MET* amplification in circulating tumor DNA. Interestingly, in the case of *MET* amplification, the patients with *MET* amplification identified in circulating tumor DNA had a poorer PFS (2.76 months) and OS (4.08 months) compared to the patients with *MET* amplification identified in tumor tissue (PFS 4.85 months, OS 7.04 months). The patients with *MET* activating mutations in tumor tissue had a longer OS of 16.32 months compared to all other study arms, but not PFS (3.95 months). The last treatment group with *MET* activating mutations in circulating tumor DNA had a similar PFS to the other three study arms (3.39 months). No results were available regarding OS in this fourth study arm [44]. This data is similar to results from earlier studies on *MET* TKIs presented in this systematic review. A possible explanation to why the patients with *MET* amplification detected in circulating tumor DNA had a poorer outcome is the greater tumor burden associated with circulating tumor cells compared to that of a localized tumor.

One other study had available preliminary results at the time of writing this systematic review (trial ID: NCT02648724). In this phase I/II study, the monoclonal anti *MET* antibody mixture entitled Sym015 was investigated. The study included 57 patients divided into three different treatment arms, one with KRAS wild type patients with *MET* amplifications, one with *MET* amplified NSCLC, and one where the patients harbored *MET*ex14 deletions. The outcome measure presented was the ORR and the results showed that the KRAS wild type cohort had 0.0% ORR while the *MET* amplified and the *MET*ex14 deletion cohort both had an ORR of 25.0% [45]. A vast difference between the already existing and published clinical trials and the ongoing ones is the greatly improved consistence in genotypes included in the newer trials. The majority of the current trials on pipe-line targeted therapies consist of cohorts of patients that are all genetically altered in the *MET* gene rather than cohorts of mixed cancer genotypes. This will make future results more easily interpreted and clinically useful.

Table 4. Ongoing clinical trials. Overview of currently ongoing clinical trials investigating potential pipe-line targeted therapies for the treatment of lung cancer with aberrant *MET* expression. NSCLC: non-small cell lung cancer, MET: mesenchymal epithelial transition factor, METex14: MET exon 14 skipping mutations, RTK: receptor tyrosine kinase, ATP: adenosine triphosphate, EGFR: epidermal growth factor receptor.

Clinical Trial ID	Study Design	Study Type	Total Participants	Cancer Type	Mutational Status	Active Drug and Effect
NCT02544633	Phase II	Non-randomized	$n = 68$	NSCLC	*MET* activating mutation *MET* amplification	MGCD265: oral RTK inhibitor targeting *MET*
NCT02920996	Phase II	Single arm	$n = 12$	NSCLC	*MET*ex14 mutation	Merestinib: reversible type II ATP-competitive *MET* inhibitor
NCT02896231	Phase I	Dose escalation	$n = 37$	NSCLC	*MET*+	PLB1001: selective *MET* inhibitor
NCT04270591	Phase Ib/II	Single arm	$n = 183$	NSCLC	*MET*ex14 mutation *MET* amplification *MET* overexpression	Glumetinib: selective *MET* inhibitor
NCT02648724	Phase I/II	Non-randomized	$n = 57$	NSCLC	*MET* amplification *MET*ex14 deletion	Sym015: monoclonal antibody mixture targeting *MET*
NCT03539536	Phase II	Single arm	$n = 275$	NSCLC	*MET*+	Telisotuzumab vedotin: antibody-drug conjugate targeting *MET*
NCT02609776	Phase I	Non-randomized	$n = 780$	Advanced NSCLC	Varying	Amivantamab: human bispecific antibody targeting *EGFR* and *MET*
NCT04077099	Phase I/II	Single arm	$n = 82$	NSCLC	Any *MET* alteration	REGN5093: human bispecific antibody targeting *MET*, inducing internalization and degradation

6. Strengths and Limitations

The most apparent limitation of the articles included in this review is the lack of control groups in most of the phase I trials, leading to difficulties in interpreting the results. Furthermore, some articles have a low number of included patients and overall, the number of participants varies greatly. Since the included studies have a high grade of variation in patient characteristics, such as age, gender, and smoking history, the comparison between the studies' results need to be interpreted with some caution. As the targeted therapies are tested on different patient categories, with varying cancer genotypes, it is difficult to compare the presented results and draw conclusions on the efficacy of the investigated drugs. On the same note, the definition of the criteria for what is considered aberrant *MET* expression varies between the included trials, with some research groups applying IHC, some performing NGS, and some using different methods. A strength of this review is the variation in location of the included trials, leading to more generalizable and widely applicable conclusions. The exclusion criterion of a publication year prior to 2013 might be considered a limitation; however, this also ensured that only current results were included in this systematic review.

7. Conclusions

It remains difficult to compare different targeted therapies and to draw conclusions regarding their potential place in the future treatment panorama. The inter and intra study variation in cohort composition and included cancer genotypes is large. Onartuzumab has shown prolonged PFS and OS among *MET* positive NSCLC patients but no convincing results in cohorts with mixed *MET* mutational statuses. The studies on crizotinib lacked control groups for comparison of the outcome measures. Many ongoing trials on pipe-line targeted therapies exist, which are investigating anti-*MET* agents on clearly defined cohorts of patients with aberrant *MET* expression. This systematic review summarized the current status of publications on the *MET* gene's implications in lung cancer development and the status of existing and up and coming targeted therapy options. More research is needed and should be encouraged to fully understand how, when, and to whom these drugs should be recommended in order to improve patient outcomes.

Author Contributions: Conceptualization, E.B., F.S. and S.L.; methodology, E.B., F.S. and S.L.; formal analysis, E.B. and F.S.; investigation, E.B. and F.S.; resources, S.L.; data curation, E.B. and F.S.; writing—original draft preparation, E.B., F.S. and F.O.; writing—review and editing, E.B., F.S., F.O. and S.L.; visualization, E.B. and F.S.; supervision, F.O. and S.L.; project administration, S.L.; funding acquisition, S.L. All authors have read and agreed to the published version of the manuscript.

Funding: The study was funded by the Swedish Cancer Foundation and the Sjögren Foundation. The APC was funded by Lund University.

Conflicts of Interest: The authors declare no conflict of interest.

Appendix A

Search Queries

Search query 1 PubMed (*n* = 183):
(((lung cancer[MeSH Terms]) OR (lung cancer)) AND ((((((("met"[All Fields]) OR ("mesenchymal epithelial transition factor"[All Fields])) OR (hgfr)) OR (hepatocyte growth factor receptor)) OR (hgf receptor[MeSH Terms])) OR (hepatocyte growth factor receptor[MeSH Terms]))) AND ((((("prognosis"[All Fields]) OR ("mortality"[All Fields])) OR ("morbidity"[All Fields])) OR ("diagnosis"[All Fields])) OR (morbidity[MeSH Terms])) AND (clinicaltrial[Filter] OR randomizedcontrolledtrial[Filter])

Search query 2 PubMed (*n* = 157):
((((lung cancer[MeSH Terms]) OR (lung cancer)) AND ((((((("met"[All Fields]) OR ("mesenchymal epithelial transition factor"[All Fields])) OR (hgfr)) OR (hepatocyte growth factor receptor)) OR (hgf receptor[MeSH Terms])) OR (hepatocyte growth factor receptor[MeSH

Terms]))) AND ((((((targeted therapy) OR (cancer therapy)) OR (targeted treatment)) OR (molecular targeted therapies[MeSH Terms])) OR (drug targeting[MeSH Terms])) OR (cancer treatment protocol[MeSH Terms]))) AND ((((("prognosis"[All Fields]) OR ("mortality"[All Fields])) OR ("morbidity"[All Fields])) OR ("diagnosis"[All Fields])) OR (morbidity[MeSH Terms])) AND (clinicaltrial[Filter] OR randomizedcontrolledtrial[Filter])

Search query 1 Embase (*n* = 384):
('lung cancer' OR 'lung cancer'/exp) AND ('met'/exp OR met OR 'mesenchymal epithelial transition factor' OR hgfr OR 'scatter factor receptor') AND ('prognosis'/exp OR prognosis OR mortality OR morbidity OR diagnosis) AND ([controlled clinical trial]/lim OR [randomized controlled trial]/lim)

Search query 2 Embase (*n* = 62):
('lung cancer' OR 'lung cancer'/exp) AND ('met'/exp OR met OR 'mesenchymal epithelial transition factor' OR hgfr OR 'scatter factor receptor') AND ('prognosis'/exp OR prognosis OR mortality OR morbidity OR diagnosis) AND ('molecularly targeted therapy'/exp OR 'molecularly targeted therapy' OR 'cancer therapy' OR 'antineoplastic protocol' OR 'drug targeting') AND ([controlled clinical trial]/lim OR [randomized controlled trial]/lim)

Table A1. Full PRISMA 2020 Checklist [27]. N/A: not applicable.

Section and Topic	Item #	Checklist Item	Location Where Item Is Reported
TITLE			
Title	1	Identify the report as a systematic review.	Page 1
ABSTRACT			
Abstract	2	See the PRISMA 2020 for Abstracts checklist.	Page 1, paragraph 2
INTRODUCTION			
Rationale	3	Describe the rationale for the review in the context of existing knowledge.	Page 2, paragraph 2
Objectives	4	Provide an explicit statement of the objective(s) or question(s) the review addresses.	Page 4, paragraph 2
METHODS			
Eligibility criteria	5	Specify the inclusion and exclusion criteria for the review and how studies were grouped for the syntheses.	Page 5, paragraph 2
Information sources	6	Specify all databases, registers, websites, organisations, reference lists and other sources searched or consulted to identify studies. Specify the date when each source was last searched or consulted.	Page 5, paragraph 1
Search strategy	7	Present the full search strategies for all databases, registers and websites, including any filters and limits used.	Page 22
Selection process	8	Specify the methods used to decide whether a study met the inclusion criteria of the review, including how many reviewers screened each record and each report retrieved, whether they worked independently, and if applicable, details of automation tools used in the process.	Page 5, paragraph 2
Data collection process	9	Specify the methods used to collect data from reports, including how many reviewers collected data from each report, whether they worked independently, any processes for obtaining or confirming data from study investigators, and if applicable, details of automation tools used in the process.	Page 5, paragraph 2

Table A1. *Cont.*

Section and Topic	Item #	Checklist Item	Location Where Item Is Reported
Data items	10a	List and define all outcomes for which data were sought. Specify whether all results that were compatible with each outcome domain in each study were sought (e.g., for all measures, time points, analyses), and if not, the methods used to decide which results to collect.	Page 6, paragraph 1
	10b	List and define all other variables for which data were sought (e.g., participant and intervention characteristics, funding sources). Describe any assumptions made about any missing or unclear information.	Page 6, paragraph 1
Study risk of bias assessment	11	Specify the methods used to assess risk of bias in the included studies, including details of the tool(s) used, how many reviewers assessed each study and whether they worked independently, and if applicable, details of automation tools used in the process.	Page 11, paragraph 1
Effect measures	12	Specify for each outcome the effect measure(s) (e.g., risk ratio, mean difference) used in the synthesis or presentation of results.	N/A
Synthesis methods	13a	Describe the processes used to decide which studies were eligible for each synthesis (e.g., tabulating the study intervention characteristics and comparing against the planned groups for each synthesis (item #5)).	Page 8, Table 3
	13b	Describe any methods required to prepare the data for presentation or synthesis, such as handling of missing summary statistics, or data conversions.	N/A
	13c	Describe any methods used to tabulate or visually display results of individual studies and syntheses.	Page 8, Table 3
	13d	Describe any methods used to synthesize results and provide a rationale for the choice(s). If meta-analysis was performed, describe the model(s), method(s) to identify the presence and extent of statistical heterogeneity, and software package(s) used.	N/A
	13e	Describe any methods used to explore possible causes of heterogeneity among study results (e.g., subgroup analysis, meta-regression).	N/A
	13f	Describe any sensitivity analyses conducted to assess robustness of the synthesized results.	N/A
Reporting bias assessment	14	Describe any methods used to assess risk of bias due to missing results in a synthesis (arising from reporting biases).	Page 11, paragraph 1
Certainty assessment	15	Describe any methods used to assess certainty (or confidence) in the body of evidence for an outcome.	N/A

Table A1. *Cont.*

Section and Topic	Item #	Checklist Item	Location Where Item Is Reported
RESULTS			
Study selection	16a	Describe the results of the search and selection process, from the number of records identified in the search to the number of studies included in the review, ideally using a flow diagram.	Page 6, Figure 3
	16b	Cite studies that might appear to meet the inclusion criteria, but which were excluded, and explain why they were excluded.	N/A
Study characteristics	17	Cite each included study and present its characteristics.	Page 11, paragraph 4
Risk of bias in studies	18	Present assessments of risk of bias for each included study.	Pages 27–28
Results of individual studies	19	For all outcomes, present, for each study: (a) summary statistics for each group (where appropriate) and (b) an effect estimate and its precision (e.g., confidence/credible interval), ideally using structured tables or plots.	Pages 29–30
Results of syntheses	20a	For each synthesis, briefly summarise the characteristics and risk of bias among contributing studies.	Page 21, paragraph 2
	20b	Present results of all statistical syntheses conducted. If meta-analysis was done, present for each the summary estimate and its precision (e.g., confidence/credible interval) and measures of statistical heterogeneity. If comparing groups, describe the direction of the effect.	N/A
	20c	Present results of all investigations of possible causes of heterogeneity among study results.	Page 21, paragraph 2
	20d	Present results of all sensitivity analyses conducted to assess the robustness of the synthesized results.	N/A
Reporting biases	21	Present assessments of risk of bias due to missing results (arising from reporting biases) for each synthesis assessed.	Pages 27–28
Certainty of evidence	22	Present assessments of certainty (or confidence) in the body of evidence for each outcome assessed.	N/A
DISCUSSION			
Discussion	23a	Provide a general interpretation of the results in the context of other evidence.	Page 18, paragraph 1
	23b	Discuss any limitations of the evidence included in the review.	Page 21, paragraph 2
	23c	Discuss any limitations of the review processes used.	Page 21, paragraph 2
	23d	Discuss implications of the results for practice, policy, and future research.	Page 18, paragraph 3

Table A1. *Cont.*

Section and Topic	Item #	Checklist Item	Location Where Item Is Reported
OTHER INFORMATION			
Registration and protocol	24a	Provide registration information for the review, including register name and registration number, or state that the review was not registered.	Page 5, paragraph 1
	24b	Indicate where the review protocol can be accessed, or state that a protocol was not prepared.	Page 5, paragraph 1
	24c	Describe and explain any amendments to information provided at registration or in the protocol.	N/A
Support	25	Describe sources of financial or non-financial support for the review, and the role of the funders or sponsors in the review.	Page 22
Competing interests	26	Declare any competing interests of review authors.	Page 22
Availability of data, code and other materials	27	Report which of the following are publicly available and where they can be found: template data collection forms; data extracted from included studies; data used for all analyses; analytic code; any other materials used in the review.	Page 22

Table A2. PRISMA Abstract Checklist [27].

Topic	No.	Item	Reported?
TITLE			
Title	1	Identify the report as a systematic review.	Yes
BACKGROUND			
Objectives	2	Provide an explicit statement of the main objective(s) or question(s) the review addresses.	Yes
METHODS			
Eligibility criteria	3	Specify the inclusion and exclusion criteria for the review.	No
Information sources	4	Specify the information sources (e.g., databases, registers) used to identify studies and the date when each was last searched.	Yes
Risk of bias	5	Specify the methods used to assess risk of bias in the included studies.	No
Synthesis of results	6	Specify the methods used to present and synthesize results.	No
RESULTS			
Included studies	7	Give the total number of included studies and participants and summarise relevant characteristics of studies.	Yes
Synthesis of results	8	Present results for main outcomes, preferably indicating the number of included studies and participants for each. If meta-analysis was done, report the summary estimate and confidence/credible interval. If comparing groups, indicate the direction of the effect (i.e., which group is favoured).	Yes
DISCUSSION			
Limitations of evidence	9	Provide a brief summary of the limitations of the evidence included in the review (e.g., study risk of bias, inconsistency and imprecision).	No
Interpretation	10	Provide a general interpretation of the results and important implications.	Yes
OTHER			
Funding	11	Specify the primary source of funding for the review.	No
Registration	12	Provide the register name and registration number.	Yes

Appendix C

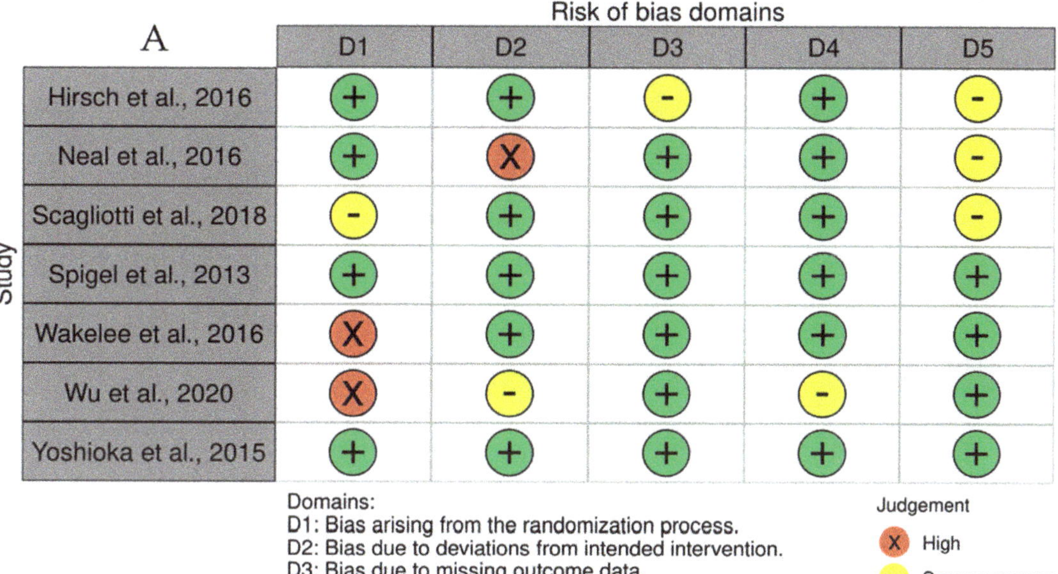

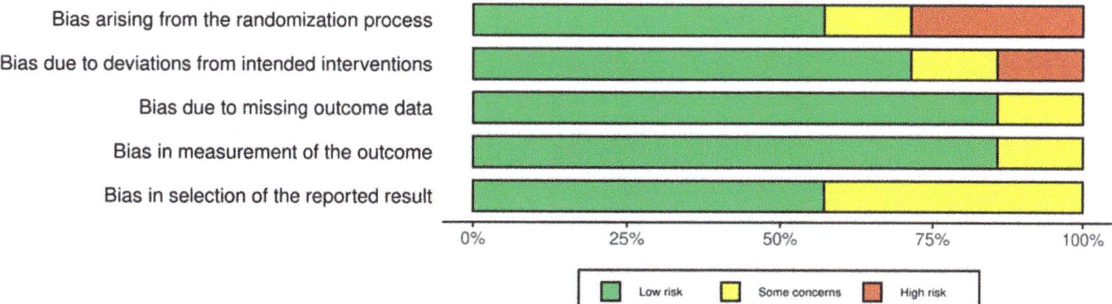

Figure A1. Risk of bias plots for all included RCTs in this systematic review, created in the Cochrane Risk of Bias Tool "robvis" [2,5,6,17,23,30,32,33]. (**A**) Traffic light plot, indicating the risk of bias for each included RCT in each of the five domains. (**B**) Summary plot, showing the summarized risk of bias in each domain, and an overall risk of bias measurement. D: domain, RCT: randomized controlled trial.

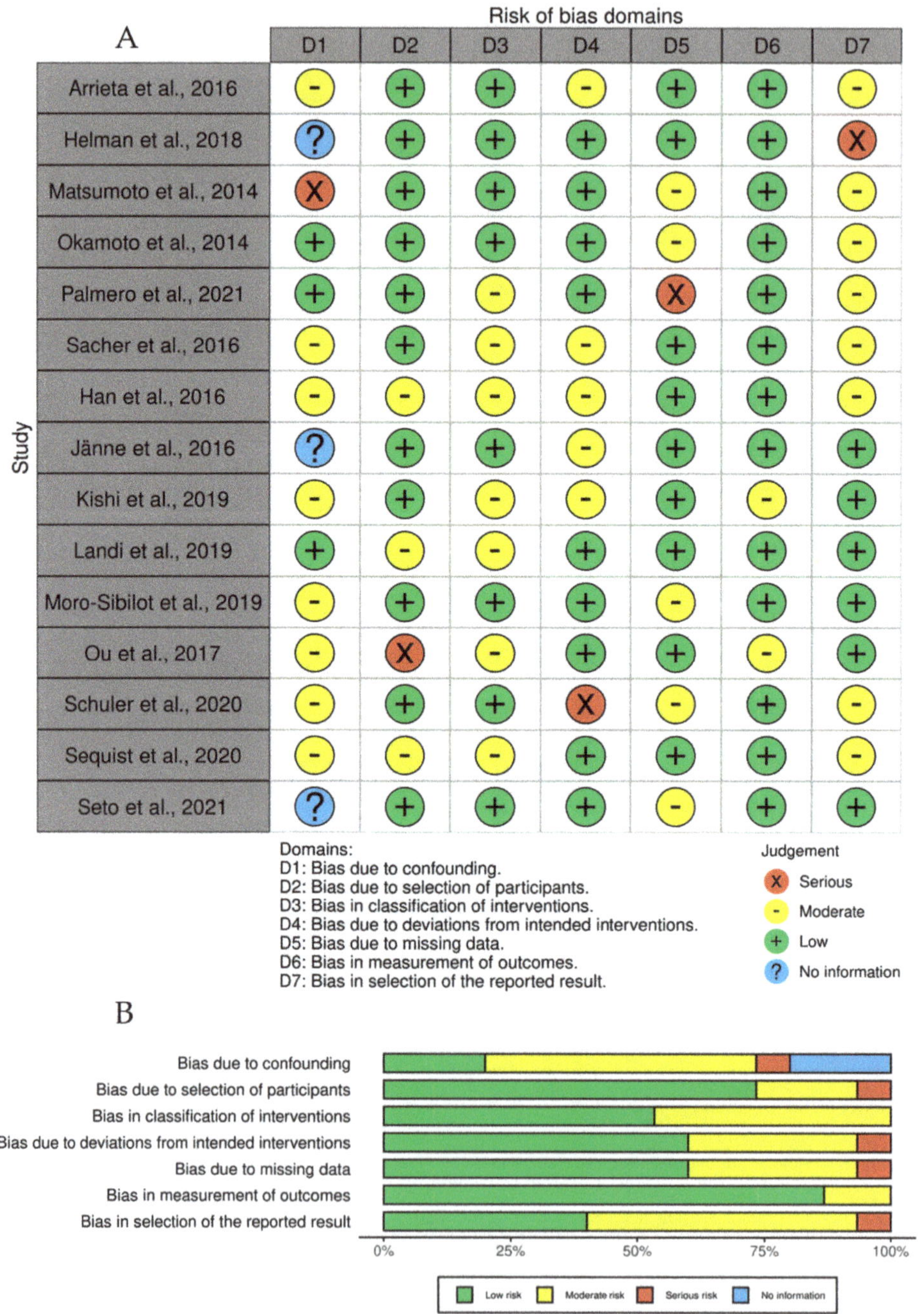

Figure A2. Risk of bias plots for all included clinical trials in this systematic review, created in the Cochrane Risk of Bias Tool "robvis" [1,3,11,14,15,18–22,24,25,28,29,31,33]. (**A**): Traffic light plot, indicating the risk of bias for each included clinical trial in each of the seven domains. (**B**): Summary plot, showing the summarized risk of bias in each domain, and an overall risk of bias measurement. D: domain.

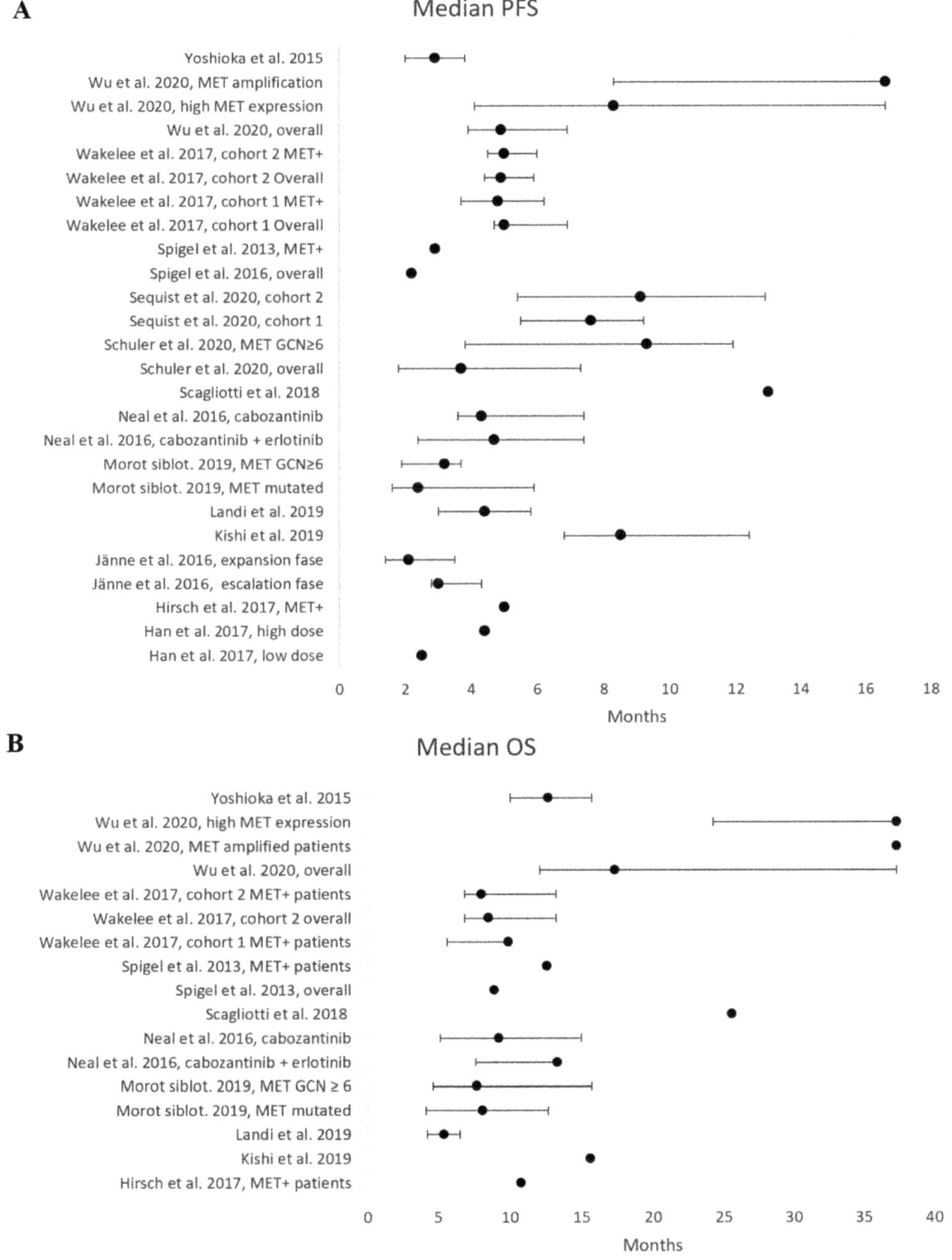

Figure A3. Forest plots presenting the median PFS and median OS with 95% confidence intervals (CI) from the included articles in this systematic review presenting results on treatment with a MET TKI. Results that are specific to a MET+ subgroup are presented separately from the overall cohort results [2,5,6,14,15,17,18,20–23,29,30,32]. (**A**) Median PFS with 95% CIs. (**B**) Median OS with 95% CIs.

Table A3. Epidemiology reported by the included studies. Median age, percentage of participants that are male gender, and smoking history, presented as percentage of participants with a history of no smoking, of the included patients in the different studies and their subgroups. EGFR: epidermal growth factor receptor, WT: wild type, MET: mesenchymal epithelial transition factor, NS: not specified.

	Mutational Status	Median Age (Years)	Male Gender (%)	History of No Smoking (%)
Arrieta et al., 2016 [1]	EGFR+/WT	60.1	33.3%	39.4%
Han et al., 2017 [14]	EGFR+/WT, MET+/−	Treatment: 63.0 Placebo: 63.0	Treatment: 56.0% Placebo: 57.0%	Treatment: 18.5% Placebo: 15.8%
Helman et al., 2018 [25]	EGFR+	61.0	28.6%	NS
Hirsch et al., 2017 [17]	MET+/−	Treatment: 68.0 Placebo: 66.0	Treatment: 70.9% Placebo: 74.1%	Treatment: 3.6% Placebo: 5.6%
Jänne et al., 2016 [21]	EGFR+/WT	59.5	39.0%	51.0%
Kishi et al., 2019 [15]	EGFR+ MET+	67.0	43%	NS
Landi et al., 2019 [18]	EGFR WT MET+	56.0	65%	23%
Matsumoto et al., 2014 [19]	EGFR WT	64.0	19.7%	89.1%
Moro-Sibilot et al., 2019 [29]	EGFR+/WT MET+/−	MET amplified: 59.0 MET mutated: 72.0	MET amplified: 56.0% MET mutated: 32.0%	MET amplified: 24.0% MET mutation: 48.0%
Neal et al., 2016 [2]	EGFR WT	65.3	45.0%	15%
Okamoto et al., 2014 [3]	NS	64.0	77.0%	18.0%
Ou et al., 2017 [24]	NS	60.0	30.0%	52.0%
Palmero et al., 2021 [11]	NS	64.3	65.0%	27.0%
Sacher et al., 2016 [28]	EGFR+/WT	64.0	45.0%	18.0%
Scagliotti et al., 2018 [30]	EGFR+ MET+/−	Treatment: 59.5 Placebo: 65.0	Treatment: 42.9% Placebo: 47.2%	Treatment: 48.2% Placebo: 60.4%
Schuler et al., 2020 [20]	EGFR WT MET+	60.0	60.0%	NS
Sequist et al., 2020 [22]	EGFR+ MET+	Cohort B: 59.0 Cohort D: 62.0	Cohort B: 59% Cohort D: 60%	NS

Table A3. *Cont.*

	Mutational Status	Median Age (Years)	Male Gender (%)	History of No Smoking (%)
Seto et al., 2021 [31]	*EGFR* WT *MET*+	68.0	66.7%	42.2%
Spigel et al., 2013 [6]	*EGFR*+/WT *MET*+/−	Treatment *MET*-: 63.0 Placebo *MET*-: 61.0 Treatment *MET*+: 66.0 Placebo *MET*+: 64.0	Treatment *MET*-: 65.0% Placebo *MET*-: 55.0% Treatment *MET*+: 51.0% Placebo *MET*+: 65.0%	Treatment *MET* -: 7.0% Placebo *MET* -: 3.0% Treatment *MET*+: 20.0% Placebo *MET*+: 19.0%
Wakelee et al., 2017 [32]	*EGFR*+/WT *MET*+/−	Cohort I treatment: 60.0 Cohort I placebo: 60.5 Cohort II treatment: 66.0 Cohort II placebo: 63.0	Cohort I treatment: 68.1% Cohort I placebo: 48.6% Cohort II treatment: 55.9% Cohort II placebo: 42.6%	Cohort I treatment: 21.7% Cohort I placebo: 25.7% Cohort II treatment: 27.1% Cohort II placebo: 13.1%
Wu et al., 2020 [23]	*EGFR*+ *MET*+	Phase Ib: 65.5 Phase II treatment: 61.0 Phase II placebo: 58.3	Phase Ib: 44.0% Phase II treatment: 35.5% Phase II placebo: 50.0%	Phase Ib: 72.0% Phase II treatment: 68.0% Phase II placebo: 67.0%
Yoshioka et al., 2015 [5]	*EGFR* WT *MET*+/−	Treatment: 63.0 Placebo: 63.0	Treatment: 66.7% Placebo: 70.8%	Treatment: 25.5% Placebo: 26.2%

References

1. Arrieta, O.; Cruz-Rico, G.; Soto-Perez-de-Celis, E.; Ramírez-Tirado, L.A.; Caballe-Perez, E.; Martínez-Hernández, J.N.; Martinez-Alvarez, I.; Soca-Chafre, G.; Macedo-Pérez, E.O.; Astudillo-de la Vega, H.; et al. Reduction in Hepatocyte Growth Factor Serum Levels is Associated with Improved Prognosis in Advanced Lung Adenocarcinoma Patients Treated with Afatinib: A Phase II Trial. *Target Oncol.* **2016**, *11*, 619–629. [CrossRef] [PubMed]
2. Neal, J.W.; Dahlberg, S.E.; Wakelee, H.A.; Aisner, S.C.; Bowden, M.; Huang, Y.; Carbone, D.P.; Gerstner, G.J.; Lerner, R.E.; Rubin, J.L.; et al. Erlotinib, cabozantinib, or erlotinib plus cabozantinib as second-line or third-line treatment of patients with *EGFR* wild-type advanced non-small-cell lung cancer (ECOG-ACRIN 1512): A randomised, controlled, open-label, multicentre, phase 2 trial. *Lancet Oncol.* **2016**, *17*, 1661–1671. [CrossRef] [PubMed]
3. Okamoto, I.; Sakai, K.; Morita, S.; Yoshioka, H.; Kaneda, H.; Takeda, K.; Hirashima, T.; Kogure, Y.; Kimura, T.; Takahashi, T.; et al. Multiplex genomic profiling of non-small cell lung cancers from the LETS phase III trial of first-line S-1/carboplatin versus paclitaxel/carboplatin: Results of a West Japan Oncology Group study. *Oncotarget* **2014**, *5*, 2293–2304. [CrossRef] [PubMed]
4. Ferlay, J.; Soerjomataram, I.; Dikshit, R.; Eser, S.; Mathers, C.; Rebelo, M.; Parkin, D.M.; Forman, D.; Bray, F. Cancer incidence and mortality worldwide: Sources, methods and major patterns in GLOBOCAN 2012. *Int. J. Cancer* **2015**, *136*, 359–386. [CrossRef] [PubMed]
5. Yoshioka, H.; Azuma, K.; Yamamoto, N.; Takahashi, T.; Nishio, M.; Katakami, N.; Ahn, M.J.; Hirashima, T.; Maemondo, M.; Kim, S.W.; et al. A randomized, double-blind, placebo-controlled, phase III trial of erlotinib with or without a c-Met inhibitor tivantinib (ARQ 197) in Asian patients with previously treated stage IIIB/IV nonsquamous nonsmall-cell lung cancer harboring wild-type epidermal growth factor receptor (ATTENTION study). *Ann. Oncol.* **2015**, *26*, 2066–2072. [PubMed]
6. Spigel, D.R.; Ervin, T.J.; Ramlau, R.A.; Daniel, D.B.; Goldschmidt, J.H., Jr.; Blumenschein, G.R., Jr.; Krzakowski, M.J.; Robinet, G.; Godbert, B.; Barlesi, F.; et al. Randomized phase II trial of Onartuzumab in combination with erlotinib in patients with advanced non-small-cell lung cancer. *J. Clin. Oncol.* **2013**, *31*, 4105–4114. [CrossRef] [PubMed]
7. Urvay, S.E.; Yucel, B.; Erdis, E.; Turan, N. Prognostic Factors in Stage III Non-Small-Cell Lung Cancer Patients. *Asian. Pac. J. Cancer Prev.* **2016**, *17*, 4693–4697.
8. Önal, Ö.; Koçer, M.; Eroğlu, H.N.; Yilmaz, S.D.; Eroğlu, I.; Karadoğan, D. Survival analysis and factors affecting survival in patients who presented to the medical oncology unit with non-small cell lung cancer. *Turk. J. Med. Sci.* **2020**, *50*, 1838–1850. [CrossRef]
9. Shields, M.D.; Marin-Acevedo, J.A.; Pellini, B. *Immunotherapy for Advanced Non-Small Cell Lung Cancer: A Decade of Progress*; American Society of Clinical Oncology Educational Book: Alexandria, VA, USA, 2021; pp. 105–127.
10. Bodén, E.; Andreasson, J.; Hirdman, G.; Malmsjö, M.; Lindstedt, S. Quantitative Proteomics Indicate Radical Removal of Non-Small Cell Lung Cancer and Predict Outcome. *Biomedicines* **2022**, *10*, 2738. [CrossRef]
11. Palmero, R.; Taus, A.; Viteri, S.; Majem, M.; Carcereny, E.; Garde-Noguera, J.; Felip, E.; Nadal, E.; Malfettone, A.; Sampayo, M.; et al. Biomarker Discovery and Outcomes for Comprehensive Cell-Free Circulating Tumor DNA Versus Standard-of-Care Tissue Testing in Advanced Non-Small-Cell Lung Cancer. *JCO Precis. Oncol.* **2021**, *5*, 93–102. [CrossRef]
12. Bethune, G.; Bethune, D.; Ridgway, N.; Xu, Z. Epidermal growth factor receptor (*EGFR*) in lung cancer: An overview and update. *J. Thorac. Dis.* **2010**, *2*, 48–51. [PubMed]
13. Coleman, N.; Hong, L.; Zhang, J.; Heymach, J.; Hong, D.; Le, X. Beyond epidermal growth factor receptor: MET amplification as a general resistance driver to targeted therapy in oncogene-driven non-small-cell lung cancer. *ESMO Open* **2021**, *6*, 100319. [CrossRef]
14. Han, K.; Chanu, P.; Jonsson, F.; Winter, H.; Bruno, R.; Jin, J.; Stroh, M. Exposure-Response and Tumor Growth Inhibition Analyses of the Monovalent Anti-c-MET Antibody Onartuzumab (MetMAb) in the Second- and Third-Line Non-Small Cell Lung Cancer. *AAPS J.* **2017**, *19*, 527–533. [CrossRef] [PubMed]
15. Kishi, K.; Sakai, H.; Seto, T.; Kozuki, T.; Nishio, M.; Imamura, F.; Hiroshi, N.; Miyako, S.; Shintaro, N.; Takashi, T.; et al. First-line onartuzumab plus erlotinib treatment for patients with MET-positive and *EGFR* mutation-positive non-small-cell lung cancer. *Cancer Treat Res. Commun.* **2019**, *18*, 100113. [CrossRef] [PubMed]
16. Organ, S.L.; Tsao, M. S An overview of the c-MET signaling pathway. *Ther. Adv. Med. Oncol.* **2011**, *3*, S7–S19. [CrossRef]
17. Hirsch, F.R.; Govindan, R.; Zvirbule, Z.; Braiteh, F.; Rittmeyer, A.; Belda-Iniesta, C.; Isla, D.; Cosgriff, T.; Boyer, M.; Ueda, M.; et al. Efficacy and Safety Results From a Phase II, Placebo-Controlled Study of Onartuzumab Plus First-Line Platinum-Doublet Chemotherapy for Advanced Squamous Cell Non-Small-Cell Lung Cancer. *Clin. Lung Cancer* **2017**, *18*, 43–49. [CrossRef]
18. Landi, L.; Chiari, R.; Tiseo, M.; D'Incà, F.; Dazzi, C.; Chella, A.; Delmonte, A.; Bonanno, L.; Giannarelli, D.; Cortinovis, D.L.; et al. Crizotinib in *MET*-Deregulated or *ROS1*-Rearranged Pretreated Non-Small Cell Lung Cancer (METROS): A Phase II, Prospective, Multicenter, Two-Arms Trial. *Clin. Cancer Res.* **2019**, *25*, 7312–7319. [CrossRef]
19. Matsumoto, Y.; Maemondo, M.; Ishii, Y.; Okudera, K.; Demura, Y.; Takamura, K.; Kobayashi, K.; Morikawa, N.; Gemm, A.; Ishimoto, O.; et al. A phase II study of erlotinib monotherapy in pre-treated non-small cell lung cancer without *EGFR* gene mutation who have never/light smoking history: Re-evaluation of *EGFR* gene status (NEJ006/TCOG0903). *Lung Cancer* **2014**, *86*, 195–200. [CrossRef]
20. Schuler, M.; Berardi, R.; Lim, W.T.; de Jonge, M.; Bauer, T.M.; Azaro, A.; Gottfried, M.; Han, J.-Y.; Lee, D.; Wollner, M.; et al. Molecular correlates of response to capmatinib in advanced non-small-cell lung cancer: Clinical and biomarker results from a phase I trial. *Ann. Oncol.* **2020**, *31*, 789–797. [CrossRef] [PubMed]

21. Jänne, P.A.; Shaw, A.T.; Camidge, D.R.; Giaccone, G.; Shreeve, S.M.; Tang, Y.; Goldberg, Z.; Martini, J.-F.; Xu, H.; James, L.P.; et al. Combined Pan-HER and ALK/ROS1/MET Inhibition with Dacomitinib and Crizotinib in Advanced Non-Small Cell Lung Cancer: Results of a Phase I Study. *J. Thorac. Oncol.* **2016**, *11*, 737–747. [CrossRef]

22. Sequist, L.V.; Han, J.Y.; Ahn, M.J.; Cho, B.C.; Yu, H.; Kim, S.W.; Yang, J.C.-H.; Lee, J.S.; Su, W.-C.; Kowalski, D.; et al. Osimertinib plus savolitinib in patients with *EGFR* mutation-positive, MET-amplified, non-small-cell lung cancer after progression on *EGFR* tyrosine kinase inhibitors: Interim results from a multicentre, open-label, phase 1b study. *Lancet Oncol.* **2020**, *21*, 373–386. [CrossRef] [PubMed]

23. Wu, Y.L.; Cheng, Y.; Zhou, J.; Lu, S.; Zhang, Y.; Zhao, J.; Kim, D.-W.; Soo, R.A.; Kim, S.-W.; Pan, H.; et al. Tepotinib plus gefitinib in patients with *EGFR*-mutant non-small-cell lung cancer with MET overexpression or MET amplification and acquired resistance to previous *EGFR* inhibitor (INSIGHT study): An open-label, phase 1b/2, multicentre, randomised trial. *Lancet Respir. Med.* **2020**, *8*, 1132–1143. [CrossRef]

24. Ou, S.I.; Govindan, R.; Eaton, K.D.; Otterson, G.A.; Gutierrez, M.E.; Mita, A.C.; Argiris, A.; Brega, N.M.; Usari, T.; Tan, W.; et al. Phase I Results from a Study of Crizotinib in Combination with Erlotinib in Patients with Advanced Nonsquamous Non-Small Cell Lung Cancer. *J. Thorac. Oncol.* **2017**, *12*, 145–151. [CrossRef]

25. Helman, E.; Nguyen, M.; Karlovich, C.A.; Despain, D.; Choquette, A.K.; Spira, A.I.; Yu, H.A.; Camidge, D.R.; Harding, T.C.; Lanman, R.B.; et al. Cell-Free DNA Next-Generation Sequencing Prediction of Response and Resistance to Third-Generation *EGFR* Inhibitor. *Clin. Lung Cancer* **2018**, *19*, 518–530. [CrossRef]

26. Morganti, S.; Tarantino, P.; Ferraro, E.; D'Amico, P.; Duso, B.A.; Curigliano, G. *Next Generation Sequencing (NGS): A Revolutionary Technology in Pharmacogenomics and Personalized Medicine in Cancer*; Springer International Publishing: Berlin/Heidelberg, Germany, 2019; pp. 9–30.

27. Page, M.J.; McKenzie, J.E.; Bossuyt, P.M.; Boutron, I.; Hoffmann, T.C.; Mulrow, C.D.; Shamseer, L.; Jennifer, M.; Elie, A.; Brennan, S.E.; et al. The PRISMA 2020 statement: An updated guideline for reporting systematic reviews. *Syst. Rev.* **2021**, *88*, 105906.

28. Sacher, A.G.; Le, L.W.; Lara-Guerra, H.; Waddell, T.K.; Sakashita, S.; Chen, Z.; Kim, L.; Zhang, T.; Kamel-Reid, S.; Salvarrey, A.; et al. A window of opportunity study of potential tumor and soluble biomarkers of response to preoperative erlotinib in early stage non-small cell lung cancer. *Oncotarget* **2016**, *7*, 25632. [CrossRef] [PubMed]

29. Moro-Sibilot, D.; Cozic, N.; Pérol, M.; Mazières, J.; Otto, J.; Souquet, P.J.; Bahleda, R.; Wislez, M.; Zalcman, G.; Guibert, S.D.; et al. Crizotinib in c-MET- or ROS1-positive NSCLC: Results of the AcSé phase II trial. *Ann. Oncol.* **2019**, *30*, 1985–1991. [CrossRef] [PubMed]

30. Scagliotti, G.V.; Shuster, D.; Orlov, S.; von Pawel, J.; Shepherd, F.A.; Ross, J.S.; Wang, Q.; Schwartz, B.; Akerley, W. Tivantinib in Combination with Erlotinib versus Erlotinib Alone for *EGFR*-Mutant NSCLC: An Exploratory Analysis of the Phase 3 MARQUEE Study. *J. Thorac. Oncol.* **2018**, *13*, 849–854. [CrossRef]

31. Seto, T.; Ohashi, K.; Sugawara, S.; Nishio, M.; Takeda, M.; Aoe, K.; Moizumi, S.; Nomura, S.; Tajima, T.; Hida, T.; et al. Capmatinib in Japanese patients with MET exon 14 skipping-mutated or MET-amplified advanced NSCLC: GEOMETRY mono-1 study. *Cancer Sci.* **2021**, *112*, 1556–1566. [CrossRef] [PubMed]

32. Wakelee, H.; Zvirbule, Z.; De Braud, F.; Kingsley, C.D.; Mekhail, T.; Lowe, T.; Schütte, W.; Lena, H.; Lawler, W.; Braiteh, F.; et al. Efficacy and Safety of Onartuzumab in Combination With First-Line Bevacizumab- or Pemetrexed-Based Chemotherapy Regimens in Advanced Non-Squamous Non-Small-Cell Lung Cancer. *Clin. Lung Cancer* **2017**, *18*, 50–59. [CrossRef]

33. McGuinness, L.A.; Higgins, J.P.T. Risk-of-bias VISualization (robvis): An R package and Shiny web app for visualizing risk-of-bias assessments. *Res. Synth. Methods* **2021**, *12*, 55–61. [CrossRef] [PubMed]

34. Andreasson, J.; Bodén, E.; Fakhro, M.; von Wachter, C.; Olm, F.; Malmsjö, M.; Hallgren, O.; Lindstedt, S. Exhaled phospholipid transfer protein and hepatocyte growth factor receptor in lung adenocarcinoma. *Respir. Res.* **2022**, *23*, 369. [CrossRef] [PubMed]

35. Liguori, N.R.; Lee, Y.; Borges, W.; Zhou, L.; Azzoli, C.; El-Deiry, W.S. Absence of Biomarker-Driven Treatment Options in Small Cell Lung Cancer, and Selected Preclinical Candidates for Next Generation Combination Therapies. *Front. Pharmacol.* **2021**, *12*, 747180. [CrossRef] [PubMed]

36. Toumazis, I.; Bastani, M.; Han, S.S.; Plevritis, S.K. Risk-Based lung cancer screening: A systematic review. *Lung Cancer* **2020**, *147*, 154–186. [CrossRef] [PubMed]

37. Scott, A.; Salgia, R. Biomarkers in lung cancer: From early detection to novel therapeutics and decision making. *Biomark. Med.* **2008**, *2*, 577–586. [CrossRef] [PubMed]

38. Smyth, E.C.; Sclafani, F.; Cunningham, D. Emerging molecular targets in oncology: Clinical potential of MET/hepatocyte growth-factor inhibitors. *OncoTargets Ther.* **2014**, *7*, 1001–1014. [CrossRef]

39. Beau-Faller, M.; Ruppert, A.M.; Voegeli, A.C.; Neuville, A.; Meyer, N.; Guerin, E.; Legrain, M.; Mennecier, B.; Wihlm, J.-M.; Massard, G.; et al. MET gene copy number in non-small cell lung cancer: Molecular analysis in a targeted tyrosine kinase inhibitor naïve cohort. *J. Thorac. Oncol.* **2008**, *3*, 331–339. [CrossRef]

40. Onozato, R.; Kosaka, T.; Kuwano, H.; Sekido, Y.; Yatabe, Y.; Mitsudomi, T. Activation of MET by gene amplification or by splice mutations deleting the juxtamembrane domain in primary resected lung cancers. *J. Thorac. Oncol.* **2009**, *4*, 5–11. [CrossRef]

41. Kent, D.G.; Green, A.R. Order Matters: The Order of Somatic Mutations Influences Cancer Evolution. *Cold Spring Harb. Perspect. Med.* **2017**, *7*, a027060. [CrossRef]

42. Jolly, C.; Van Loo, P. Timing somatic events in the evolution of cancer. *Genome Biol.* **2018**, *19*, 95. [CrossRef]

43. Dong, Y.; Xu, J.; Sun, B.; Wang, J.; Wang, Z. MET-Targeted Therapies and Clinical Outcomes: A Systematic Literature Review. *Mol. Diagn. Ther.* **2022**, *26*, 203–227. [CrossRef] [PubMed]

44. Kollmannsberger, C.; Hurwitz, H.; Bazhenova, L.; Cho, B.C.; Hong, D.; Park, K.; Sharma, S.; Der-Torossian, H.; Christensen, J.G.; Faltaos, D.; et al. Phase I Study Evaluating Glesatinib (MGCD265), An Inhibitor of MET and AXL, in Patients with Non-small Cell Lung Cancer and Other Advanced Solid Tumors. *Target Oncol.* **2023**, *18*, 105–118. [CrossRef] [PubMed]

45. Camidge, D.R.; Janku, F.; Bueno, A.M.; Catenacci, D.V.; Lee, J.; Lee, S.H.; Chung, H.C.; Dowlati, A.; Rohrberg, K.S.; Font, F.E.; et al. 1490PD—A phase Ia/IIa trial of Sym015, a MET antibody mixture, in patients with advanced solid tumours. *Ann. Oncol.* **2019**, *30*, V610–V611. [CrossRef]

Review

The Role of cMET in Gastric Cancer—A Review of the Literature

Filip Van Herpe * and Eric Van Cutsem

Department of Digestive Oncology, University Hospitals Gasthuisberg, 3000 Leuven, Belgium
* Correspondence: filip.vanherpe@uzleuven.be; Tel.: +32-16344727

Simple Summary: cMET is a proto-oncogene that has been extensively studied in gastric cancer. Gastric cancer (GC) is a heterogenous disease with varied histology and molecular profiling. It still implies a poor prognosis in stage IV. New targeted therapeutic options are being investigated. In this review, we analyzed all studies performed on gastric cancer with MET-inhibitors. In first-line therapy, the addition of MET-inhibition to chemotherapy did not show any benefit in allcomers. Different tyrosine kinase inhibitors (TKI) have been investigated in small cohorts with different diagnostic assays added to the inclusion criteria. Determining patients with gastric cancer who benefit from cMET inhibitors remains difficult. Potentially only *MET* amplification detected by comprehensive genomic testing could be a good targeted option, although the prevalence is limited to less than 5% of all patients with gastric cancer.

Abstract: Gastric cancer (GC) is an important cause of cancer worldwide with over one million new cases yearly. The vast majority of cases present in stage IV disease, and it still bears a poor prognosis. However, since 2010, progress has been made with the introduction of targeted therapies against HER2 and with checkpoint inhibitors (PDL1). More agents interfering with other targets (FGFR2B, CLDN18.2) are being investigated. cMET is a less frequent molecular target that has been studied for gastric cancer. It is a proto-oncogene that leads to activation of the MAPK pathway and the PI3K pathway, which is responsible for activating the MTOR pathway. The prevalence of cMET is strongly debated as different techniques are being used to detect MET-driven tumors. Because of the difference in diagnostic assays, selecting patients who benefit from cMET inhibitors is difficult. In this review, we discuss the pathway of cMET, its clinical significance and the different diagnostic assays that are currently used, such as immunohistochemy (IHC), fluorescence in situ hybridization (FISH), the H-score and next-generation sequencing (NGS). Next, we discuss all the current data on cMET inhibitors in gastric cancer. Since the data on cMET inhibitors are very heterogenous, it is difficult to provide a general consensus on the outcome, as inclusion criteria differ between trials. Diagnosing cMET-driven gastric tumors is difficult, and potentially the only accurate determination of cMET overexpression/amplification may be next-generation sequencing (NGS).

Keywords: cMET; gastro-oesophageal cancer; gastric cancer; review

Citation: Van Herpe, F.; Van Cutsem, E. The Role of cMET in Gastric Cancer—A Review of the Literature. *Cancers* **2023**, *15*, 1976. https://doi.org/10.3390/cancers15071976

Academic Editors: Jan Trøst Jørgensen and Jens Mollerup

Received: 22 November 2022
Revised: 10 March 2023
Accepted: 23 March 2023
Published: 26 March 2023

1. Introduction

Gastric cancer (GC) is an important cause of cancer worldwide with over one million new cases per year and, in 2020, an estimated 769,000 deaths [1].

Gastric adenocarcinoma is a heterogenous disease and is categorized in different subgroups based on histology (diffuse vs. intestinal), as well as on molecular profiling, microsatellite instability (MSI), Epstein–Barr virus (EBV), genomic stability (GS) and chromosomal instability (CIN). Apart from immunotherapy-sensitive MSI or EBV-driven gastric cancer, the prognosis in stage IV disease remains poor.

The first-line standard of care therapy remains the doublet chemotherapeutic combination of platinum-based therapy (cisplatin or oxaliplatin) in combination with a fluoropyrimidine (5-fluorouracil, capecitabine or S-1). Occasionally, a third cytotoxic agent is added (usually docetaxel). Over the last 15 years, molecular-driven targeted therapy has evolved rapidly. Patients with metastatic GC can be subdivided into two groups based on HER2 expression status: HER2 positive disease, meaning a 3+ score on protein immunohistochemistry (IHC) and a 2+ score on IHC followed by fluorescence in situ hybridization (FISH) of $\geq$2. For patients with HER2 positive disease, trastuzumab can be added to standard of care chemotherapy as the ToGA trial showed improved progression-free survival (PFS) and overall survival (OS) in HER2-positive patients [2].

In 2015, VEGFR-2-inhibition was added in the second line with ramucirumab for gastric cancer in combination with paclitaxel [3]. It showed an improved OS compared to placebo as well as paclitaxel monotherapy. Recently, in 2021, the addition of a checkpoint inhibitor (the PD-1 antibody nivolumab) was added in HER2 negative patients harboring a PDL1 CPS score > 5 on tumor samples in first-line gastric cancer in combination with platinum-based doublet chemotherapy [4]. A combined treatment of another checkpoint inhibitor, pembrolizumab, with trastuzumab, in HER2 positive patients, showed a very high response rate (ORR) in the recent findings of the Keynote 811 trial [5]. However, in PDL1 and HER2 negative populations, there is still a need for improvement. Due to increased comprehensive genomic analysis of tumor DNA and the development of new targets, more targeted therapies are being found in gastric cancer. This includes targets such as CLDN18.2 and FGFR2b overexpression, and the landscape is currently still expanding. One of the potential targets in gastric cancer is cMET. *MET* is a proto-oncogene which encodes for a transmembrane receptor with tyrosine kinase activity. In this review, we discuss how diagnostic testing of *MET* overexpression/amplification is performed in gastric cancer, and, next, provide an overview of all clinical studies that have been published with cMET inhibitors in gastric cancer.

2. Materials and Methods

A literature search was performed of all studies published from 2008 until July 2022 on Pubmed (https://pubmed.ncbi.nlm.nih.gov/, accessed on 21 November 2022) as well as on clinical trials.gov (https://clinicaltrials.gov/, accessed on 21 November 2022) including cMET in gastric cancer. Search terms such as "Gastric cancer", "cMET", "gastro-esophageal cancer" and "c-MET" were used in different combinations. All background information regarding cMET was found by selecting articles from the search term: "cMET in gastric cancer" on Pubmed. From the 322 articles found, 51 were selected to include in this review.

3. Results

3.1. General Background and Preclinical Data on cMET in Gastric Cancer

MET, also known as the N-methyl-NO-nitroso-guanidine human osteosarcoma transforming gene, was originally discovered in 1984 by Cooper et al. working on osteosarcoma [6]. The *MET-gene* is located on chromosome 7q21-31 and consists of a heterodimer with a small extracellular alpha-chain subunit (50 kDa) and a larger single-pass transmembrane beta-chain subunit (145 kDa). The extracellular domain contains three functional domains: the SEMA domain and the plexin-semaphoring-integrin (PSI) domain, together with four immunoglobulin-like regions in plexins and transcription factors (IPT 1–4). The intra-cellular part is subdivided into three domains: a juxtamembrane (JM) domain, a tyrosine kinase (TK) domain and a C-terminal multi-functional docking site (MFDS). These are all regulated by phosphorylation. Phosphorylation of the JM domain results in inhibition of the kinase domain and degradation of cMET, whereas phosphorylation of the TK domain results in upregulation of the kinase activity of cMET [7]. Phosphorylation in the MFDS domain directly mediates recruitment of downstream signaling molecules, such as SHIP2, PIR3K, GRB2, GAB1, etc. [8–11].

cMET can become activated through homodimerization upon binding of HGF or "hepatocyte growth factor" through its HGF-ligand binding sites (IPT3–4 and SEMA domains), which leads to phosphorylation and activation of downstream signaling. This process is also called the *canonical activation* of cMET. Alternatively, cMET can be activated in an HGF-independent manner, so-called *non-canonical activation*. For example, des-gamma-carboxyl prothrombin (DCP) is shown to induce cell proliferation via cMET-Janus kinase 1-STAT3 signaling by causing auto-phosphorylation in the cMET-PK domain in hepatocellular carcinoma [12]. cMET activity can also be monitored by the interaction of several *signal modifiers*. Integrin $\alpha6\beta4$ potentiates HGF-triggered activation of RAS and PI3K [13]. Class B pexin transactivates cMET in response to stimulation of semaphoring and induces the execution of cMET-dependent biological responses [14]. Transmembrane cell adhesion molecules of the CD44 family link the cMET cytoplasmatic domain to actin microfilaments via growth factor receptor bound protein 2 (GRB2), also facilitating cMET induced activation of RAS via the son of sevenless protein (SOS). The FAS receptor (FAS-R) also interacts with the cMET extracellular domain, thereby preventing FAS-R and FAS ligand recognition, FAS self-aggregation, and limiting apoptosis through the extrinsic pathway [15]. Functional interactions have also been described with epidermal growth factor receptors (EGFR) enabling activation of cMET after cellular stimulation by EGFR. cMET can even be stimulated by EGFR in the absence of HGF and the simultaneous activation of cMET and EGFR is synergistic. cMET can, conversely, also upregulate EGFR ligands. There is also evidence that ERBB2 and ERBB3 receptors can cause transactivation through cMET, which could help induce resistance to targeted therapies and strengthen downstream pathways, such as Akt and ERK/MAP kinase [16–20]. Apart from receptor- or ligand-driven activation, cMET can also be activated by hypoxia, inactivation of tumor suppressor genes, microRNAs, autocrine cMET induced activation, etc. [21].

The first relation between cMET and gastric cancer (GC) was described in 1992 in eleven gastric cell lines that showed the presence of *MET* amplification on chromosome 7 in mostly diffuse type gastric cancer and was indicative of poor prognosis [22]. In a separate study, about 18% of 154 gastric cancers included stained positive for cMET and showed more prevalent MET expression in more advanced GC [23]. Other studies in gastric cancer cell lines showed that anti-HGF inhibited further cell growth and that HGF could be produced by gastric fibroblasts, the invasiveness of which could be promoted by MET expression on gastric cancer cells [24,25]. In 2011, Toiyama et al. investigated the co-expression of HGF and MET as a predictor for peritoneal dissemination. In 100 patients with gastric cancer, the expression of HGF and MET was higher in patients with more advanced disease as well as peritoneal metastasis showing that it plays an important role in epithelial-mesenchymal transition [26]. Comprehensive molecular characterization in gastric cancer revealed by mRNA sequencing alternative splicing events of the *MET 2* exon showed skipping in 30% of cases, which also resulted in MET overexpression. In 17% (47/212) of gastric cancers, new variants of MET were skipped in exon 18 and 19. The removed exons encoded regions of the kinase domain. In up to 8% of cases, *MET* amplification could be detected [27]. Peng et al. published a systematic review with meta-analysis that included 2258 patients with gastric cancer in a total of 16 studies that provided data on *MET* expression and amplification. Although these data were subject to a publication and selection bias, overexpression and/or amplification of MET was associated with poorer survival [28]. The combination of this data together led to the further development of cMET inhibitors in gastric cancer as new potential molecular drivers [29]. An overview of the cMET-pathway and its therapeutic landscape is given in Figure 1.

The cMET tyrosine receptor kinase can be activated through protein overexpression, gene amplification, increased HGF ligand autocrine expression, enhanced paracrine ligand-mediated stimulation, inadequate cMET degradation, ligand-independent activation and, rarely, gene mutation. In gastric cancer, *HGF/cMET mutations* are exceedingly rare. Activation of MET in gastric cancer is thought to be primarily a result of receptor overexpression and/or genomic upregulation (amplification/fusion) [30].

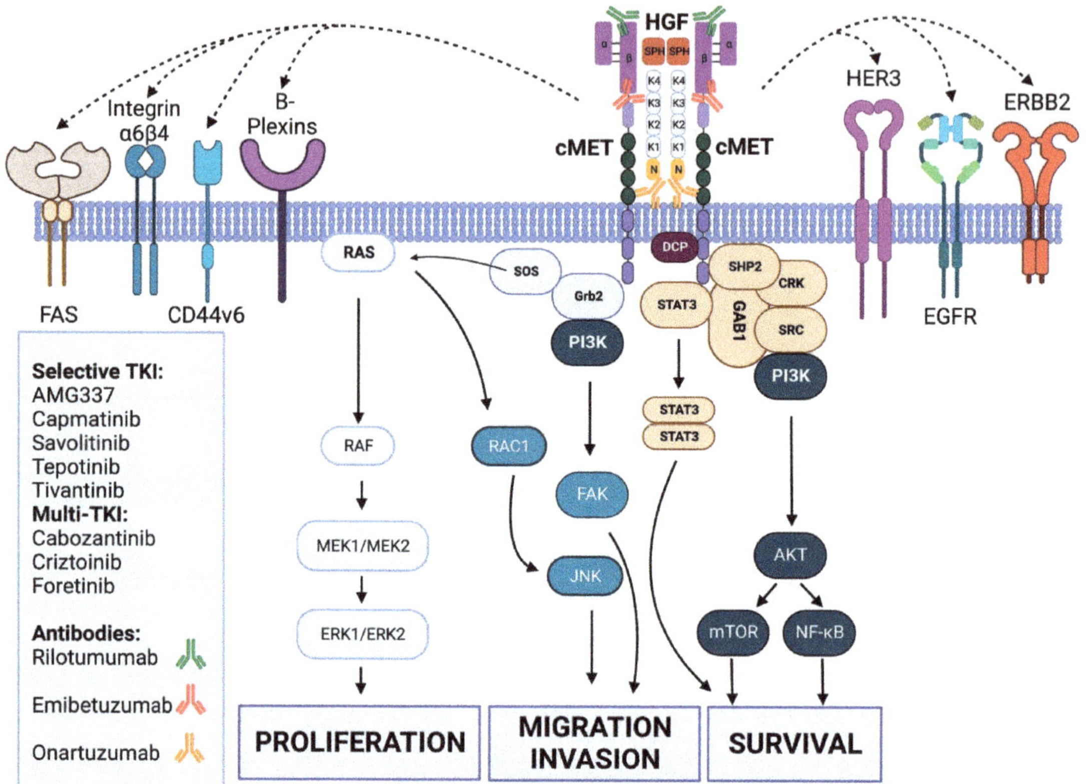

Figure 1. An overview of all the cMET inhibitor compounds with corresponding activity to the cMET pathway. HGF (hepatocyte growth factor) binds to cMET (mesenchymal-epithelial transition factor) with low- and high-affinity binding sites that induces cMET homodimerization and autophosphorylation by canonical activation. This activates GAB1, SRC, CRK, SHP2 and STAT3, where, together with PI3K, this leads to activation of AKT, the mTOR and NF-kB signaling cascades that control cell survival. Via Grb2 and SOS, the RAS/RAF and MAPK pathway is activated, inducing cell proliferation. Grb2 also activates PI3K, which, via FAK, controls migration/invasion. RAS by itself activates RAC1 that activates JNK, also responsible for cell migration and invasion. cMET can functionally be influenced by ERBB2, HER3 and EGFR receptors and launch synergistic activation. cMET is also regulated by different cell remodelers, such as FAS, integrin alpha 6 beta 4, CD44v6 and B-plexins, that facilitate the activation of the cMET tyrosine kinase domains and downstream cascade. (Created with BioRender.com, accessed on 21 November 2022).

Over the years, different methods have been developed to detect MET overexpression and/or amplification in gastric cancer, all of which have their advantages/disadvantages. *MET* protein expression can be evaluated by immunohistochemistry (IHC), with confirmation of in situ hybridization (ISH), or by next generation sequencing (NGS). The most successful clinical experience comes from non-small-cell lung cancer (NSCLC) in which *cMET* alterations are found in 3–4% of all patients. The *MET* exon 14 skipping mutation is most common and, more rarely, also MET amplification can be found. In the past, protein immunohistochemistry (IHC) and mass spectrometry have been used to detect *MET* exon 14 mutations; however, it has been shown to be unreliable as a screening tool for *MET*-exon-14-positive patients. Sanger sequencing or RT-PCR can be used to detect *MET* alterations, and, in practice, next generation sequencing (NGS) is performed [31].

3.1.1. Immunohistochemistry (IHC) and Fluorescence In Situ Hybridization (FISH)

Diagnostic determination of protein overexpression by immunohistochemistry (IHC) and fluorescence in situ hybridization (FISH) has a long history in HER2-positive breast and gastric cancer [32].

Several commercial MET antibodies exist to determine protein overexpression by IHC, although comparison of the performance has not yet been performed as of today. To detect MET protein expression by IHC, a slide with hematoxylin and eosin-stained sections is selected. The most commonly used antibody is SP44, a rabbit monoclonal anti-total MET antibody clone. The first test for scoring MET overexpression is performed by analysis of the percentage of positive tumor cells (scale 0–100%) with a staining intensity of 0 to 3+: negative (0), weak (1+), moderate (2+) or strong (3+). A cut point of >50% of tumor cells staining moderately or strongly has been associated with treatment benefit in NSCLC and is often used as a standardized cut-off of MET overexpression.

The second immunohistochemical technique for scoring MET overexpression is evaluated using the H-score. The H-score multiplies the percentage of cells with 1+, 2+ or 3+ staining by the percentage of positive cells (from 0% to 100%). The H-score ranges from 0–300 with $\geq$200 indicating overexpression, although the cut points do vary between studies [33–36]. However, MET IHC overexpression does not strongly correlate with *MET* amplification. This can be explained by the inclusion of lower levels of *MET* amplification not causing substantial protein expression, or expression that is being controlled post-transcriptionally [37].

MET gene amplification is commonly assessed with an in situ hybridization technique (ISH) by a MET/CEP7 dual color probe set. This technique was developed to distinguish polysomy from true amplification, as polysomy typically does not result in response to targeted therapy. In *MET* polysomy every additional chromosome 7 will have a corresponding MET-gene location and this will not result in increase in the MET/chromosome 8 centromere ratio (MET/CEP7). In *MET* amplification, there will be an increase in MET copies on chromosome 7, resulting in an increase in the MET/CEP7 ratio. *MET* gene amplification can be defined using FISH or by gene copy number (GCN) > 5 based on the Cappuzzo criteria [38]. Alternative criteria include a MET GCN of $\geq$6 and a MET GCN of $\geq$15, although when determining only gene copy numbers, these do not enable differentiation between polysomy and true focal amplification because other zones of the chromosome are not searched for, and the absolute number of MET-containing chromosomes cannot be determined. For that reason, a ratio between MET and CEP7 can, therefore, enable focal amplification. Depending on the literature, a MET/CEP7 ratio $\geq$ 2.0 or 2.2 is chosen. In a different study, a categorization of the degree of amplification was performed in three groups based on MET/CEP7 ratios: low $\geq$ 1.8 to $\leq$2.2; intermediate > 2.2 to <5; and high $\geq$ 5. In NSCLC, *MET* amplification (MET/CEP7 ratio > 2.2) was only detected in 1% of patients with MET overexpression (H score $\geq$ 200). In gastric cancer, however, a MET IHC H-score of 150 had a 75% sensitivity and 78% specificity to detect *MET* amplification (MET/CEP7 ratio > 2.0 and GCN > 4.0) showing the difficulty of establishing a correlation between protein overexpression and its ability to detect *MET* amplification [39].

3.1.2. Next-Generation Sequencing (NGS)/Comprehensive Genomic Profiling (CGP)

No standardized copy number nor cutoff has been determined as of today for next-generation sequencing. *MET* amplification can be detected by next-generation sequencing (NGS) but strongly depends on the quality of the DNA (e.g., DNA from old samples or DNA mixed-up with normal cells) as it increases the background noise and makes detecting gene copy numbers more difficult [13]. As of today, NGS testing has been expanded to comprehensive genomic profiling (CGP), and to different tumor types, as *MET* amplification is a rare oncogenic driver across all solid tumor types. The only pitfall is that *MET* polysomy and *MET* amplification may not be distinguished by some NGS assays and do not control for CEP7. Therefore, if possible *MET* amplification is detected, it preferably needs to be confirmed by in situ hybridization (ISH) [40].

For instance, in NSCLC, there is a wide diversity of alterations leading to *MET* exon 14 skipping; therefore, enrichment for NGS is necessary. Amplicon or hybrid-capture-based DNA NGS was initially used so as not to miss low-frequency alleles in a broad area of interest around *MET* exon 14, increasing the detection rate by up to 2.6%. Additionally, RNA testing increased the positive testing ratio to 3.9%. The reason RNA testing resulted in higher positive samples was that it only needed to detect exon 13–15 fusion mRNA creating the MET-driven phenotype of NSCLC [41]. Most of this experience was based on exon-14-skipping mutations and cannot be generalized to gastric cancer.

3.2. Clinical Exposure of cMET-Driven Therapies in Gastric Cancer
3.2.1. Monoclonal Antibodies

Rilotumumab is a monoclonal antibody that binds to the hepatocyte growth factor (HGF) and prevents it binding to the cMET receptor. In 2014, a randomized phase 2 study with rilotumumab was investigated in combination with epirubicin-cisplatin and capecitabine (ECX). Rilotumumab improved PFS in the combined arm with ECX to a median of 5.7 months compared to 4.2 months compared to placebo. The objective response rate (ORR) was 39% and the disease control rate (DCR) was 80% in the combined rilotumumab group [42]. Because of a distinct difference in effect between patients with high MET expression compared to patients with low MET expression, a subsequent phase 3 trial was set up: RILOMET-1 and RILOMET-2 (in Asia). In 2017, a first-line phase 3 study with rilotumumab was investigated (RILOMET-1) in standard of care therapy in patients overexpressing MET. Patients were screened with MET immunohistochemy; $\geq$25% of tumor cells with membrane staining of $\geq$1+ staining intensity were eligible. Different MET-detection tools were analyzed, e.g., the previously mentioned H-score, *MET* amplification via the MET/CEP ratio of 2 or more, as well as the average of MET copies < or >5. A total of 1477 patients were screened, of which 1043 (81%) were deemed MET positive. A total of 608 patients were randomly assigned to receive rilotumumab plus ECX or placebo plus ECX [43].

The study protocol was stopped early because of a higher number of deaths in the rilotumumab group. In an additional analysis, no biomarker (MET IHC, amplification, MET copies) could show a distinctive effect of the investigational drug. Preliminary analysis showed that rilotumumab was ineffective, with a median OS of 9.6 months compared to 11 months in the chemotherapy-alone group. Because of the high number of deaths, RILOMET II was closed shortly afterwards [43].

Another first-line phase 2 randomized clinical trial was performed in 2014 in the MEGA trial comparing standard of care chemotherapy folfox to folfox + panitumumab or folfox + rilotumumab. No pre-set diagnostic cMET-assay was included, although cMET staining and *MET* amplification were reported within all patients. IHC was positive in 60% of all patients, though no further details were given about the threshold for MET positivity. A total of 162 patients were included in the study. *MET* amplification was detected in 3 out of a total of 100 patients. Progression-free survival and overall survival was comparable between all arms; so, no added benefit was shown of rilotumumab or panitumumab in the first-line treatment of gastro-esophageal cancer [44].

In the METGastric trial, onartuzumab was evaluated in a randomized phase 3 study in a first-line combination with a backbone of mFOLFOX6. Onartuzumab is a recombinant, fully humanized, monovalent monoclonal antibody that binds with the extracellular domain of cMET. It prevents HGF from binding to the cMET-receptor and, therefore, restricts intracellular signaling. In an earlier phase II trial, the combination onartuzumab-erlotinib (EGFR inhibitor) resulted in an improved overall survival in patients with non-small-cell lung cancer (NSCLC) who were cMET positive, defined as 50% of tumor cells staining with an IHC intensity of 2+/3+. Therefore, the same screening with immunohistochemistry (IHC) and a 50% cell ratio was used in this phase 3 trial in gastric cancer. The proportion of patients with higher expression intensity was almost doubled in this study (38% vs. 21%) compared to the RILOMET-1 study. Unfortunately, the phase 3 part of the study was terminated early as the parallel phase 2 part could not ascertain the right patient selection. From the patients eligible for analysis, no difference in ORR could be found between standard of care and addition of onartuzumab [45].

A phase 2 non-randomized single-arm trial included 65 patients with advanced gastric cancer treated with emibetuzumab, an immunoglobulin G4 monoclonal bivalent anti-cMET antibody that blocks cMET signaling by blocking ligand-dependent cMET activation, as well as internalizing the cMET-receptor to be degraded in a ligand-independent manner. Patients were included beyond progression in second-line chemotherapy and screened for MET protein expression by immunohistochemistry (IHC) to be 2+ or 3+ positive in more than 60% of tumor cells. Of the 15 patients that were included, no patients had a partial response, apart from one patient with a −22% reduction in the target lesion, although he developed ascites and was considered progressive disease. *MET* amplification by ISH, defined by a MET/CEP7 ratio of ≥2, was found in 3 out of 4 patients with high IHC expression, although no relation could be found between any diagnostic marker and outcome [46].

3.2.2. Tyrosine Kinase Inhibitors

A phase 1b study with AMG337 confirmed an overall response ratio (ORR) of 29.7% in *MET*-amplified patients with acceptable toxicity. AMG337 is a highly selective and potent small molecule inhibitor of cMET receptor signaling. A subsequent phase 2 study in patients with advanced esophagogastric cancer and other solid tumors who had received prior therapy was set up. Screening for *MET* amplification was performed by a central laboratory defined as a MET/CEP-7 ratio > 2.0. Over 2000 patients were screened of which 132 (6%) had a *MET* amplification; finally, 55 patients with measurable disease were enrolled. An ORR of 19% was reached in the cohort of esophago-gastric cancer, though not in other tumor types, but the study was terminated early as the study product could not uphold the earlier seen ORR of up to 62% in a small cohort of 13 patients. Among all the patients included in the analysis, the mean MET/CEP7 ratio was 7.7 (2.4–12.0) in the 8 responders and 7.1 (2.0–20.4) in the 39 non-responders; therefore, biomarker analysis did not show an association between the level of *MET* gene amplification and response to treatment. Possibly, the full potential of AMG337 could not have been investigated as this was a single-arm study and early termination likely influenced the final evaluation [47].

Crizotinib is a small molecule oral inhibitor of the anaplastic lymphoma kinase (ALK), c-MET/hepatocyte growth factor receptor (HGFR), and ROS receptor tyrosine kinases. Crizotinib is approved for *ALK* and *ROS1* gene rearrangement in non-small-cell lung cancer (NSCLC). Several cases with *MET* amplification in esophagogastric cancer showed efficacy for crizotinib for which a phase II trial was designed. MET overexpression was determined by central testing and initial screening was based on an IHC of 2+ or 3+; *MET* amplification was assessed by FISH, and the number of cMET gene copies for inclusion was set at ≥6 copies. cMET was prospectively analyzed in 570 patients with esophageal/junction or gastric adenocarcinoma and *MET* amplification was found in 35 patients (=6.1%). Finally, 11 patients were enrolled, of which 9 patients started therapy with crizotinib. The median copy number of MET was 7 (range 6–11). A tumor response rate was achieved in 2/3 patients (67%) with a low tumor MET amplification, and in 3/6 (50%) with an intermediate tumor MET amplification. Unfortunately, the trial was prematurely stopped due to insufficient accrual [48].

Capmatinib is an oral-type Ib cMET inhibitor that was studied within solid tumors of which nine patients with gastric cancer were treated. Patients were eligible after third-line treatment and inclusion was allowed based on immunohistochemistry (IHC), H-score > 150, MET/CEP7 ratio of ≥2 or a gene copy number of ≥5. Only two out of nine patients reached stable disease and no clear correlation between the different diagnostic techniques of MET overexpression and response was observed [49].

Foretinib, is a small-molecule multikinase inhibitor that targets MET, RON, AXL, TIE-2 and VEGFR2 receptors. It binds in the adenosine triphosphate pocket of its targets, resulting in conformational change and kinase inhibition. Foretinib has been evaluated in a phase 2 single-arm multicentric study with two dose cohorts (intermittent vs. continuous dosing). No diagnostic cMET biomarker was mandatory to enter the study. Patients eligible for the study had progressed beyond first-line chemotherapy. Only three patients had *MET* amplification and an additional 22% increased copy number due to polysomy. Across both arms, no patients experienced a partial response (PR) and 15 patients had stable disease (SD). There was no difference in response rate between patients with *MET* amplification and/or polysomy compared to patients without a cMET-driven biomarker. The median PFS was 1.7 months, while the median OS was 7.4 months with intermittent dosing and 4.3 months with daily dosing. This suggested that cMET signaling may not be critical in patients with gastric cancer without *MET* amplification [50].

Tivantinib is a low-molecular-weight, orally available selective inhibitor of cMET. It disrupts cMET phosphorylation in a non-ATP competitive manner. It was studied in an open-label phase 2 single-arm multicenter trial. A total of 30 patients were included without the necessity of a pre-diagnostic cMET assay. Pretreatment of one or two systemic therapies was allowed. A total of 11 patients achieved disease control, although only stable disease was reached. The median PFS was 43 days, and the median survival time was 344 days. As for the earlier studies, although IHC and FISH was performed, no biomarker was correlated with clinical benefit [51].

In 2017, a phase 1 run-in trial in solid tumors, and later phase 2, specifically for first-line esophago-gastric cancer, evaluated tivantinib in combination with folfox. A total of 34 patients was included in the phase 2 part and no pre-study diagnostic biomarker was necessary. The overall response ratio was similar to historic cohorts of standard chemotherapy folfox and no additional effect could be attributed to the addition of tivantinib. Moreover, no relation could be seen between IHC or MET protein expression [52].

The VIKTORY UMBRELLA trial is a basket study of patients with gastric cancer based on clinical sequencing and focuses on eight different biomarker groups of which *MET* amplification and MET overexpression (IHC 3+) are two. In this study savolitinib, a class I cMET inhibitor and small molecule receptor tyrosine kinase inhibitor was used. The study investigated the targeted therapy in second-line and compared it to second-line paclitaxel/ramucirumab. The incidence of MET overexpression by IHC (3+) was 8.8% (42/479) in this group, while 17 (40.5%) of 42 MET-overexpressed tumors had *MET*-amplified tumors by next-generation sequencing (NGS) or FISH, and 25 (59.5%) patients had no *MET* amplification. The overall response ratio in the *MET*-amplified arm was 50% (10 of 20). Patients with high MET copy number ($\geq$10 MET gene copies by tissue NGS) had high response rates to savolitinib. One patient with upfront peritoneal metastases could be curatively resected after downstaging with savolitinib and was still alive one year after surgery. This shows that truly *MET*-amplified gastric cancer patients can gain an additional benefit from targeted therapy and even show better overall survival compared to patients receiving standard of care second-line chemotherapy [53].

Other TKIs have been studied in different tumor types but no extensive data in gastric cancer can be found. Cabozantinib has been investigated thoroughly in renal cell carcinoma (RCC) and hepatocellular carcinoma (HCC). It is a blocker of VEGFR-1 to 3, and the TAM family (TYRO3, AXL, MER), as well as cMET [54]. In the diffuse type gastric cancer cell line, one study investigated cabozantinib and showed strong targeting of cMET and VEGFR-2, suggesting its pivotal role. A study in all-comers 3d-line gastroesophageal cancer is now open, combining pembrolizumab-cabozantinib (NCT04164979), as well as the CAMILLA trial evaluating cabozantinib/durvalumab, with or without tremelimumab (NCT03539822).

Tepotinib is a selective cMET inhibitor that interrupts the cMET signal transduction pathway. In the VISION trial in NSCLC, tepotinib 500 mg in 152 patients showed a partial response in 50% of patients with the *MET* exon-14-skipping mutation resulting in an ORR of 46%. EMA recently approved the use of tepotinib in *MET* exon-14-skipping non-small-cell lung cancer [55].

One small interventional trial was performed in a cohort of solid tumors refractory to standard therapy. No MET diagnostic assay was necessary to enter the study. IHC and FISH analysis were determined throughout the study. Two patients with gastric cancer entered the study and one patient reached a PFS of 4, 6 months into therapy after four prior lines of chemotherapy. Therefore, the efficacy of tepotinib in gastric cancer is still unknown [56]. An overview of all clinical studies in gastric cancer with cMET inhibitors is given in Table 1.

Table 1. Overview of all studies that have been performed in gastric cancer with MET-inhibitors.

Type	Target	Name	Mechanism of Action	Trial (Ref.)	Phase	N° of Patients Included	Inclusion Diagnostic Marker	Effect
Monoclonal antibody	HGF	mFolfox + Rilotumumab	Blocks HGF	Rilomet I [19]	III	608	IHC ≥ 1 ≤	No benefit of mAb
		mFolfox + Rilotumumab or EGFRi		MEGA [21]	III	162	IHC 2+ or 3+	No benefit of mAb
	MET	Onartuzumab + mFolfox	Blocks MET	METGastric [22]	III	562	none	No benefit of mAb
	HGF and MET	Emibetuzumab	Blocks HGF binding and internalization of MET	NA [23]	II	15	IHC 2+ or 3+	No patients with PR
Tyrosine kinase inhibitor		Crizotinib	Multi TKI	AcSé Crizotinib Program [25]	II	9	IHC 2+ or 3+	ORR 33%
		Foretinib		NA [27]	II	74	None	No PR, 15 patiënts SD
		Capmatinib	Selective MET TKI	NA [26]	II	9	IHC 2+ or 3+ or Hscore > 150 MET/CEN7 ≥ 2 or MET GCN ≥ 5	2 patients with SD
		Tivantinib		NA [28]	II	30	None	No ORR, 36 DCR
		Savolitinib		Viktory Umbrella Trial [30]	II	20	Amplification ≥ 10 on NGS	ORR 50%
		AMG337		NA [24]	II	55	MET/CEN7 ≥ 2	ORR 19%

ORR: objective response ratio; SD: stable disease; PR: partial response; mAb: monoclonal antibody; TKI: tyrosine kinase inhibitor; DCR: disease control rate; IHC: immunohistochemistry; GCN: gene copy number; NGS: next-generation sequencing.

4. Discussion

Among all the studies performed within cMET in gastric cancer, not many successes have been achieved. The RILOMET-1, MEGA and METGastric study all included patients in first-line gastroesophageal cancer and did not show a significant benefit in addition to chemotherapy.

One of the possible reasons why rilotumumab and onartuzumab did not show an increased benefit in first line is explained by the mechanism of action. It is a ligand-dependent antibody and in *MET* amplification the cMET pathway is autonomously active regardless of the effect on the binding ligand.

Secondly, screening by positive immunohistochemistry for cMET is not an appropriate biomarker. Overexpression based on immunohistochemistry did not show the good correlation with cMET-activity as an oncogenic driver typically seen in *MET* amplification. Moreover, in gastric cancer *MET* amplifications are still limited to around 5% of all patients, and, therefore, using cMET-targeted therapy for all first-line patients with gastric cancer results in overtreatment in patients without a *MET* amplification. This also means that the efficacy of these cMET-specific treatments could potentially be underrated [57–59].

In the study with AMG337, a potential benefit in patients with gastric cancer with an ORR of 19% could be shown. Patients were included based on a FISH MET/CEP7 ratio $\geq$ 2.0, but no distinct difference was found in the MET/CEP7 ratio in the responder and non-responder groups. In the study with crizotinib and capmatinib, screening was also based on IHC, and prospective analysis of *MET* gene amplification by GCN $\geq$ 5 or the MET/CEP7 ratio $\geq$ 2 could not distinguish responders from non-responders. However, this study had a low number of patients, and as mentioned earlier, the gene copy number is possibly not a good marker to detect *MET* amplification.

Foretinib was studied without cMET-specific inclusion criteria based on a diagnostic assay but was monitored in the study. Three patients had a *MET* amplification, but, across all patients, no partial responses were seen. A total of 15 patients showed stable disease. Again, a low number of patients was included in the study to determine a clinically significant benefit. Tivantinb was combined with a folfox regimen in a phase 1, and later in a phase 2, open-label trial. Although, in total, 49 patients were treated, no difference was seen compared to a historical cohort of folfox alone. No specific biomarker analysis could distinguish patients in terms of outcome.

The VIKTORY umbrella trial showed the most successful response for savolitinib on cMET-inhibitors for patients with gastric cancer and compared this to paclitaxel/ramucirumab second-line therapy. *MET* amplification was determined by NGS or FISH and an ORR of 50% was reached, showing the possible potential for cMET inhibitors in gastric cancer in a well-defined population.

5. Conclusions

All the forementioned studies had a different inclusion strategy, e.g., *MET* overexpression, rarely, a FISH MET/CEP7 ratio $\geq$ 2, or no prescreening assay was needed. Studies without an inclusion strategy could not find an overt biomarker predicting cMET sensitivity. Diagnostic assays of *MET* overexpression/amplification are not as homogeneous in gastric cancer compared to non-small-cell lung cancer (NSCLC), where the *MET* exon-14-skipping mutation has been extensively investigated and a clear strategy has been set out.

No beneficial effect has been seen in addition to chemotherapy for allcomers in first-line gastric cancer. In the cohorts with TKIs, all patient groups remain limited, and the inclusion strategies all remain different, apart from the clear analysis of the VIKTORY umbrella trial, which detected most patients based on NGS analysis with a good ORR of 50%.

The most important problem remains the need for clinically meaningful cut-off points, including the level of *MET* amplification as well as *MET* overexpression, to determine treatment-related decision-making.

Determining *MET* amplification by NGS or whole-exome sequencing could be the most accurate technique to predict *MET*-inhibitor sensitivity; however, confirmation by ISH will be important to distinguish true *MET* amplification from polysomy, depending on the type of NGS assay. More data on comprehensive genomic testing and *MET* inhibitors will be needed in other (basket) studies and further investigation will be needed to identify optimal predictive biomarkers under targeted therapy.

Author Contributions: Conceptualization, F.V.H. and E.V.C.; methodology, F.V.H.; investigation, F.V.H.; writing—original draft preparation, F.V.H.; writing—review and editing, F.V.H. and E.V.C.; visualization, F.V.H.; supervision, E.V.C.; project administration, F.V.H. All authors have read and agreed to the published version of the manuscript.

Funding: This research received no external funding.

Informed Consent Statement: Informed consent was obtained from all subjects involved in the cited studies.

Data Availability Statement: The data can be shared up on request.

Acknowledgments: Thanks are due to biorender.com for providing a platform for creating this illustration (Figure 1).

Conflicts of Interest: The authors declare no conflict of interest.

References

1. Sung, H.; Ferlay, J.; Siegel, R.L.; Laversanne, M.; Soerjomataram, I.; Jemal, A.; Bray, F. Global Cancer Statistics 2020: GLOBOCAN Estimates of Incidence and Mortality Worldwide for 36 Cancers in 185 Countries. *CA Cancer J. Clin.* **2021**, *71*, 209–249. [CrossRef]

2. Bang, Y.J.; Van Cutsem, E.; Feyereislova, A.; Chung, H.C.; Shen, L.; Sawaki, A.; Lordick, F.; Ohtsu, A.; Omuro, Y.; Satoh, T.; et al. Trastuzumab in combination with chemotherapy versus chemotherapy alone for treatment of HER2-positive advanced gastric or gastro-oesophageal junction cancer (ToGA): A phase 3, open-label, randomised controlled trial. *Lancet* **2010**, *376*, 687–697. [CrossRef]

3. Muro, K.; Van Cutsem, E.; Narita, Y.; Pentheroudakis, G.; Baba, E.; Li, J.; Ryu, M.H.; Zamaniah, W.I.W.; Yong, W.P.; Yeh, K.H.; et al. Pan-Asian adapted ESMO Clinical Practice Guidelines for the management of patients with metastatic gastric cancer: A JSMO-ESMO initiative endorsed by CSCO, KSMO, MOS, SSO and TOS. *Ann. Oncol.* **2019**, *30*, 19–33. [CrossRef]

4. Janjigian, Y.Y.; Shitara, K.; Moehler, M.; Garrido, M.; Salman, P.; Shen, L.; Wyrwicz, L.; Yamaguchi, K.; Skoczylas, T.; Campos Bragagnoli, A.; et al. First-line nivolumab plus chemotherapy versus chemotherapy alone for advanced gastric, gastro-oesophageal junction, and oesophageal adenocarcinoma (CheckMate 649): A randomised, open-label, phase 3 trial. *Lancet* **2021**, *398*, 27–40. [CrossRef] [PubMed]

5. Janjigian, Y.Y.; Kawazoe, A.; Yañez, P.; Li, N.; Lonardi, S.; Kolesnik, O.; Barajas, O.; Bai, Y.; Shen, L.; Tang, Y.; et al. The KEYNOTE-811 trial of dual PD-1 and HER2 blockade in HER2-positive gastric cancer. *Nature* **2021**, *600*, 727–730. [CrossRef] [PubMed]

6. Zeng, Y.; Jin, R.U. Molecular pathogenesis, targeted therapies, and future perspectives for gastric cancer. *Semin. Cancer Biol.* **2021**, *86 Pt 3*, 566–582. [CrossRef] [PubMed]

7. Fu, J.; Su, X.; Li, Z.; Deng, L.; Liu, X.; Feng, X.; Peng, J. HGF/c-MET pathway in cancer: From molecular characterization to clinical evidence. *Oncogene* **2021**, *40*, 4625–4651. [CrossRef]

8. Gandino, L.; Longati, P.; Medico, E.; Prat, M.; Comoglio, P.M. Phosphorylation of ser 985 negatively regulates the hepatocyte growth factor receptor kinase. *J. Biol. Chem.* **1994**, *269*, 1815–1820. [CrossRef]

9. Kadoyama, K.; Funakoshi, H.; Ohya-Shimada, W.; Nakamura, T.; Matsumoto, K.; Matsuyama, S.; Nakamura, T. Disease-dependent reciprocal phosphorylation of serine and tyrosine residues of c-Met/HGF receptor contributes disease retardation of a transgenic mouse model of ALS. *Neurosci. Res.* **2009**, *65*, 194–200. [CrossRef]

10. Nakayama, M.; Sakai, K.; Yamashita, A.; Nakamura, T.; Suzuki, Y.; Matsumoto, K. Met/HGF receptor activation is regulated by juxtamembrane Ser985 phosphorylation in hepatocytes. *Cytokine* **2013**, *62*, 446–452. [CrossRef]

11. Peschard, P.; Fournier, T.M.; Lamorte, L.; Naujokas, M.A.; Band, H.; Langdon, W.Y.; Park, M. Mutation of the c-Cbl TKB domain binding site on the Met receptor tyrosine kinase converts it into a transforming protein. *Mol. Cell.* **2001**, *8*, 995–1004. [CrossRef] [PubMed]

12. Suzuki, M.; Shiraha, H.; Fujikawa, T.; Takaoka, N.; Ueda, N.; Nakanishi, Y.; Koike, K.; Takaki, A.; Shiratori, Y. Des-gamma-carboxy prothrombin is a potential autologous growth factor for hepatocellular carcinoma. *J. Biol. Chem.* **2005**, *280*, 6409–6415. [CrossRef] [PubMed]

13. Trusolino, L.; Bertotti, A.; Comoglio, P.M. A signaling adapter function for alpha6beta4 integrin in the control of HGF-dependent invasive growth. *Cell* **2001**, *107*, 643–654. [CrossRef] [PubMed]

14. Conrotto, P.; Corso, S.; Gamberini, S.; Comoglio, P.M.; Giordano, S. Interplay between scatter factor receptors and B plexins controls invasive growth. *Oncogene* **2004**, *23*, 5131–5137. [CrossRef] [PubMed]

15. Wang, X.; DeFrances, M.C.; Dai, Y.; Pediaditakis, P.; Johnson, C.; Bell, A.; Michalopoulos, G.K.; Zarnegar, R. A mechanism of cell survival: Sequestration of Fas by the HGF receptor. *Met. Mol. Cell.* **2002**, *9*, 411–421. [CrossRef]

16. Jo, M.; Stolz, D.B.; Esplen, J.E.; Dorko, K.; Michalopoulos, G.K.; Strom, S.C. Cross-talk between epidermal growth factor receptor and c-Met signal pathways in transformed cells. *J. Biol. Chem.* **2000**, *275*, 8806–8811. [CrossRef]

17. Puri, N.; Salgia, R. Synergism of EGFR and c-Met pathways, cross-talk and inhibition, in non-small cell lung cancer. *J. Carcinog.* **2008**, *7*, 9. [CrossRef]

18. Bachleitner-Hofmann, T.; Sun, M.Y.; Chen, C.T.; Tang, L.; Song, L.; Zeng, Z.; Shah, M.; Christensen, J.G.; Rosen, N.; Solit, D.B.; et al. HER kinase activation confers resistance to MET tyrosine kinase inhibition in MET oncogene-addicted gastric cancer cells. *Mol. Cancer Ther.* **2008**, *7*, 3499–3508. [CrossRef]

19. Khoury, H.; Naujokas, M.A.; Zuo, D.; Sangwan, V.; Frigault, M.M.; Petkiewicz, S.; Dankort, D.L.; Muller, W.J.; Park, M. HGF converts ErbB2/Neu epithelial morphogenesis to cell invasion. *Mol. Biol. Cell.* **2005**, *16*, 550–561. [CrossRef]

20. Benvenuti, S.; Lazzari, L.; Arnesano, A.; Li Chiavi, G.; Gentile, A.; Comoglio, P.M. Ron kinase transphosphorylation sustains MET oncogene addiction. *Cancer Res.* **2011**, *71*, 1945–1955. [CrossRef]

21. Wang, H.; Rao, B.; Lou, J.; Li, J.; Liu, Z.; Li, A.; Cui, G.; Ren, Z.; Yu, Z. The function of the HGFc-Met axis in hepatocellular carcinoma. *Front Cell Dev. Biol.* **2020**, *8*, 55. [CrossRef] [PubMed]

22. Cooper, C.S.; Park, M.; Blair, D.G.; Tainsky, M.A.; Huebner, K.; Croce, C.M.; Vande Woude, G.F. Molecular cloning of a new transforming gene from a chemically transformed human cell line. *Nature* **1984**, *311*, 29–33. [CrossRef] [PubMed]

23. Kuniyasu, H.; Yasui, W.; Kitadai, Y.; Yokozaki, H.; Ito, H.; Tahara, E. Frequent amplification of the c-met gene in scirrhous type stomach cancer. *Biochem. Biophys. Res. Commun.* **1992**, *189*, 227–232. [CrossRef]

24. Kaji, M.; Yonemura, Y.; Harada, S.; Liu, X.; Terada, I.; Yamamoto, H. Participation of c-met in the progression of human gastric cancers: Anti-c-met oligonucleotides inhibit proliferation or invasiveness of gastric cancer cells. *Cancer Gene Ther.* **1996**, *3*, 393–404.

25. Inoue, T.; Chung, Y.S.; Yashiro, M.; Nishimura, S.; Hasuma, T.; Otani, S.; Sowa, M. Transforming growth factor-beta and hepatocyte growth factor produced by gastric fibroblasts stimulate the invasiveness of scirrhous gastric cancer cells. *Jpn. J. Cancer Res.* **1997**, *88*, 152–159. [CrossRef] [PubMed]

26. Toiyama, Y.; Yasuda, H.; Saigusa, S.; Matushita, K.; Fujikawa, H.; Tanaka, K.; Tanaka, K.; Mohri, Y.; Inoue, Y.; Goel, A.; et al. Co-expression of hepatocyte growth factor and c-Met predicts peritoneal dissemination established by autocrine hepatocyte growth factor/c-Met signaling in gastric cancer. *Int. J. Cancer.* **2012**, *130*, 2912–2921. [CrossRef]

27. Cancer Genome Atlas Research Network. Comprehensive molecular characterization of gastric adenocarcinoma. *Nature* **2014**, *513*, 202–209. [CrossRef]

28. Peng, Z.; Zhu, Y.; Wang, Q.; Gao, J.; Li, Y.; Li, Y.; Ge, S.; Shen, L. Prognostic significance of MET amplification and expression in gastric cancer: A systematic review with meta-analysis. *PLoS ONE* **2014**, *9*, e84502. [CrossRef]

29. Bradley, C.A.; Salto-Tellez, M.; Laurent-Puig, P.; Bardelli, A.; Rolfo, C.; Tabernero, J.; Khawaja, H.A.; Lawler, M.; Johnston, P.G.; Van Schaeybroeck, S. MErCuRIC consortium. Targeting c-MET in gastrointestinal tumours: Rationale, opportunities and challenges. *Nat. Rev. Clin. Oncol.* **2017**, *14*, 562–576. [CrossRef]

30. Ma, P.C.; Maulik, G.; Christensen, J.; Salgia, R. c-Met: Structure, functions and potential for therapeutic inhibition. *Cancer Metastasis Rev.* **2003**, *22*, 309–325. [CrossRef]

31. Socinski, M.A.; Pennell, N.A.; Davies, K.D. Exon 14 Skipping Mutations in Non-Small-Cell Lung Cancer: An Overview of Biology, Clinical Outcomes, and Testing Considerations. *JCO Precis. Oncol.* **2021**, *5*.

32. Mo, H.N.; Liu, P. Targeting MET in cancer therapy. *Chronic Dis. Transl. Med.* **2017**, *3*, 148–153. [CrossRef] [PubMed]

33. Benvenuti, S.; Comoglio, P.M. The MET receptor tyrosine kinase in invasion and metastasis. *J. Cell Physiol.* **2007**, *213*, 316–325. [CrossRef] [PubMed]

34. Hack, S.P.; Bruey, J.M.; Koeppen, H. HGF/MET-directed therapeutics in gastroesophageal cancer: A review of clinical and biomarker development. *Oncotarget* **2014**, *5*, 2866–2880. [CrossRef] [PubMed]

35. Guo, R.; Luo, J.; Chang, J.; Rekhtman, N.; Arcila, M.; Drilon, A. MET-dependent solid tumours—Molecular diagnosis and targeted therapy. *Nat. Rev. Clin. Oncol.* **2020**, *17*, 569–587. [CrossRef]

36. Janjigian, Y.Y.; Tang, L.H.; Coit, D.G.; Kelsen, D.P.; Francone, T.D.; Weiser, M.R.; Jhanwar, S.C.; Shah, M.A. MET expression and amplification in patients with localized gastric cancer. *Cancer Epidemiol. Biomarkers Prev.* **2011**, *20*, 1021–1027. [CrossRef]

37. Spigel, D.R.; Ervin, T.J.; Ramlau, R.A.; Daniel, D.B.; Goldschmidt, J.H.; Blumenschein, G.R.; Krzakowski, M.J.; Robinet, G.; Godbert, B.; Barlesi, F.; et al. Randomized phase II trial of Onartuzumab in combination with erlotinib in patients with advanced non-small-cell lung cancer. *J. Clin. Oncol.* **2013**, *31*, 4105–4114. [CrossRef] [PubMed]

38. Cappuzzo, F.; Marchetti, A.; Skokan, M.; Rossi, E.; Gajapathy, S.; Felicioni, L.; Del Grammastro, M.; Sciarrotta, M.G.; Buttitta, F.; Incarbone, M.; et al. Increased MET gene copy number negatively affects survival of surgically resected non-small-cell lung cancer patients. *J. Clin. Oncol.* **2009**, *27*, 1667–1674. [CrossRef]

39. Mignard, X.; Ruppert, A.M.; Antoine, M.; Vasseur, J.; Girard, N.; Mazières, J.; Moro-Sibilot, D.; Fallet, V.; Rabbe, N.; Thivolet-Bejui, F.; et al. c-MET Overexpression as a Poor Predictor of MET Amplifications or Exon 14 Mutations in Lung Sarcomatoid Carcinomas. *J. Thorac. Oncol.* **2018**, *13*, 1962–1967. [CrossRef]

40. Coleman, N.; Hong, L.; Zhang, J.; Heymach, J.; Hong, D.; Le, X. Beyond epidermal growth factor receptor: MET amplification as a general resistance driver to targeted therapy in oncogene-driven non-small-cell lung cancer. *ESMO Open* **2021**, *6*, 100319. [CrossRef]

41. Bubendorf, L.; Dafni, U.; Schöbel, M.; Finn, S.P.; Tischler, V.; Sejda, A.; Marchetti, A.; Thunnissen, E.; Verbeken, E.K.; Warth, A.; et al. Prevalence and clinical association of MET gene overexpression and amplification in patients with NSCLC: Results from the European Thoracic Oncology Platform (ETOP) Lungscape project. *Lung Cancer* **2017**, *111*, 143–149. [CrossRef] [PubMed]

42. Catenacci, D.V.T.; Tebbutt, N.C.; Davidenko, I.; Murad, A.M.; Al-Batran, S.E.; Ilson, D.H.; Tjulandin, S.; Gotovkin, E.; Karaszewska, B.; Bondarenko, I.; et al. Rilotumumab plus epirubicin, cisplatin, and capecitabine as first-line therapy in advanced MET-positive gastric or gastro-oesophageal junction cancer (RILOMET-1): A randomised, double-blind, placebo-controlled, phase 3 trial. *Lancet Oncol.* **2017**, *18*, 1467–1482. [CrossRef] [PubMed]

43. Iveson, T.; Donehower, R.C.; Davidenko, I.; Tjulandin, S.; Deptala, A.; Harrison, M.; Nirni, S.; Lakshmaiah, K.; Thomas, A.; Jiang, Y.; et al. Rilotumumab in combination with epirubicin, cisplatin, and capecitabine as first-line treatment for gastric or oesophagogastric junction adenocarcinoma: An open-label, dose de-escalation phase 1b study and a double-blind, randomised phase 2 study. *Lancet Oncol.* **2014**, *15*, 1007–1018. [CrossRef]

44. Malka, D.; François, E.; Penault-Llorca, F.; Castan, F.; Bouché, O.; Bennouna, J.; Ghiringhelli, F.; de la Fouchardière, C.; Borg, C.; Samalin, E.; et al. FOLFOX alone or combined with rilotumumab or panitumumab as first-line treatment for patients with advanced gastroesophageal adenocarcinoma (PRODIGE 17-ACCORD 20-MEGA): A randomised, open-label, three-arm phase II trial. *Eur. J Cancer.* **2019**, *115*, 97–106. [CrossRef]

45. Shah, M.A.; Bang, Y.J.; Lordick, F.; Alsina, M.; Chen, M.; Hack, S.P.; Bruey, J.M.; Smith, D.; McCaffery, I.; Shames, D.S.; et al. Effect of Fluorouracil, Leucovorin, and Oxaliplatin with or Without Onartuzumab in HER2-Negative, MET-Positive Gastroesophageal Adenocarcinoma: The METGastric Randomized Clinical Trial. *JAMA Oncol.* **2017**, *3*, 620–627. [CrossRef] [PubMed]

46. Sakai, D.; Chung, H.C.; Oh, D.Y.; Park, S.H.; Kadowaki, S.; Kim, Y.H.; Tsuji, A.; Komatsu, Y.; Kang, Y.K.; Uenaka, K.; et al. A non-randomized, open-label, single-arm, Phase 2 study of emibetuzumab in Asian patients with MET diagnostic positive, advanced gastric cancer. *Cancer Chemother. Pharmacol.* **2017**, *80*, 1197–1207. [CrossRef] [PubMed]

47. Van Cutsem, E.; Karaszewska, B.; Kang, Y.K.; Chung, H.C.; Shankaran, V.; Siena, S.; Go, N.F.; Yang, H.; Schupp, M.; Cunningham, D. A Multicenter Phase II Study of AMG 337 in Patients with. *Clin. Cancer Res.* **2019**, *25*, 2414–2423. [CrossRef]

48. Aparicio, T.; Cozic, N.; de la Fouchardière, C.; Meriaux, E.; Plaza, J.; Mineur, L.; Guimbaud, R.; Samalin, E.; Mary, F.; Lecomte, T.; et al. The Activity of Crizotinib in Chemo-Refractory MET-Amplified Esophageal and Gastric Adenocarcinomas: Results from the AcSé-Crizotinib Program. *Target. Oncol.* **2021**, *16*, 381–388. [CrossRef]

49. Bang, Y.J.; Su, W.C.; Schuler, M.; Nam, D.H.; Lim, W.T.; Bauer, T.M.; Azaro, A.; Poon, R.T.P.; Hong, D.; Lin, C.C.; et al. Phase 1 study of capmatinib in MET-positive solid tumor patients: Dose escalation and expansion of selected cohorts. *Cancer Sci.* **2020**, *111*, 536–547. [CrossRef]

50. Shah, M.A.; Wainberg, Z.A.; Catenacci, D.V.; Hochster, H.S.; Ford, J.; Kunz, P.; Lee, F.C.; Kallender, H.; Cecchi, F.; Rabe, D.C.; et al. Phase II study evaluating 2 dosing schedules of oral foretinib (GSK1363089), cMET/VEGFR2 inhibitor, in patients with metastatic gastric cancer. *PLoS ONE* **2013**, *8*, e54014. [CrossRef]

51. Kang, Y.K.; Muro, K.; Ryu, M.H.; Yasui, H.; Nishina, T.; Ryoo, B.Y.; Kamiya, Y.; Akinaga, S.; Boku, N. A phase II trial of a selective c-Met inhibitor tivantinib (ARQ 197) monotherapy as a second- or third-line therapy in the patients with metastatic gastric cancer. *Investig. New Drugs.* **2014**, *32*, 355–361. [CrossRef] [PubMed]

52. Pant, S.; Patel, M.; Kurkjian, C.; Hemphill, B.; Flores, M.; Thompson, D.; Bendell, J. A Phase II Study of the c-Met Inhibitor Tivantinib in Combination with FOLFOX for the Treatment of Patients with Previously Untreated Metastatic Adenocarcinoma of the Distal Esophagus, Gastroesophageal Junction, or Stomach. *Cancer Investig.* **2017**, *35*, 463–472. [CrossRef]

53. Lee, J.; Kim, S.T.; Kim, K.; Lee, H.; Kozarewa, I.; Mortimer, P.G.S.; Odegaard, J.I.; Harrington, E.A.; Lee, J.; Lee, T.; et al. Tumor Genomic Profiling Guides Patients with Metastatic Gastric Cancer to Targeted Treatment: The VIKTORY Umbrella Trial. *Cancer Discov.* **2019**, *9*, 1388–1405. [CrossRef] [PubMed]

54. El-Khoueiry, A.B.; Hanna, D.L.; Llovet, J.; Kelley, R.K. Cabozantinib: An evolving therapy for hepatocellular carcinoma. *Cancer Treat. Rev.* **2021**, *98*, 102221. [CrossRef] [PubMed]

55. Paik, P.K.; Felip, E.; Veillon, R.; Sakai, H.; Cortot, A.B.; Garassino, M.C.; Mazieres, J.; Viteri, S.; Senellart, H.; Van Meerbeeck, J.; et al. Tepotinib in Non-Small-Cell Lung Cancer with. *N. Engl. J. Med.* **2020**, *383*, 931–943. [CrossRef] [PubMed]

56. Shitara, K.; Yamazaki, K.; Tsushima, T.; Naito, T.; Matsubara, N.; Watanabe, M.; Sarholz, B.; Johne, A.; Doi, T. Phase I trial of the MET inhibitor tepotinib in Japanese patients with solid tumors. *Jpn. J. Clin. Oncol.* **2020**, *50*, 859–866. [CrossRef]

57. Nakajima, M.; Sawada, H.; Yamada, Y.; Watanabe, A.; Tatsumi, M.; Yamashita, J.; Matsuda, M.; Sakaguchi, T.; Hirao, T.; Nakano, H. The prognostic significance of amplification and overexpression of c-met and c-erb B-2 in human gastric carcinomas. *Cancer* **1999**, *85*, 1894–1902. [CrossRef]

58. Lee, H.E.; Kim, M.A.; Lee, H.S.; Jung, E.J.; Yang, H.K.; Lee, B.L.; Bang, Y.J.; Kim, W.H. MET in gastric carcinomas: Comparison between protein expression and gene copy number and impact on clinical outcome. *Br. J. Cancer* **2012**, *107*, 325–333. [CrossRef]

59. Eder, J.P.; Vande Woude, G.F.; Boerner, S.A.; LoRusso, P.M. Novel therapeutic inhibitors of the c-Met signaling pathway in cancer. *Clin. Cancer Res.* **2009**, *15*, 2207–2214. [CrossRef]

Review

MET Amplification as a Resistance Driver to TKI Therapies in Lung Cancer: Clinical Challenges and Opportunities

Kang Qin [1], Lingzhi Hong [1,2], Jianjun Zhang [1] and Xiuning Le [1,*]

1   Department of Thoracic/Head and Neck Medical Oncology, The University of Texas MD Anderson Cancer Center, Houston, TX 77030, USA

2   Department of Imaging Physics, The University of Texas MD Anderson Cancer Center, Houston, TX 77030, USA

*   Correspondence: xle1@mdanderson.org

Simple Summary: Secondary amplifications/copy number changes of the gene *MET* (*MET* protocol oncogene) play a significant role in the development of resistance to targeted drugs in advanced non-small cell lung cancer (NSCLC). In this review, we aim to clarify the biological mechanisms of *MET* amplification-mediated resistance to tyrosine kinase inhibitors, discuss the challenges of commonly used assays for the identification of *MET* amplifications. We also summarize the latest findings on combined strategies to overcome acquired *MET* amplification-mediated resistance, especially the combinatory regimens with EGFR-TKIs and MET-TKIs.

Abstract: Targeted therapy has emerged as an important pillar for the standard of care in oncogene-driven non-small cell lung cancer (NSCLC), which significantly improved outcomes of patients whose tumors harbor oncogenic driver mutations. However, tumors eventually develop resistance to targeted drugs, and mechanisms of resistance can be diverse. *MET* amplification has been proven to be a driver of resistance to tyrosine kinase inhibitor (TKI)-treated advanced NSCLC with its activation of *EGFR*, *ALK*, *RET*, and *ROS-1* alterations. The combined therapy of MET-TKIs and EGFR-TKIs has shown outstanding clinical efficacy in *EGFR*-mutated NSCLC with secondary *MET* amplification-mediated resistance in a series of clinical trials. In this review, we aimed to clarify the underlying mechanisms of *MET* amplification-mediated resistance to tyrosine kinase inhibitors, discuss the ways and challenges in the detection and diagnosis of *MET* amplifications in patients with metastatic NSCLC, and summarize the recently published clinical data as well as ongoing trials of new combination strategies to overcome *MET* amplification-mediated TKI resistance.

Keywords: NSCLC; MET amplification; TKIs; resistance mechanism; detection; diagnosis; combined-therapy

Citation: Qin, K.; Hong, L.; Zhang, J.; Le, X. MET Amplification as a Resistance Driver to TKI Therapies in Lung Cancer: Clinical Challenges and Opportunities. *Cancers* **2023**, *15*, 612. https://doi.org/10.3390/cancers15030612

Academic Editors: Jan Trøst Jørgensen and Jens Mollerup

Received: 7 December 2022
Revised: 13 January 2023
Accepted: 14 January 2023
Published: 18 January 2023

1. Introduction

Non-small cell lung cancer (NSCLC) constitutes around 85% of lung cancer, which has been the leading cause of cancer-related deaths worldwide [1]. In the past decade, ground-breaking progress in personalized therapy and targeted agents has led to unprecedented clinical improvements in the subgroup of patients with advanced or metastatic NSCLC carrying active oncogenic driver alterations. Molecular profiling to identify actionable oncogenic drivers is now recommended as part of the initial clinical work-up for patients with metastatic NSCLC. Currently, In NSCLC, especially of the non-squamous histology, predictive biomarkers recommended for testing by the NCCN profiling panel are *EGFR*, *KRAS*, and *BRAF* mutations; *ALK*, *RET*, and *ROS1* gene rearrangements; *MET* alterations including *MET* exon 14 skipping mutations and *MET* amplifications; *ERBB2 (HER2)* mutations; and *NTRK* 1/2/3 gene fusions [2]. Although the well-established targeted drugs show outstanding efficacy in initial disease control, drug resistance always inevitably develops. Clarifying and overcoming the resistance to targeted drugs with novel strategies is one of the major challenges in the era of personalized therapy.

The MET proto-oncogene (hereafter referred to as *MET*) encodes the receptor tyrosine kinase or hepatocyte growth factor (HGF) receptor, which, along with its ligand HGF (HGF/MET axis), functions as an essential regulator of cell survival, proliferation, motility and migration. Dysregulation of MET signaling has been found in a variety of cancers through different mechanisms, such as activating point mutations of the *MET* gene, overexpression of the ligand HGF, *MET* gene copy number gain (*MET*-CNG)/amplification, and *MET* gene fusions [3,4].

MET amplification occurs in 1–6% of NSCLC cases and was considered as a negative prognostic factor [5–7]. In recent years, increasing evidence has implicated that *MET* amplification was a key driver of acquired resistance to these aforementioned targeted therapies such as EGFR-TKIs and ALK-TKIs. Although an increasing number of drugs acting on MET signaling is currently achievable, for example, the MET/ALK/ROS tyrosine-kinase inhibitor crizotinib or selective MET-TKIs (capmatinib, savolitinib, tepotinib) [7], more strategies are needed to overcome *MET* amplification-mediated acquired resistance to TKIs. Therefore, it is important and necessary to clarify the underlying molecular mechanisms of *MET* amplification-mediated resistance, and to find out appropriate ways to identify *MET* copy number gains and amplifications, so that researchers can develop effective therapeutic strategies to overcome this resistance and prolong the life of NSCLC patients. Our review focuses on the molecular mechanisms of acquired resistance to targeted therapies mediated by *MET* amplifications, and the ways and challenges in detection and diagnosis of *MET* amplifications in NSCLC. We also summarize the recently published clinical data as well as the ongoing trials focusing on new combination strategies to overcome *MET* amplification-mediated TKI resistance.

2. MET Biology, Structure, Function, and Pathways

The receptor tyrosine kinases (RTKs) are encoded by a family of proto-oncogenes with more than 75 members that regulate cellular growth, oncogenesis, tumor metastasis, and progression through downstream signaling pathways such as the RAS-RAF-MEK-ERK and PI3K-AKT-mTOR pathways [8]. *MET*, together with *EGFR*, *ALK*, *BRAF*, etc. are all members of this family, which were found to be frequently mutated in advanced NSCLC [9]. Tyrosine kinase inhibitors bind and act on these RTKs, and lead to the inhibition of downstream signaling pathways which would otherwise induce tumor cell growth and proliferation [10]. In patients with advanced NSCLC undergoing TKI treatments, acquired resistance always develops and limits the long-term application of these targeted agents. Bypassing the activation of *MET*-related pathways has proven to be one of the underlying reasons [11].

The human *MET* gene is a 120 kb proto-oncogene that is located on chromosome 7 band 7q21–q31. Hepatocyte growth factor receptor (HGFR) or MET protein is the product of the *MET* proto-oncogene, and its ligand HGF is a disulfide-linked a-b heterodimeric molecule, also known as plasminogen-related growth factor-1(PRGF-1) [12]. MET protein is normally expressed in various epithelial and mesenchymal cell types. Upon HGF binding, the HGF/MET signaling pathway is activated, then MET undergoes homodimerization and autophosphorylation of a series of tyrosine residues within the intracellular region, including Y1230, Y1234, Y1235, Y1313, Y1349, and Y1356, etc., which lead to the activation of multiple intracellular signaling pathways including the RAS-RAF-MAPK, JAK-STAT and PI3K-AKT/mTOR, and phospholipase C pathways [13] (Figure 1). The signalings have been shown to trigger a variety of cellular responses, including cell proliferation, tissue regeneration, angiogenesis, and cellular invasion, etc. [14]. Oncogenic *MET* alterations, including the overexpression of MET protein or *MET* gene alterations, such as mutations, amplifications, or fusions, cause dysregulation of the HGF/MET signaling pathway, and lead to a wide range of human cancers, including papillary renal cell carcinoma, gastric cancer, and non-small cell lung cancer, etc. [3,15].

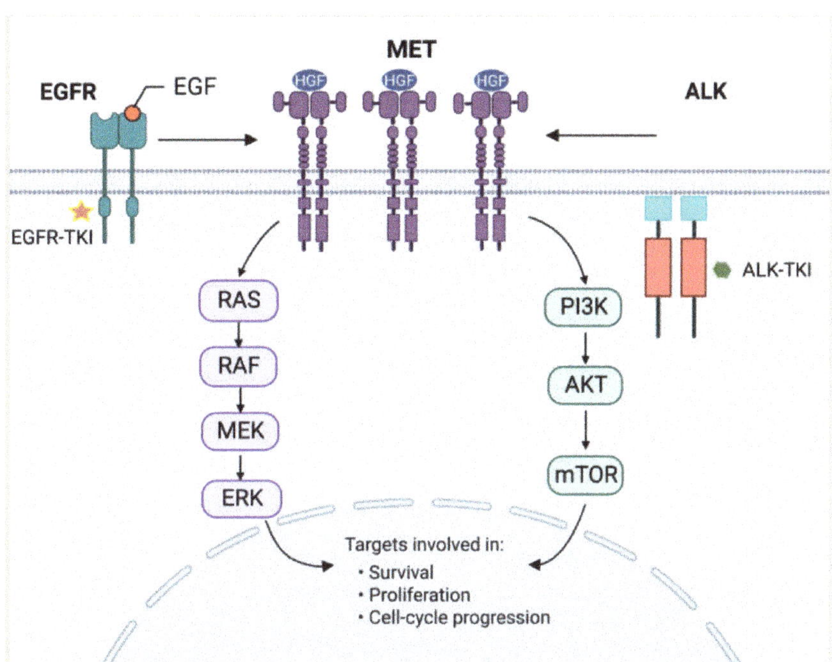

Figure 1. Mechanisms of *MET* amplification-mediated resistance to molecularly targeted therapies in non-small cell lung cancer (NSCLC). *EGFR* mutation or *ALK*-rearrangement as the primary driver oncogene shown. MET, EGFR, and ALK are all members of the RTK family, which regulates cellular proliferation and survival through common downstream pathways such as the PI3K-AKT-mTOR and RAS-RAF-MEK-ERK pathways.

3. *MET* Amplification as a Mediator of Resistance to Targeted Agents in NSCLC

Increased gene copy numbers (GCN) of the *MET* gene could be observed in approximately 1–3% of NSCLC, either due to de novo amplification or as a secondary resistance mechanism in response to targeted therapies [4]. Acquired resistance to EGFR-TKIs can develop via both *EGFR*-dependent and *EGFR*-independent mechanisms. Acquisition of the Exon20 T790M mutation has been proven to be the most common *EGFR*-dependent cause, with MET signaling dysregulation as the most common *EGFR*-independent cause [16,17]. *MET* amplification-mediated resistance has a prevalence of 5–21% after first/second generation EGFR-TKI treatment, 7–15% after first-line osimertinib therapy, and 5–50% of osimertinib resistance after secondary and/or further-line osimertinib treatment [18–20].

The underlying mechanism by which *MET* amplification leads to EGFR-TKI resistance may be associated with phosphorylation of ErbB3 (HER3), which functions as a key activator of the PI3K/AKT and MEK/MAPK pathways, providing bypass signaling in the presence of EGFR-TKIs [17,21,22]. A study by Y.Yarden and colleagues showed that a combination of mAb33 (an anti-HER3 antibody) with cetuximab and third-generation EGFR-TKI osimertinib markedly reduced HER3, and also downregulated MET expression [21]. In another study, inhibition of MET through an inhibitor or knockdown of the *MET* gene restored the effects of osimertinib on ErbB3 inactivation and ErbB3 phosphorylation suppression [22]. Taken together, these findings suggested that phosphorylation of ErbB3 was involved in *MET* amplification-mediated acquired resistance to EGFR-TKIs in advanced NSCLC. In addition, an upregulation of mTOR and Wnt signaling proteins was observed in MET-TKIs/EGFR-TKI-resistant NSCLC cell lines, implying the role of alternative cell signaling pathways in TKI resistance [23]. Furthermore, MET-TKIs and

EGFR-TKIs showed a synergistic inhibitory effect on cell proliferation and downstream activation of signal transduction. Therefore, a combination of HGF and EGF tyrosine kinase inhibitors could potentially be targeted in a synergistic fashion to overcome *MET* amplification-mediated resistance to EGFR-TKIs [24,25].

ALK-rearranged NSCLC is another major subtype of lung cancer, which occurs in around 3–5% of lung adenocarcinomas [26]. As a member of the RTKs family, ALK also regulates cellular proliferation and survival through pathways such as the PI3K-AKT-mTOR, RAS-RAF-MEK-ERK, and JAK-STAT pathways [27,28]. Around 50% of resistance to second-generation ALK-TKIs (ceritinib, alectinib, and brigatinib, etc.) is caused by *ALK*-independent resistance mechanisms, most often due to activation of bypass signaling pathways, including activation of MET, EGFR, and IGF-1R (insulin-like growth factor 1 receptor), etc. [29,30]. MET overactivation was shown to be involved in the development of acquired resistance to alectinib, but not to crizotinib in NSCLC cell lines [31–33]. MET activation-mediated resistance was found to be overcome by crizotinib, which was initially developed as a MET receptor TKI [32–34]. However, more evidence is needed to fully clarify the mechanism and functions of *MET* amplification in ALK downstream signaling and ALK-TKI resistance development.

KRAS is the most frequently mutated cancer-related driver in non-small cell lung cancer, which could be observed in over 30% of NSCLC patients. *KRAS G12C* variants are the most commonly found subtype of oncogenic *KRAS* alterations, which have been identified in around 10% of NSCLC cases [35]. Acquired focal *MET* amplification in a patient with *KRAS G12C*-mutant lung adenocarcinoma treated with sotorasib was also documented [36].

A preclinical study reported that constitutive activation of *KRAS* could lead to the persistent stimulation of downstream signaling pathways, for example, the PI3K/AKT/mTOR cascade and the overexpression of MET protein [37], which indicated that MET amplification was one of the acquired bypass mechanisms of resistance to *KRAS* inhibitors. Lito et al. demonstrated that the inhibition of SHP2, which functions as a valuable co-inhibitory target in *KRAS G12C* signaling and is also a central node in RTK and RAS inhibition signaling, was able to overcome *KRAS G12C* inhibitor resistance in vitro [38,39].

MET amplification is also a known resistance mechanism in *RET*-rearranged NSCLC. Data on acquired resistance to *RET*-specific inhibitors, such as selpercatinib and pralsetinib, have suggested that on-target mutations at non-gatekeeper sites or the emergence of off-target alterations such as *MET* amplification or *NTRK* fusion are potential mechanisms of acquired resistance [40–42]. Furthermore, combinational therapy with crizotinib, which is a MET/ALK/ROS1 TKI, with selpercatinib in patients who had *RET* fusion-positive and *MET*-amplified NSCLC showed clinical efficacy in selpercatinib-resistant tumors [40].

Although there have been many studies that investigated and identified the underlying mechanism of acquired resistance to the targeted agents mediated by *MET* amplification, this issue requires further elucidation through more preclinical and clinical studies.

4. Detection of *MET* Amplification and Overexpression

Since *MET* amplification is a common resistance mechanism to different TKI resistances in lung cancer and inhibitors are available to be used, it is then critical to detect *MET* amplification with appropriate methods and cut-offs so that patients can be identified to be offered potential anti-MET treatment. *MET* copy number gains can occur either as polysomy (multiple copies of chromosome 7) or true amplification (regional or focal copy number gains without chromosome 7 duplication) [43]. True amplification is more likely to lead to oncogene addiction [44]. Various assays have been developed for the detection of *MET* copy number changes. Fluorescence in situ hybridization (FISH) is the gold standard method for *MET* amplification detection. Next-generation sequencing (NGS) is becoming more popular clinically, as the results cover multiple oncogenes, and NGS profiling can be utilized for tissue or liquid biopsy/circulating tumor DNA, either DNA- or RNA-based. Immunochemistry (IHC) is mainly used for the identification of MET overexpression.

Quantitative real-time polymerase chain reaction (qRT-PCR) is less commonly used. Each assay has its advantages and disadvantages.

4.1. Fluorescence In Situ Hybridization (FISH)

Fluorescence in situ hybridization (FISH) is the standard method, and is also the way that is mostly used clinically for identification of *MET* amplification. *MET* amplification can be defined by FISH, either by determining gene copy number or by taking the ratio of *MET* to CEP7 (centromere 7 enumeration probe). *MET* amplification is defined as *MET* GCN $\geq$ 5 with the Cappuzzo criteria, which means five or more copies of *MET* are detected per tumor cell [45–47]. Cut-off points such as a *MET* GCN of $\geq$6 or 10 or 15 are also used in some studies [48–52]. However, GCN itself cannot distinguish true amplification from polysomy. *MET* amplification can also be determined by the MET/CEP7 ratio, and a cut-off value of *MET*/CEP7 ratio round 2 is commonly used to define *MET* amplification [46,47,53–57].

In some studies, *MET* amplification was categorized into three degrees using the *MET*/CEP7 ratio: low amplification $1.8 \leq MET/CEP7 \leq 2.2$; intermediate amplification $2.2 < MET/CEP7 < 5$; and high amplification $MET/CEP7 \geq 5$ [44]. Compared with GCN, the *MET*/CEP7 ratio identifies *MET* amplification more accurately when there is no concurrent chromosome 7 polysomy [58]. However, there is no consensus on a single definition cut-off value with the FISH assay; other cut-off values may also be used. For instance, in a study by Buckingham et al., tumor cells with CEN7 signals on average ≥ 3.6 were categorized as polysomic *MET* amplification [59].

Tumors harboring de novo *MET* amplifications (high level, i.e., *MET* to CEP7 ratio ≥ 5) are thought to be primarily dependent on the MET signaling pathway for growth, as there are often no other concurrent oncogenic drivers. These amplifications are identified in <1–5% of NSCLCs, and indicate a poor prognosis [60–63].

Furthermore, the literature suggests that, compared with other assays, FISH is unique in that it can capture the various levels of *MET* gene amplification, including "true" high-level *MET* gene amplified cases characterized by a high *MET* GCN ($\geq$6 per cell) without concomitant polysomy (i.e., a high *MET*/CEN7 ratio) [64]. However, FISH only detects tissue samples, and it is inapplicable when tissue samples are not available, which limits its clinical use [65].

4.2. Next-Generation Sequencing (NGS)

Simultaneous targeted DNA- and RNA-based next-generation sequencing (NGS) offers the most straightforward and comprehensive profiling for not only *MET* amplifications, but for all treatment relevant genetic alterations, including the *MET*14-skipping aberrations, which cannot be identified by FISH but are of high clinical significance; therefore, NGS has also been widely applied in clinical practice for detection of MET copy number gains [66].

Two methods are commonly used for NGS-targeted approaches: capture hybridization-based sequencing and amplicon-based sequencing, and each has its own advantages and disadvantages. A head-to-head study compared these two types of methods, and indicated that amplicon-based approaches have a much-simplified workflow, and require smaller amounts of DNA for assessment. By contrast, hybridization-based NGS profiling was less likely to miss mutations, and performed better with respect to sequencing complexity and uniformity of coverage [67–70].

However, *MET* amplification detected via NGS is reported as continuous variables, and there is a lack of consensus on a single cut-off value. Normally, the cut-off value ranges from GCN 2.3–10. For example, in the TATTON study, *MET* amplification used a cut-off value as GCN ≥ 5 [71]; in the INC280 study [72], *MET* amplification was determined with a GCN ≥ 2.3; in the ongoing phase 2 INSIGHT 2 (NCT03940703) study, *MET* amplification was defined as GCN ≥ 6 [73].

Now that NGS is increasingly used to optimize precision oncology therapy in NSCLC, the question is whether NGS assays can replace the FISH method regarding the classification of *MET* copy number status. Copious studies have investigated this question. Heydt C. et al. [74] compared 35 *MET*-amplified NSCLC samples (including 5 samples showing a low-level *MET* amplification, 10 samples with an intermediate-level *MET* amplification, and 10 samples with a high-level *MET* amplification), and found that MET-IHC had the best agreement with MET-FISH. Furthermore, only high-level *MET*-amplified cases (GCN $\geq$ 6), showed better concordance between NGS and FISH detections than those in intermediate- or low-level *MET*-amplified patients. This was confirmed through a study by Schubart C. et al., which compared detection results of 205 consecutive NSCLC cases with *MET* alterations, using either an amplicon-based, 15-gene NGS panel, or the standard FISH method. Among the 205 patients detected, 9 cases were classified as *MET*-amplified by NGS, and 16 cases were classified as high-level *MET*-amplification by FISH, yielding a discrepancy of 43.7% (7/16); only cases harboring a *MET* GCN > 10 showed the best concordance when comparing FISH versus NGS (80%, 4/5) [64]. In a study by Peng et al. [75], the concordance rate among FISH and NGS was only 62.5% (25/40). In addition, amplification identified by NGS was found to be an ineffective predictive biomarker, and failed to distinguish significant clinical outcomes. The PR rate was 60.0% (6/10, with *MET* GCN $\geq$ 5) vs. 40.0% (12/30, with *MET* GCN < 5); the median PFS was 4.8 months vs. 2.2 months (p = 0.357). A study by Lai et al. also demonstrated a low concordance between the FISH assay and NGS profiling; among samples with FISH-positive results with GCN $\geq$ 8, only one-third were identified as *MET* amplification with NGS [76]. Of the 18/39 patients identified as MET-high (two amplifications and 16 polysomies), only 8/18 were deemed to have *MET* CNG by NGS. Of the two *MET*-amplified tumors (3.4 and 2 by ratio), the latter was reported as non-*MET*-amplified on NGS. In addition, only 1/3 tumors with a *MET* CNG greater than 8 by FISH were identified as *MET*-amplified with NGS. The result of the TATTON study also showed low consistency between NGS and FISH for *MET* amplification; among all 47 FISH-positive patients, only 12 had *MET* amplification by NGS [71]. Taken together, FISH is the standard method for the detection of various levels of *MET* amplifications during routine diagnostics; NGS is widely used, but is not yet able to replace FISH for the detection of *MET* gene copy number gains.

In recent years, the use of liquid biopsy for genomic profiling has made multi-gene sequencing more easily accessible to patients. NGS of circulating cell-free DNA (cfDNA) has also been used to detect *MET* alterations in clinical studies, including the VISION study [77] and the INSIGHT 2 study [73]. Both studies used liquid biopsy to prospectively screen patients for enrollment and establish the role of liquid biopsy as a tissue-sparing, less-invasive and more easily accessible method for the detection of *MET* alterations.

4.3. Immunohistochemistry (IHC)

MET can be transcriptionally induced in cancer cells in the setting of hypoxia/inflammation to activate proliferation, decrease apoptosis, and promote migration. Thus, tumors can rely on MET signaling, even in the absence of a genomic driver such as *MET* amplification, mutation, or fusion [78]. MET can also be overexpressed in cancers that harbor an activating genomic signature, including those with primary/secondary *MET* amplification, or *MET* exon 14 alterations. Therefore, MET protein overexpression detected by IHC is also commonly used for screening of *MET* gene amplification. Various scoring systems are currently in clinical use to define MET protein expression and overexpression. The most common way is categorizing the MET expression based on a 0–3+ scale into four degrees: negative (0), weak (1+), moderate (2+), or strong (3+). By the MetMab criteria, the cut-off for MET overexpression should be 2+ in at least 50% of the cells [79].

The H-score system multiplies the percentage of cells with 1+, 2+, or 3+ staining by the SI (staining intensity) score [80]. H-scores range from 0–300, and over 200 usually defines MET overexpression. However, cut points vary as well [81,82]. Investigators also used a median H-score (of the range of H-scores obtained from samples exclusively within a given study) as a cut point for overexpression; this approach makes standardization across studies difficult [50,83]. The H-scoring system multiplies the percentage of cells with 1+, 2+, or 3+ staining by the staining intensity score [80]. H-scores range from 0–300; ≥200 usually denotes overexpression, but cut points vary [82,84].

Whether IHC screening for MET overexpression can be used for *MET* amplification detection remains controversial, and attempts to take MET IHC as a marker of MET dependency have largely been unsuccessful [63,82,84]. MET IHC demonstrated poor correlation with the *MET*/CEP7 ratio in sarcomatoid lung cancer, regardless of the stage [82]. In a tri-institutional cohort of patients with metastatic lung adenocarcinoma, more than 30% of cases were MET IHC-positive, but only 2% were *MET*-amplified. MET IHC even failed to detect MET in two of the three *MET*-amplified patients [63]. MET IHC was not an effective predictive marker for MET-directed therapies in some clinical trials [85,86]. For example, in a study by Spigel D.R. et al. [85], the HR of PFS in patients with MET IHC 3+ status was 0.86 (95% CI, 0.58 to 1.29), compared with 1.06 (95% CI, 0.85 to 1.32) in patients with MET IHC 2+ status. Moreover, no statistically significant differences in OS, PFS, or ORR between the onartuzumab and placebo arms were observed when analyzed using MET FISH status. Coupling the reports from the growing literature, there is a strong challenge in taking MET IHC as an effective way of screening for MET dependency. Therefore, MET IHC is not viewed as an effective way of screening for MET dependency, and it is less commonly used for clinical *MET* amplification detection.

4.4. Reverse-Transcription Polymerase Chain Reaction—qRT-PCR

Real-time PCR(RT-PCR), also known as quantitative PCR (qPCR), is the gold-standard for sensitive, specific detection and quantification of nucleic acid targets, and is a valid method for *MET* exon 14-skipping mutation detection [87]. However, unlike gene point mutation, *MET* amplification is difficult to test with qPCR or qRT-PCR; therefore, it is less commonly used in clinical settings for *MET* amplification detection compared to FISH/NGS [16,88–91].

Although seldomly used, the detection of *MET* amplification using ddPCR shows very high concordance rates with FISH, either in tissue samples only (100%, 102/102) or among both peripheral blood and tissue samples (94.17%, 97/103). This indicates that ddPCR is an optional non-invasive method for detecting of *MET* CNG in blood samples as compared with the FISH method in tissue samples; thus, it may be an alternative method for MET amplification detection when FISH is not applicable, especially when tumor tissue is not available [65]. Again, cut-off values vary, and consensus on the standard definition for *MET* amplification by PCR remains to be proposed.

5. Drug Combination Strategies to Overcome Secondary *MET* Amplification Resistance

Since acquired *MET* amplification can bypass the initial oncogene driver to mediate resistance, it is reasonable to hypothesize that inhibition of MET signaling, together with continued inhibition of the initial oncogene driver, can overcome resistance. In the last a few years, there has also been much progression in the development of new agents that act on the HGF/MET pathways. MET-targeting drugs that are currently used in clinics and trials can be divided into three general categories: small molecule inhibitors (e.g., crizotinib, savolitinib, tepotinib, and foretinib), antibodies against the MET receptor (e.g., onartuzumab or amivantamab), and antibody-drug conjugates (e.g., telisotuzumab, vedotin) [92]. In some of the preclinical studies, it has been shown that adding a MET inhibitor to *MET*-amplified *EGFR*-mutant-resistant NSCLC cells can overcome resistance [93].

Therefore, numerous clinical studies have shown preliminary efficacy using this approach. In patients with *EGFR*-mutant NSCLC and *MET* amplification with disease progression on EGFR TKI treatment, subsequent treatment with an MET inhibitor and EGFR-TKI combination rendered clinical benefits in a series of phase I/II studies (Table 1) [71–73,94–101].

Table 1. Summary of key clinical studies on combined therapies to overcome acquired MET-amplification-mediated resistance to EGFR-TKIs in EGFR-mutated NSCLC.

Study (Author, Year, NCT ID)	Treatment	Phase of Study (Number of Patients)	MET Diagnostic Assays and Criteria	Concurrent EGFR Mutations and Prior EGFR-TKIs	Lines of Therapies (Prior EGFR TKIs)	mPFS, Months	mOS, Months	ORR%
Combined therapies with first-generation EGFR-TKIs and MET inhibitors								
Yang et al. (2021) NCT02374645 [94]	Savolitinib plus Gefitinib	phase Ib $n = 64$ safety run-in $n = 13$ (savolitinib + gefitinib $n = 6$; savolitinib + gefitinib $n = 7$); expansion savolitinib+ gefitinib $n = 51$	MET GCN ≥ 5 or MET/CEP7 ratio ≥ 2 by FISH	EGFR-mutated advanced NSCLC	≥ 1 (A prior EGFR-TKI)	4.2 (95% CI: 3.5, 8.5)	NR	NR; In EGFR T790M-negative: ORR: 52% (12/23)
McCoach CE, et al. (2021) NCT01911507 [95]	Capmatinib + Erlotinib	Phase I/II $n = 17$ Cohort A (EGFR mutant $n = 12$) cohort B (EGFR wildtype, $n = 5$)	CNG or MET/CEN7 ratio outside of normal range by FISH; MET IHC 2-3+;	Cohort A: EGFR Mutant; cohort B: EGFR wildtype	≥ 1 prior EGFR TKI	NR	NR	Cohort A: 50%; Cohort B: 75%
Wu et al. (2020). INSIGHT study NCT01982955 [96]	Tepotinib + Gefitinib vs. Chemotherapy (pemetrexed + cisplatin or carboplatin);	Phase Ib (18)/Phase II (55)	MET OE (IHC 2+ or IHC3+) or MET amp (FISH, mean GCN ≥ 5, and/or MET/CEP7 ratio of ≥ 2)	EGFR-mutant, T790M-negative	≥ 2	Overall: 4.9 (90% CI: 3.9–6.9) vs. 4.4 (90% CI: 4.2–6.8)HR 0.67 (90% CI:0.35–1.28) In the high MET subgroup (IHC3+): mPFS: 8.3 (90% CI: 4.1–16.6) vs. 4.4 (90% CI: 4.1–6.8), HR 0.35, 90% CI: 0.17–0.74 In the MET amplification subgroup:16.6 (90% CI: 8.3–not estimable) vs. 4.2 (90% CI: 1.4–7.0); HR 0.13, 90% CI: 0.04–0.43	Phase II Overall: 17.3 (90% CI: 12.1–37.3) vs. 18.7 (90% CI:15.9–20.7); HR 0.69, (90% CI: 0.34–1.41) In the high (IHC3+) MET subgroup: 37.3 (90% CI 24·2–37·3) vs. 17·9 (12.0–20.7); HR 0.33, 90% CI: 0.14–0.76. In the MET amplification subgroup 37.3 months (90% CI not estimable) vs. 13.1 [3.25–not estimable]; HR 0.08, 90% CI: 0.01–0.51)	Phase II Overall: 45% (29.7–61.3) vs. 33% (17.8–52.1) In the high (IHC3+) MET subgroup: 68% (47.0–85.3) vs. 33% (14.2–57.7) In the MET amplification subgroup: 67% (39.1–87.7) vs. 43% (12.9–77.5)
Camidge et al. (2022) [97]	Telisotuzumab Vedotin + erlotinib	phase 1b 42 NSCLC pts received T + E; 37 were c-MET+ (36 evaluable; 35 H-score ≥ 150, 1 MET amplified)	c-Met+ (central lab IHC H-score ≥ 150 or local lab MET amplification)		≥ 1	NR 95%CI: 2.8–NE	5.9 m 95CI: 1.2–NE	EGFR mut+: 34.5 (95%CI: 17.9–54.3) EGFR wildtype: 28.6% (95%CI: 3.7–71.0)
Wu et al. (2018). NCT01610336 [72]	Capmatinib (INC280) + Gefitinib	Phase Ib(61)/phaseII (100)(GCN < 4: $n = 41$ $4 \leq$ GCN < 6: $N = 18$; GCN ≥ 6: $n = 36$)	IHC, MET OE 2+ or 3+; FISH, MET Amp GCN ≥ 5, MET/(CEP7) ratio of ≥ 2:1 50% of tumor cells with IHC 3+ or MET GCN < 4)	EGFR-mutated advanced NSCLC	≥ 2 (≥ 1 prior EGFR-TKI)	Overall: 5.5 (95% CI, 3.8 to 5.6; mPFS in GCN ≥ 6 subgroup: 5.49 (95% CI, 4.21 to 7.29), mPFS in the $4 \leq$ GCN < 6 subgroup: 5.39 (95% CI, 3.65 to 7.46); mPFS in the GCN < 4 subgroup: 3.91(95% CI, 3.65 to 5.55) mPFS in the IHC2+/ GCN ≥ 5 subgroup: 7.29 (95%CI, 1.81 to 9.07) mPFS in the IHC3+ subgroup: 5.45 (95% CI, 3.71 to 7.10)	NR	Phase Ib/II overall: 43%; Phase II overall: 29%; GCN ≥ 6: 47%; $4 \leq$ GCN < 6: 22%; GCN < 4: 12%; IHC 3+: 32%
Camidge et al. (2022) NCT01900652 [98]	emibetuzumab + erlotinib vs. emibetuzumab monotherapy	Phase II emibetuzumab + erlotinib ($n = 83$); emibetuzumab monotherapy ($n = 28$)	≥ 10% of cells expressing MET at ≥ 2+ by IHC	EGFR-mt NSCLC	≥ 1	3.3 vs. 1.6	NR	3.0 for emibetuzumab + erlotinib (95% CI: 0.4, 10.5) vs. 4.3% for emibetuzumab (95% CI: 0.1, 21.9)
Combined therapies with third-generation EGFR-TKIs and MET-TKIs								
Yu et al. (2021) ORCHARD Study [99]	Osimertinib + Savolitinib	phase II ($n = 17$)	NGS (criteria NR; GCN ranged from 7 to 68)	EGFRm	2 (progressed after prior first-line Osimertinib)	NR	NR	ORR: 41% (7/17)
Sequist et al. (2020) TATTON study NCT02143466 [71]	Osimertinib + Savolitinib	phase 1B; Part B $n = 138$ (osimertinib 80 mg psavolitinib 600 mg or 300 mg): (Part B1: previously received third generation EGFR TKI $n = 69$; part B2: no previous third-generation EGFR TKI, Thr790Met negative, $n = 51$; Part B3: no previous third-generation EGFR TKI, Thr790Met positive $n = 18$); Part D $n = 42$ (osimertinib plus savolitinib; no previous third-generation EGFR TKI, Thr790Met negative)	MET gene copy number gain ≥ 5 or MET/CEP7 ratio ≥ 2 by FISH; MET + 3 expression in ≥ 50% of tumor cells by IHC; ≥ 20% tumor cells, coverage of $\geq 200\times$ sequencing depth and ≥ 5 copies of MET over tumor ploidy by NGS	EGFR mutation-positive(with or without T790M mutation)	≥ 2 (≥ 1 prior EGFR-TKI)	Part B overall: 5.5–11.1; Part D: 9.0 (95%CI: 5.4–12.9)	NR	part B: 33–67%; part D: 62%

Table 1. *Cont.*

Study (Author, Year, NCT ID)	Treatment	Phase of Study (Number of Patients)	MET Diagnostic Assays and Criteria	Concurrent EGFR Mutations and Prior EGFR-TKIs	Lines of Therapies (Prior EGFR TKIs)	mPFS, Months	mOS, Months	ORR%
E. Felip et al. (2019) NCT02335944 [100]	Capmatinib + Nazartinib n = 68 (66 had known MET status: 23 MET+, 43 MET−)	Phase 1b/II study	MET+: IHC 3+ and/or GCN ≥4	EGFR-mutant stage IIIB/IV NSCLC	≥1	7.7 (95% CI: 5.4–12.2)	18.8 (95% CI:14.0–21.3)	43.5 (95% CI:23.2–65.5)
NCT03940703 INSIGHT 2 study [73]	Tepotinib plus Osimertinib vs. chemotherapy	Phase II (n = 425)	METamp by FISH testing (GCN ≥ 5 and/or MET/CEP7 ratio ≥ 2) or METamp determined by using NGS (GCN ≥ 2.3)	EGFR-mutated NSCLC	2	NR	NR	54.5% among the 22 patients with FISH detected MET amplification and at least 9 months of follow-up; 45.8% among the 48 participants with follow-up of 3 months or more; 50.0% for the 16 patients who were followed up for 9 months or more and 56.5% for the 23 followed up for 3 months or more
NCT03778229 SAVANNAH Study [101]	Osimertinib + Savolitinib	Phase II (n = 193)	High levels of MET overexpression and/or amplification, defined as IHC90+ and/or FISH10+, (IHC50+ and/or FISH5+; n = 193)	EGFRm+, MET+, progressed on prior Osimertinib	≥2	All patients (IHC50+ and/or FISH5+; n = 193): 5.3 (4.2, 5.8); Patients with high levels of MET (IHC90+ and/or FISH10+): (n = 108) 7.1 (5.3, 8.0) Patients with high levels of MET (IHC90+ and/or FISH10+) No prior chemo (n = 87): 7.2 (4.7, 9.2) Patients with lower levels of MET (n = 77): 2.8 (2.6, 4.3)	NR	Overall: All patients (IHC50+ and/or FISH5+; n = 193): 5.3 (4.2, 5.8) Patients with high levels of MET (IHC90+ and/or FISH10+) (n = 108): 7.1 (5.3, 8.0) Patients with high levels of MET (IHC90+ and/or FISH10+) No prior chemo (n = 87): 7.2 (4.7, 9.2) Patients with lower levels of MET (n = 77): 9 (4, 18)

Abbreviations: NSCLC, non-small cell lung cancer; ORR, objective response rate; mPFS, median progression-free survival; OS, median overall survival; MET, mesenchymal-epithelial transition factor; amp, amplification; EGFR, epidermal growth factor receptor; IHC, immunohistochemistry; FISH, fluorescence in situ hybridization; GCN, gene copy number; Pem, pemetrexed; Dox, docetaxel; Gem, gemcitabine; LBx, liquid biopsy; TBx, tissue biopsy; [CI], confidence interval; NR, not reported.

The TATTON trial [71] demonstrated the clinical benefits of osimertinib plus savolitinib in patients with previously treated *EGFR*-mutant *MET*-amplified NSCLC, with an objective response rate (ORR) of 44%. Among patients progressed on a third-generation EGFR-TKI, the ORR was 30%. Notably, an ORR of 64% was observed among 23 patients with *EGFR*-mutant T790M-negative NSCLC without prior third-generation EGFR-TKI treatments. Another study confirmed that the combination of capmatinib with geftinib demonstrated a PFS of 3.3 months, and an ORR up to 47% in patients with *EGFR* mutation and *MET* amplification (defined by CGN ≥ 6) [5]. In the INSIGHT study, the combination of tepotinib and gefitinib showed significantly a better PFS (16.6 months vs. 4.2 months) and OS (37.3 months vs. 17.9 months, respectively) than chemotherapy in patients with resistant *EGFR*-mutant NSCLC, especially in patients with high MET over-expression [96]. The ongoing ORCHARD study (NCT03944772) included 20 patients with *MET* amplification who progressed on first-line osimertininb monotherapy, and received second-line combinatory treatment with osimertinib and savolitinib [99]. Initial benefits for the patients were presented with good tolerance: among 17 patients who were evaluable for confirmed response analysis at data cut-off (DCO), 7 patients had confirmed partial response (ORR 41%, 7/17) and 7 patients had stable disease (DCR 41%, 7/17). The ORCHARD study is still ongoing, and more results are expected to be released in the future.

With those successes, pivotal trials were designed to further evaluate this combination approach to overcome *MET* amplification-medicated resistance in *EGFR*-mutant NSCLC. The data in the INSIGHT study led to the investigation of tepotinib plus osimertinib in the INSIGHT 2 trial [73]. Preliminary data from the INSIGHT 2 study suggested that the combination of tepotinib and osimertinib has activity in patients with *EGFR*-mutated advanced NSCLC with *MET* amplification who progressed on first-line osimertinib. In the first 48 patients who had over 3 months follow up, the ORR was 45.8% [95% CI, 31–61%], with duration of response not reached [73]. Similarly, results of the TATTON trial led to further development of the savolitnib plus osimertinib combination in the SAVANNAH trial [101]. SAVANNAH is a global, randomized, single-arm phase II trial that is studying the efficacy of savolitinib with osimertinib in patients with *EGFR*-mutant, locally advanced or metastatic NSCLC with MET overexpression and/or amplification, who progressed following treatment with osmiertinib. Patients were treated with savolitinib with osimertinib. Preliminary results demonstrated an ORR of 32% in the total population. In patients with high levels of MET overexpression and/or *MET* amplification (defined as IHC90+ and/or FISH10+), the ORR was high at 49% [95% CI, 39–59%] [101].

Other studies investigating the clinical evidence of dual inhibition of EGFR and MET with small molecule inhibitors or monoclonal antibodies, such as capmatinib plus geftinib, telisotuzumab plus erlotinib, savolitinib plus geftinib, onartuzumab plus erlotinib, capmatinib plus erlotinib, and emibetuzumab plus erlotinib within patients with NSCLC with EGFR-mutant and *MET* alterations, are summarized in Table 1 [71–73,94–101].

6. Conclusions

Acquired *MET* amplification functions as a mechanism of resistance to targeted therapies in NSCLC, and now has been proven to be a pharmaceutical target to overcome this resistance. Although evidence was most abundant and convincing for *MET* amplification mediated resistance to EGFR TKIs, its role in mediating resistance to other targeted therapies, such as ALK, ROS1, or RET TKIs, is being recognized. Case reports and case series have indicated that dual inhibition of MET and the initial oncogene driver pathway may be a valid clinical approach to overcome resistance. With *MET* amplification serving as a general resistance mechanism to targeted therapies in lung cancer, it is crucial to be able to detect *MET* amplification in a reliable manner, in order to identify the appropriate patients who can benefit from MET-targeting therapy.

MET amplification or MET overexpression could be detected by multiple clinical pathology laboratory tests, including FISH, NGS, and IHC. However, clinically meaningful cut-offs need to be standardized for continuous variables, including the copy number gain of *MET* amplification and MET overexpression. As clinical diagnostic methods migrate towards more comprehensive and technically sophisticated NGS assays, further understanding of NGS assays for the detection of *MET* amplification is needed, both in tumors and plasma, and ideally both in DNA and RNA. The effective detection of *MET*-dependent cancers is critical, given that *MET*-directed targeted therapy is active in many of these cancers. Importantly, the level of activity of *MET*-targeted therapies is associated with the degree of oncogenic addiction to *MET* pathway signaling.

To overcome *MET* amplification-mediated acquired resistance to TKIs, combination therapies to inhibit both MET and the primary driver oncogene are necessary. EGFR-MET TKI combination therapy has shown promising clinical efficacy in this setting. While awaiting the maturation of the large clinical trial results and regulatory approval of this approach, further studies are necessary that aim to standardize the *MET* amplification detection assay and the cut-off values.

Author Contributions: Conceptualization, K.Q. and X.L.; methodology, K.Q.; software, K.Q. and L.H.; data curation, K.Q.; writing—original draft preparation, K.Q.; writing—review and editing, K.Q., J.Z. and X.L.; visualization, L.H.; supervision, X.L.; All authors have read and agreed to the published version of the manuscript.

Funding: X.L. is supported by Damon Runyon Cancer Research Foundation, V Foundation for Cancer Research, and Lung Cancer Research Foundation.

Conflicts of Interest: There is no conflict of interest related to this work. Outside of this work, X.L. receives consulting/advisory fees from EMD Serono (Merck KGaA), AstraZeneca, Spectrum Pharmaceutics, Novartis, Eli Lilly, Boehringer Ingelheim, Hengrui Therapeutics, Janssen, Blueprint Medicines, Sensei Biotherapeutics, and Abbvie, and Research Funding from Eli Lilly, EMD Serono, Teligene, ArriVent, Regeneron, and Boehringer Ingelheim. Outside of this work, J.Z. reports grants from Merck, grants and personal fees from Johnson and Johnson and Novartis, personal fees from Bristol Myers Squibb, AstraZeneca, GenePlus, Innovent and Hengrui.

Abbreviations

NSCLC	non-small cell lung cancer
RTKs	receptor tyrosine kinases
MET	mesenchymal-epithelial transition factor
EGFR	epidermal growth factor receptor
ALK	anaplastic lymphoma kinase

References

1. Jemal, A.; Siegel, R.; Ward, E.; Murray, T.; Xu, J.; Thun, M.J. Cancer statistics, 2007. *CA Cancer J. Clin.* **2007**, *57*, 43–66. [CrossRef] [PubMed]
2. Ettinger, D.S.; Wood, D.E.; Aisner, D.L.; Akerley, W.; Bauman, J.R.; Bharat, A.; Shapiro, T.A.; Singh, A.P.; Stevenson, J.; Hughes, M.; et al. NCCN Guidelines Insights: Non-Small Cell Lung Cancer, Version 2.2021. *J. Natl. Compr. Cancer Netw.* **2021**, *19*, 254–266. [CrossRef] [PubMed]
3. Comoglio, P.M.; Trusolino, L.; Boccaccio, C. Known and novel roles of the MET oncogene in cancer: A coherent approach to targeted therapy. *Nat. Rev. Cancer* **2018**, *18*, 341–358. [CrossRef] [PubMed]
4. Lee, M.; Jain, P.; Wang, F.; Ma, P.C.; Borczuk, A.; Halmos, B. MET alterations and their impact on the future of non-small cell lung cancer (NSCLC) targeted therapies. *Expert Opin. Ther. Targets* **2021**, *25*, 249–268. [CrossRef]
5. Miranda, O.; Farooqui, M.; Siegfried, J.M. Status of Agents Targeting the HGF/c-Met Axis in Lung Cancer. *Cancers* **2018**, *10*, 280. [CrossRef]
6. Planchard, D. Have We Really MET a New Target? *J. Clin. Oncol.* **2018**, *36*, JCO2018793455. [CrossRef]
7. Garon, E.B.; Brodrick, P. Targeted Therapy Approaches for MET Abnormalities in Non-Small Cell Lung Cancer. *Drugs* **2021**, *81*, 547–554. [CrossRef]
8. Ciuffreda, L.; Incani, U.; Steelman, L.; Abrams, S.; Falcone, I.; DEL Curatolo, A.; Chappell, W.; Franklin, R.A.; Vari, S.; Cognetti, F.; et al. Signaling Intermediates (MAPK and PI3K) as Therapeutic Targets in NSCLC. *Curr. Pharm. Des.* **2014**, *20*, 3944–3957. [CrossRef]
9. Cardarella, S.; Ogino, A.; Nishino, M.; Butaney, M.; Shen, J.; Lydon, C.; Yeap, B.Y.; Sholl, L.M.; Johnson, B.E.; Jänne, P.A. Clinical, pathologic, and biologic features associated with BRAF mutations in non-small cell lung cancer. *Clin. Cancer Res.* **2013**, *19*, 4532–4540. [CrossRef]
10. Domvri, K.; Zarogoulidis, P.; Darwiche, K.; Browning, R.F.; Li, Q.; Turner, J.F.; Kioumis, I.; Spyratos, D.; Porpodis, K.; Papaiwannou, A.; et al. Molecular Targeted Drugs and Biomarkers in NSCLC, the Evolving Role of Individualized Therapy. *J. Cancer* **2013**, *4*, 736–754. [CrossRef]
11. Soucheray, M.; Capelletti, M.; Pulido, I.; Kuang, Y.; Paweletz, C.P.; Becker, J.H.; Kikuchi, E.; Xu, C.; Patel, T.B.; Al-Shahrour, F.; et al. Intratumoral Heterogeneity in *EGFR*-Mutant NSCLC Results in Divergent Resistance Mechanisms in Response to EGFR Tyrosine Kinase Inhibition. *Cancer Res.* **2015**, *75*, 4372–4383. [CrossRef] [PubMed]
12. Seid, T.; Hagiya, M.; Shimonishi, M.; Nakamura, T.; Shimizu, S. Organization of the human hepatocyte growth factor-encoding gene. *Gene* **1991**, *102*, 213–219. [CrossRef] [PubMed]
13. Mars, W.M.; Zarnegar, R.; Michalopoulos, G.K. Activation of hepatocyte growth factor by the plasminogen activators uPA and tPA. *Am. J. Pathol.* **1993**, *143*, 949–958.
14. Ma, P.C.; Maulik, G.; Christensen, J.; Salgia, R. c-Met: Structure, functions and potential for therapeutic inhibition. *Cancer Metastasis Rev.* **2003**, *22*, 309–325. [CrossRef] [PubMed]
15. Dong, Y.; Xu, J.; Sun, B.; Wang, J.; Wang, Z. MET-Targeted Therapies and Clinical Outcomes: A Systematic Literature Review. *Mol. Diagn. Ther.* **2022**, *26*, 203–227. [CrossRef] [PubMed]
16. Bean, J.; Brennan, C.; Shih, J.-Y.; Riely, G.; Viale, A.; Wang, L.; Chitale, D.; Motoi, N.; Szoke, J.; Broderick, S.; et al. *MET* amplification occurs with or without *T790M* mutations in *EGFR* mutant lung tumors with acquired resistance to gefitinib or erlotinib. *Proc. Natl. Acad. Sci. USA* **2007**, *104*, 20932–20937. [CrossRef]
17. Engelman, J.A.; Zejnullahu, K.; Mitsudomi, T.; Song, Y.; Hyland, C.; Park, J.O.; Lindeman, N.; Gale, C.-M.; Zhao, X.; Christensen, J.; et al. MET Amplification Leads to Gefitinib Resistance in Lung Cancer by Activating ERBB3 Signaling. *Science* **2007**, *316*, 1039–1043. [CrossRef]
18. Wu, Y.-L.; Soo, R.A.; Locatelli, G.; Stammberger, U.; Scagliotti, G.; Park, K. Does c-Met remain a rational target for therapy in patients with EGFR TKI-resistant non-small cell lung cancer? *Cancer Treat. Rev.* **2017**, *61*, 70–81. [CrossRef]

19. Ramalingam, S.S.; Cheng, Y.; Zhou, C.; Ohe, Y.; Imamura, F.; Cho, B.C.; Lin, M.-C.; Majem, M.; Shah, R.; Gray, J.; et al. Mechanisms of acquired resistance to first-line osimertinib: Preliminary data from the phase III FLAURA study. *Ann. Oncol.* **2018**, *29* (Suppl. S8), LBA50. [CrossRef]

20. Papadimitrakopoulou, V.A.; Wu, Y.L.; Han, J.Y.; Ahn, M.J.; Ramalingam, S.S.; John, T.; Okamoto, I.; Yang, J.H.; Bulusu, K.C.; Laus, G.; et al. Analysis of resistance mechanisms to osimertinib in patients with EGFR T790M advanced NSCLC from the AURA3 study. *Ann. Oncol.* **2018**, *29*, viii741. [CrossRef]

21. Romaniello, D.; Marrocco, I.; Nataraj, N.B.; Ferrer, I.; Drago-Garcia, D.; Vaknin, I.; Oren, R.; Lindzen, M.; Ghosh, S.; Kreitman, M.; et al. Targeting HER3, a Catalytically Defective Receptor Tyrosine Kinase, Prevents Resistance of Lung Cancer to a Third-Generation EGFR Kinase Inhibitor. *Cancers* **2020**, *12*, 2394. [CrossRef] [PubMed]

22. Shi, P.; Oh, Y.-T.; Zhang, G.; Yao, W.; Yue, P.; Li, Y.; Kanteti, R.; Riehm, J.; Salgia, R.; Owonikoko, T.K.; et al. Met gene amplification and protein hyperactivation is a mechanism of resistance to both first and third generation EGFR inhibitors in lung cancer treatment. *Cancer Lett.* **2016**, *380*, 494–504. [CrossRef] [PubMed]

23. Fong, J.T.; Jacobs, R.J.; Moravec, D.N.; Uppada, S.B.; Botting, G.M.; Nlend, M.; Puri, N. Alternative Signaling Pathways as Potential Therapeutic Targets for Overcoming EGFR and c-Met Inhibitor Resistance in Non-Small Cell Lung Cancer. *PLoS ONE* **2013**, *8*, e78398. [CrossRef] [PubMed]

24. Puri, N.; Salgia, R. Synergism of EGFR and c-Met pathways, cross-talk and inhibition, in non-small cell lung cancer. *J. Carcinog.* **2008**, *7*, 9. [CrossRef] [PubMed]

25. Nakagawa, T.; Takeuchi, S.; Yamada, T.; Nanjo, S.; Ishikawa, D.; Sano, T.; Kita, K.; Nakamura, T.; Matsumoto, K.; Suda, K.; et al. Combined Therapy with Mutant-Selective EGFR Inhibitor and Met Kinase Inhibitor for Overcoming Erlotinib Resistance in EGFR-Mutant Lung Cancer. *Mol. Cancer Ther.* **2012**, *11*, 2149–2157. [CrossRef]

26. Proietti, A.; Alì, G.; Pelliccioni, S.; Lupi, C.; Sensi, E.; Boldrini, L.; Servadio, A.; Chella, A.; Ribechini, A.; Cappuzzo, F.; et al. Anaplastic lymphoma kinase gene rearrangements in cytological samples of non-small cell lung cancer: Comparison with histological assessment. *Cancer Cytopathol.* **2014**, *122*, 445–453. [CrossRef]

27. Morales La Madrid, A.; Campbell, N.; Smith, S.; Cohn, S.L.; Salgia, R. Targeting ALK: A promising strategy for the treatment of non-small cell lung cancer, non-Hodgkin's lymphoma, and neuroblastoma. *Target Oncol.* **2012**, *7*, 199–210; Erratum in *Target Oncol.* **2013**, *8*, 301. [CrossRef]

28. Shaw, A.T.; Solomon, B. Targeting Anaplastic Lymphoma Kinase in Lung Cancer. *Clin. Cancer Res.* **2011**, *17*, 2081–2086. [CrossRef]

29. Katayama, R. Drug resistance in anaplastic lymphoma kinase-rearranged lung cancer. *Cancer Sci.* **2018**, *109*, 572–580. [CrossRef]

30. Choi, S.H.; Kim, D.H.; Choi, Y.J.; Kim, S.Y.; Lee, J.-E.; Sung, K.J.; Kim, W.S.; Choi, C.-M.; Rho, J.K.; Lee, J.C. Multiple receptor tyrosine kinase activation related to ALK inhibitor resistance in lung cancer cells with ALK rearrangement. *Oncotarget* **2017**, *8*, 58771–58780. [CrossRef]

31. Gouji, T.; Takashi, S.; Mitsuhiro, T.; Yukito, I. Crizotinib Can Overcome Acquired Resistance to CH5424802: Is Amplification of the MET Gene a Key Factor? *J. Thorac. Oncol.* **2014**, *9*, e27–e28. [CrossRef] [PubMed]

32. Kogita, A.; Togashi, Y.; Hayashi, H.; Banno, E.; Terashima, M.; DE Velasco, M.A.; Sakai, K.; Fujita, Y.; Tomida, S.; Takeyama, Y.; et al. Activated MET acts as a salvage signal after treatment with alectinib, a selective ALK inhibitor, in ALK-positive non-small cell lung cancer. *Int. J. Oncol.* **2015**, *46*, 1025–1030. [CrossRef] [PubMed]

33. Tanimoto, A.; Yamada, T.; Nanjo, S.; Takeuchi, S.; Ebi, H.; Kita, K.; Matsumoto, K.; Yano, S. Receptor ligand-triggered resistance to alectinib and its circumvention by Hsp90 inhibition in EML4-ALK lung cancer cells. *Oncotarget* **2014**, *5*, 4920–4928. [CrossRef] [PubMed]

34. Isozaki, H.; Ichihara, E.; Takigawa, N.; Ohashi, K.; Ochi, N.; Yasugi, M.; Ninomiya, T.; Yamane, H.; Hotta, K.; Sakai, K.; et al. Non–Small Cell Lung Cancer Cells Acquire Resistance to the ALK Inhibitor Alectinib by Activating Alternative Receptor Tyrosine Kinases. *Cancer Res.* **2016**, *76*, 1506–1516. [CrossRef]

35. Hong, D.S.; Fakih, M.G.; Strickler, J.H.; Desai, J.; Durm, G.A.; Shapiro, G.I.; Falchook, G.S.; Price, T.J.; Sacher, A.; Li, B.T.; et al. KRAS G12C Inhibition with Sotorasib in Advanced Solid Tumors. *N. Engl. J. Med.* **2020**, *383*, 1207–1217. [CrossRef]

36. Awad, M.; Liu, S.; Arbour, K.; Zhu, V.; Johnson, M.; Heist, R.; Patil, T.; Riely, G.; Jacobson, J.; Dilly, J.; et al. Abstract LB002: Mechanisms of acquired resistance to KRAS G12C inhibition in cancer. *Cancer Res.* **2021**, *81*, LB002. [CrossRef]

37. Zhao, Y.; Murciano-Goroff, Y.R.; Xue, J.Y.; Ang, A.; Lucas, J.; Mai, T.T.; Da Cruz Paula, A.F.; Saiki, A.Y.; Mohn, D.; Achanta, P.; et al. Diverse alterations associated with resistance to KRAS(G12C) inhibition. *Nature* **2021**, *599*, 679–683. [CrossRef]

38. Lito, P.; Solomon, M.; Li, L.-S.; Hansen, R.; Rosen, N. Allele-specific inhibitors inactivate mutant KRAS G12C by a trapping mechanism. *Science* **2016**, *351*, 604–608. [CrossRef]

39. Ahmed, T.A.; Adamopoulos, C.; Karoulia, Z.; Wu, X.; Sachidanandam, R.; Aaronson, S.A.; Poulikakos, P.I. SHP2 Drives Adaptive Resistance to ERK Signaling Inhibition in Molecularly Defined Subsets of ERK-Dependent Tumors. *Cell Rep.* **2019**, *26*, 65–78.e5. [CrossRef]

40. Rosen, E.Y.; Johnson, M.L.; Clifford, S.E.; Somwar, R.; Kherani, J.F.; Son, J.; Bertram, A.A.; Davare, M.A.; Gladstone, E.G.; Ivanova, E.V.; et al. Overcoming MET-Dependent Resistance to Selective RET Inhibition in Patients with RET Fusion–Positive Lung Cancer by Combining Selpercatinib with Crizotinib. *Clin. Cancer Res.* **2021**, *27*, 34–42. [CrossRef]

41. Subbiah, V.; Hu, M.I.-N.; Gainor, J.F.; Mansfield, A.S.; Alonso, G.; Taylor, M.H.; Zhu, V.W.; Lopez, P.G.; Amatu, A.; Doebele, R.C.; et al. Clinical activity of the RET inhibitor pralsetinib (BLU-667) in patients with RET fusion+ solid tumors. *J. Clin. Oncol.* **2020**, *38*, 109. [CrossRef]

42. Lin, J.; Liu, S.; McCoach, C.; Zhu, V.; Tan, A.; Yoda, S.; Peterson, J.; Do, A.; Prutisto-Chang, K.; Dagogo-Jack, I.; et al. Mechanisms of resistance to selective RET tyrosine kinase inhibitors in RET fusion-positive non-small-cell lung cancer. *Ann. Oncol.* **2020**, *31*, 1725–1733. [CrossRef] [PubMed]

43. Hellman, A.; Zlotorynski, E.; Scherer, S.W.; Cheung, J.; Vincent, J.B.; Smith, D.I.; Trakhtenbrot, L.; Kerem, B. A role for common fragile site induction in amplification of human oncogenes. *Cancer Cell* **2002**, *1*, 89–97. [CrossRef] [PubMed]

44. Noonan, S.A.; Berry, L.; Lu, X.; Gao, D.; Barón, A.E.; Chesnut, P.; Sheren, J.; Aisner, D.L.; Merrick, D.; Doebele, R.C.; et al. Identifying the Appropriate FISH Criteria for Defining MET Copy Number–Driven Lung Adenocarcinoma through Oncogene Overlap Analysis. *J. Thorac. Oncol.* **2016**, *11*, 1293–1304. [CrossRef]

45. Cappuzzo, F.; Marchetti, A.; Skokan, M.; Rossi, E.; Gajapathy, S.; Felicioni, L.; del Grammastro, M.; Sciarrotta, M.G.; Buttitta, F.; Incarbone, M.; et al. Increased *MET* Gene Copy Number Negatively Affects Survival of Surgically Resected Non–Small-Cell Lung Cancer Patients. *J. Clin. Oncol.* **2009**, *27*, 1667–1674. [CrossRef]

46. Kato, H.; Arao, T.; Matsumoto, K.; Fujita, Y.; Kimura, H.; Hayashi, H.; Nishiki, K.; Iwama, M.; Shiraishi, O.; Yasuda, A.; et al. Gene amplification of EGFR, HER2, FGFR2 and MET in esophageal squamous cell carcinoma. *Int. J. Oncol.* **2013**, *42*, 1151–1158. [CrossRef]

47. Graziano, F.; Galluccio, N.; Lorenzini, P.; Ruzzo, A.; Canestrari, E.; D'Emidio, S.; Catalano, V.; Sisti, V.; Ligorio, C.; Magnani, M.; et al. Genetic activation of the MET pathway and prognosis of patients with high-risk, radically resected gastric cancer. *J. Clin. Oncol.* **2011**, *29*, 4789–4795. [CrossRef]

48. Yin, X.; Zhang, T.; Su, X.; Ji, Y.; Ye, P.; Fu, H.; Fan, S.; Shen, Y.; Gavine, P.R.; Gu, Y. Relationships between Chromosome 7 Gain, MET Gene Copy Number Increase and MET Protein Overexpression in Chinese Papillary Renal Cell Carcinoma Patients. *PLoS ONE* **2015**, *10*, e0143468. [CrossRef]

49. Lee, H.E.; A Kim, M.; Jung, E.-J.; Yang, H.-K.; Lee, B.L.; Bang, Y.-J.; Kim, W.H. MET in gastric carcinomas: Comparison between protein expression and gene copy number and impact on clinical outcome. *Br. J. Cancer* **2012**, *107*, 325–333. [CrossRef]

50. Nagatsuma, A.K.; Aizawa, M.; Kuwata, T.; Doi, T.; Ohtsu, A.; Fujii, H.; Ochiai, A. Expression profiles of HER2, EGFR, MET and FGFR2 in a large cohort of patients with gastric adenocarcinoma. *Gastric Cancer* **2015**, *18*, 227–238. [CrossRef]

51. Tanaka, A.; Sueoka-Aragane, N.; Nakamura, T.; Takeda, Y.; Mitsuoka, M.; Yamasaki, F.; Hayashi, S.; Sueoka, E.; Kimura, S. Co-existence of positive MET FISH status with EGFR mutations signifies poor prognosis in lung adenocarcinoma patients. *Lung Cancer* **2012**, *75*, 89–94. [CrossRef] [PubMed]

52. Wolf, J.; Seto, T.; Han, J.Y.; Reguart, N.; Garon, E.B.; Groen, H.J.; Tan, D.S.; Hida, T.; de Jonge, M.; Heist, R.S.; et al. Capmatinib in *MET* Exon 14-Mutated or *MET*-Amplified Non-Small-Cell Lung Cancer. *N. Engl. J. Med.* **2020**, *383*, 944–957. [CrossRef] [PubMed]

53. An, X.; Wang, F.; Shao, Q.; Wang, F.-H.; Wang, Z.-Q.; Chen, C.; Li, C.; Luo, H.-Y.; Zhang, D.-S.; Xu, R.H.; et al. MET amplification is not rare and predicts unfavorable clinical outcomes in patients with recurrent/metastatic gastric cancer after chemotherapy. *Cancer* **2014**, *120*, 675–682. [CrossRef]

54. Kondo, S.; Ojima, H.; Tsuda, H.; Hashimoto, J.; Morizane, C.; Ikeda, M.; Ueno, H.; Tamura, K.; Shimada, K.; Kanai, Y.; et al. Clinical impact of c-Met expression and its gene amplification in hepatocellular carcinoma. *Int. J. Clin. Oncol.* **2013**, *18*, 207–213. [CrossRef] [PubMed]

55. Go, H.; Jeon, Y.K.; Park, H.J.; Sung, S.-W.; Seo, J.-W.; Chung, D.H. High MET Gene Copy Number Leads to Shorter Survival in Patients with Non-small Cell Lung Cancer. *J. Thorac. Oncol.* **2010**, *5*, 305–313. [CrossRef] [PubMed]

56. Lennerz, J.K.; Kwak, E.L.; Ackerman, A.; Michael, M.; Fox, S.B.; Bergethon, K.; Lauwers, G.Y.; Christensen, J.G.; Wilner, K.D.; Haber, D.A.; et al. *MET* Amplification Identifies a Small and Aggressive Subgroup of Esophagogastric Adenocarcinoma With Evidence of Responsiveness to Crizotinib. *J. Clin. Oncol.* **2011**, *29*, 4803–4810. [CrossRef]

57. Jardim, D.L.; Tang, C.; Gagliato, D.D.M.; Falchook, G.S.; Hess, K.; Janku, F.; Fu, S.; Wheler, J.J.; Zinner, R.G.; Naing, A.; et al. Analysis of 1,115 Patients Tested for *MET* Amplification and Therapy Response in the MD Anderson Phase I Clinic. *Clin. Cancer Res.* **2014**, *20*, 6336–6345. [CrossRef]

58. Reams, A.B.; Roth, J.R. Mechanisms of gene duplication and amplification. *Cold Spring Harb. Perspect. Biol.* **2015**, *7*, a016592. [CrossRef]

59. Buckingham, L.E.; Coon, J.S.; Morrison, L.E.; Jacobson, K.K.; Jewell, S.S.; Kaiser, K.A.; Mauer, A.M.; Muzzafar, T.; Polowy, C.; Basu, S.; et al. The Prognostic Value of Chromosome 7 Polysomy in Non-small Cell Lung Cancer Patients Treated with Gefitinib. *J. Thorac. Oncol.* **2007**, *2*, 414–422. [CrossRef]

60. Tong, J.H.; Yeung, S.F.; Chan, A.W.H.; Chung, L.Y.; Chau, S.L.; Lung, R.W.M.; Tong, C.Y.; Chow, C.; Tin, E.K.Y.; Yu, Y.H.; et al. MET Amplification and Exon 14 Splice Site Mutation Define Unique Molecular Subgroups of Non–Small Cell Lung Carcinoma with Poor Prognosis. *Clin. Cancer Res.* **2016**, *22*, 3048–3056. [CrossRef]

61. Le, X. Heterogeneity in MET-Aberrant NSCLC. *J. Thorac. Oncol.* **2021**, *16*, 504–506. [CrossRef]

62. Kron, A.; Scheffler, M.; Heydt, C.; Ruge, L.; Schaepers, C.; Eisert, A.-K.; Merkelbach-Bruse, S.; Riedel, R.; Nogova, L.; Fischer, R.N.; et al. Genetic Heterogeneity of MET-Aberrant NSCLC and Its Impact on the Outcome of Immunotherapy. *J. Thorac. Oncol.* **2021**, *16*, 572–582. [CrossRef] [PubMed]

63. Guo, R.; Luo, J.; Chang, J.; Rekhtman, N.; Arcila, M.; Drilon, A. MET-dependent solid tumours—Molecular diagnosis and targeted therapy. *Nat. Rev. Clin. Oncol.* **2020**, *17*, 569–587. [CrossRef] [PubMed]

64. Schubart, C.; Stöhr, R.; Tögel, L.; Fuchs, F.; Sirbu, H.; Seitz, G.; Seggewiss-Bernhardt, R.; Leistner, R.; Sterlacci, W.; Vieth, M.; et al. *MET* Amplification in Non-Small Cell Lung Cancer (NSCLC)—A Consecutive Evaluation Using Next-Generation Sequencing (NGS) in a Real-World Setting. *Cancers* **2021**, *13*, 5023. [CrossRef] [PubMed]

65. Fan, Y.; Sun, R.; Wang, Z.; Zhang, Y.; Xiao, X.; Liu, Y.; Xin, B.; Xiong, H.; Lu, D.; Ma, J. Detection of MET amplification by droplet digital PCR in peripheral blood samples of non-small cell lung cancer. *J. Cancer Res. Clin. Oncol.* **2022**, *148*, 1–11. [CrossRef] [PubMed]

66. Hu, T.; Chitnis, N.; Monos, D.; Dinh, A. Next-generation sequencing technologies: An overview. *Hum. Immunol.* **2021**, *82*, 801–811. [CrossRef] [PubMed]

67. Hung, S.S.; Meissner, B.; Chavez, E.A.; Ben-Neriah, S.; Ennishi, D.; Jones, M.R.; Shulha, H.P.; Chan, F.C.; Boyle, M.; Kridel, R.; et al. Assessment of Capture and Amplicon-Based Approaches for the Development of a Targeted Next-Generation Sequencing Pipeline to Personalize Lymphoma Management. *J. Mol. Diagn.* **2018**, *20*, 203–214. [CrossRef]

68. Samorodnitsky, E.; Jewell, B.M.; Hagopian, R.; Miya, J.; Wing, M.R.; Lyon, E.; Damodaran, S.; Bhatt, D.; Reeser, J.W.; Datta, J.; et al. Evaluation of Hybridization Capture Versus Amplicon-Based Methods for Whole-Exome Sequencing. *Hum. Mutat.* **2015**, *36*, 903–914. [CrossRef]

69. Wang, J.; Yu, H.; Zhang, V.W.; Tian, X.; Feng, Y.; Wang, G.; Gorman, E.; Wang, H.; Lutz, R.E.; Schmitt, E.S.; et al. Capture-based high-coverage NGS: A powerful tool to uncover a wide spectrum of mutation types. *Genet. Med.* **2016**, *18*, 513–521. [CrossRef]

70. Wong, S.Q.; Li, J.; Salemi, R.; Sheppard, K.E.; Do, H.; Tothill, R.W.; McArthur, G.A.; Dobrovic, A. Targeted-capture massively-parallel sequencing enables robust detection of clinically informative mutations from formalin-fixed tumours. *Sci. Rep.* **2013**, *3*, 3494. [CrossRef]

71. Hartmaier, R.J.; Markovets, A.A.; Ahn, M.J.; Sequist, L.V.; Han, J.Y.; Cho, B.C.; Yu, H.A.; Kim, S.W.; Yang, J.C.; Janne, P.A.; et al. Osimertinib + Savolitinib to Overcome Acquired MET-Mediated Resistance in Epidermal Growth Factor Receptor Mutated MET-Amplified Non-Small Cell Lung Cancer: TATTON. *Cancer Discov.* **2023**, *13*, 98–113. [CrossRef]

72. Wu, Y.L.; Zhang, L.; Kim, D.W.; Liu, X.; Lee, D.H.; Yang, J.C.; Ahn, M.-J.; Vansteenkiste, J.F.; Su, W.-C.; Tan, D.S.; et al. Phase Ib/II Study of Capmatinib (INC280) Plus Gefitinib After Failure of Epidermal Growth Factor Receptor (EGFR) Inhibitor Therapy in Patients With EGFR-Mutated, MET Factor-Dysregulated Non-Small-Cell Lung Cancer. *J. Clin. Oncol.* **2018**, *36*, 3101–3109, Erratum in *J. Clin. Oncol.* **2019**, *37*, 261. [CrossRef] [PubMed]

73. Smit, E.F.; Dooms, C.; Raskin, J.; Nadal, E.; Tho, L.M.; Le, X.; Mazieres, J.; Hin, H.; Morise, M.; Wu, Y.L.; et al. INSIGHT 2: A phase II study of tepotinib plus osimertinib in *MET*-amplified NSCLC and first-line osimertinib resistance. *Future Oncol.* **2022**, *18*, 1039–1054. [CrossRef] [PubMed]

74. Heydt, C.; Becher, A.-K.; Wagener-Ryczek, S.; Ball, M.; Schultheis, A.M.; Schallenberg, S.; Rüsseler, V.; Büttner, R.; Merkelbach-Bruse, S. Comparison of in situ and extraction-based methods for the detection of MET amplifications in solid tumors. *Comput. Struct. Biotechnol. J.* **2019**, *17*, 1339–1347. [CrossRef] [PubMed]

75. Peng, L.-X.; Jie, G.-L.; Li, A.-N.; Liu, S.-Y.; Sun, H.; Zheng, M.-M.; Zhou, J.-Y.; Zhang, J.-T.; Zhang, X.-C.; Zhou, Q.; et al. MET amplification identified by next-generation sequencing and its clinical relevance for MET inhibitors. *Exp. Hematol. Oncol.* **2021**, *10*, 52. [CrossRef]

76. Lai, G.G.Y.; Lim, T.H.; Lim, J.; Liew, P.J.R.; Kwang, X.L.; Nahar, R.; Aung, Z.W.; Takano, A.; Lee, Y.Y.; Lau, D.P.X.; et al. Clonal MET Amplification as a Determinant of Tyrosine Kinase Inhibitor Resistance in Epidermal Growth Factor Receptor–Mutant Non–Small-Cell Lung Cancer. *J. Clin. Oncol.* **2019**, *37*, 876–884. [CrossRef]

77. Sakai, H.; Morise, M.; Kato, T.; Matsumoto, S.; Sakamoto, T.; Kumagai, T.; Tokito, T.; Atagi, S.; Kozuki, T.; Tanaka, H.; et al. Tepotinib in patients with NSCLC harbouring *MET* exon 14 skipping: Japanese subset analysis from the Phase II VISION study. *Jpn. J. Clin. Oncol.* **2021**, *51*, 1261–1268. [CrossRef]

78. Drilon, A.; Cappuzzo, F.; Ou, S.-H.I.; Camidge, D.R. Targeting MET in Lung Cancer: Will Expectations Finally Be MET? *J. Thorac. Oncol.* **2017**, *12*, 15–26. [CrossRef]

79. Spigel, D.R.; Ervin, T.J.; Ramlau, R.A.; Daniel, D.B.; Goldschmidt, J.H., Jr.; Blumenschein, G.R., Jr.; Krzakowski, M.J.; Robinet, G.; Godbert, B.; Barlesi, F.; et al. Randomized Phase II Trial of Onartuzumab in Combination With Erlotinib in Patients With Advanced Non–Small-Cell Lung Cancer. *J. Clin. Oncol.* **2013**, *31*, 4105–4114. [CrossRef]

80. Stenger, M. Calculating H-Score. 2015. Available online: https://ascopost.com/issues/april-102015/calculating-h-score/ (accessed on 8 August 2022).

81. Aisner, D.L.; Sholl, L.M.; Berry, L.D.; Rossi, M.R.; Chen, H.; Fujimoto, J.; Moreira, A.L.; Ramalingam, S.S.; Villaruz, L.C.; Otterson, G.A.; et al. The Impact of Smoking and TP53 Mutations in Lung Adenocarcinoma Patients with Targetable Mutations—The Lung Cancer Mutation Consortium (LCMC2). *Clin. Cancer Res.* **2018**, *24*, 1038–1047. [CrossRef]

82. Mignard, X.; Ruppert, A.-M.; Antoine, M.; Vasseur, J.; Girard, N.; Mazières, J.; Moro-Sibilot, D.; Fallet, V.; Rabbe, N.; Thivolet-Bejui, F.; et al. c-MET Overexpression as a Poor Predictor of MET Amplifications or Exon 14 Mutations in Lung Sarcomatoid Carcinomas. *J. Thorac. Oncol.* **2018**, *13*, 1962–1967. [CrossRef] [PubMed]

83. Camidge, D.R.; Otterson, G.A.; Clark, J.W.; Ou, S.-H.I.; Weiss, J.; Ades, S.; Shapiro, G.I.; Socinski, M.A.; Murphy, D.A.; Conte, U.; et al. Crizotinib in Patients with MET-Amplified NSCLC. *J. Thorac. Oncol.* **2021**, *16*, 1017–1029. [CrossRef] [PubMed]

84. Sterlacci, W.; Fiegl, M.; Gugger, M.; Bubendorf, L.; Savic, S.; Tzankov, A. MET overexpression and gene amplification: Prevalence, clinico-pathological characteristics and prognostic significance in a large cohort of patients with surgically resected NSCLC. *Virchows Arch.* **2017**, *471*, 49–55. [CrossRef] [PubMed]

85. Spigel, D.R.; Edelman, M.; O'Byrne, K.; Paz-Ares, L.; Mocci, S.; Phan, S.; Shames, D.S.; Smith, D.; Yu, W.; Paton, V.E.; et al. Results From the Phase III Randomized Trial of Onartuzumab Plus Erlotinib Versus Erlotinib in Previously Treated Stage IIIB or IV Non–Small-Cell Lung Cancer: METLung. *J. Clin. Oncol.* **2017**, *35*, 412–420. [CrossRef]

86. Neal, J.W.; E Dahlberg, S.; A Wakelee, H.; Aisner, S.C.; Bowden, M.; Huang, Y.; Carbone, D.P.; Gerstner, G.J.; Lerner, R.E.; Rubin, J.L.; et al. Erlotinib, cabozantinib, or erlotinib plus cabozantinib as second-line or third-line treatment of patients with EGFR wild-type advanced non-small-cell lung cancer (ECOG-ACRIN 1512): A randomised, controlled, open-label, multicentre, phase 2 trial. *Lancet Oncol.* **2016**, *17*, 1661–1671. [CrossRef]

87. Cai, Y.-R.; Zhang, H.-Q.; Zhang, Z.-D.; Mu, J.; Li, Z.-H. Detection of MET and SOX2 amplification by quantitative real-time PCR in non-small cell lung carcinoma. *Oncol. Lett.* **2011**, *2*, 257–264. [CrossRef]

88. Zhang, Y.; Tang, E.-T.; Du, Z. Detection of MET Gene Copy Number in Cancer Samples Using the Droplet Digital PCR Method. *PLoS ONE* **2016**, *11*, e0146784. [CrossRef]

89. Xu, C.-W.; Wang, W.-X.; Wu, M.-J.; Zhu, Y.-C.; Zhuang, W.; Lin, G.; Du, K.-Q.; Huang, Y.-J.; Chen, Y.-P.; Chen, G.; et al. Comparison of the *c-MET* gene amplification between primary tumor and metastatic lymph nodes in non-small cell lung cancer. *Thorac. Cancer* **2017**, *8*, 417–422. [CrossRef]

90. Bardelli, A.; Corso, S.; Bertotti, A.; Hobor, S.; Valtorta, E.; Siravegna, G.; Sartore-Bianchi, A.; Scala, E.; Cassingena, A.; Zecchin, D.; et al. Amplification of the *MET* Receptor Drives Resistance to Anti-EGFR Therapies in Colorectal Cancer. *Cancer Discov.* **2013**, *3*, 658–673. [CrossRef]

91. Galimi, F.; Torti, D.; Sassi, F.; Isella, C.; Corà, D.; Gastaldi, S.; Ribero, D.; Muratore, A.; Massucco, P.; Siatis, D.; et al. Genetic and Expression Analysis of MET, MACC1, and HGF in Metastatic Colorectal Cancer: Response to Met Inhibition in Patient Xenografts and Pathologic Correlations. *Clin. Cancer Res.* **2011**, *17*, 3146–3156. [CrossRef]

92. Pasquini, G.; Giaccone, G. C-MET inhibitors for advanced non-small cell lung cancer. *Expert Opin. Investig. Drugs* **2018**, *27*, 363–375. [CrossRef] [PubMed]

93. Le, X.; Puri, S.; Negrao, M.V.; Nilsson, M.B.; Robichaux, J.; Boyle, T.; Hicks, J.K.; Lovinger, K.L.; Roarty, E.; Rinsurongkawong, W.; et al. Landscape of EGFR-Dependent and -Independent Resistance Mechanisms to Osimertinib and Continuation Therapy Beyond Progression in *EGFR*-Mutant NSCLC. *Clin. Cancer Res.* **2018**, *24*, 6195–6203. [CrossRef] [PubMed]

94. Yang, J.-J.; Fang, J.; Shu, Y.-Q.; Chang, J.-H.; Chen, G.-Y.; He, J.X.; Li, W.; Liu, X.-Q.; Yang, N.; Zhou, C.; et al. A phase Ib study of the highly selective MET-TKI savolitinib plus gefitinib in patients with EGFR-mutated, MET-amplified advanced non-small-cell lung cancer. *Investig. New Drugs* **2021**, *39*, 477–487. [CrossRef] [PubMed]

95. McCoach, C.E.; Yu, A.; Gandara, D.R.; Riess, J.W.; Vang, D.P.; Li, T.; Lara, P.N.; Gubens, M.; Lara, F.; Mack, P.C.; et al. Phase I/II Study of Capmatinib Plus Erlotinib in Patients With MET-Positive Non–Small-Cell Lung Cancer. *JCO Precis. Oncol.* **2021**, *1*, 177–190. [CrossRef] [PubMed]

96. Wu, Y.-L.; Cheng, Y.; Zhou, J.; Lu, S.; Zhang, Y.; Zhao, J.; Kim, D.-W.; Soo, R.A.; Kim, S.-W.; Pan, H.; et al. Tepotinib plus gefitinib in patients with EGFR-mutant non-small-cell lung cancer with MET overexpression or MET amplification and acquired resistance to previous EGFR inhibitor (INSIGHT study): An open-label, phase 1b/2, multicentre, randomised trial. *Lancet Respir. Med.* **2020**, *8*, 1132–1143. [CrossRef]

97. Camidge, D.R.; Barlesi, F.; Goldman, J.W.; Morgensztern, D.; Heist, R.; Vokes, E.; Spira, A.; Angevin, E.; Su, W.-C.; Kelly, K.; et al. Phase Ib Study of Telisotuzumab Vedotin in Combination With Erlotinib in Patients With c-Met Protein-Expressing Non-Small-Cell Lung Cancer. *J. Clin. Oncol.* **2022**, *3*, JCO2200739. [CrossRef]

98. Camidge, D.R.; Moran, T.; Demedts, I.; Grosch, H.; Mileham, K.; Molina, J.; Juan-Vidal, O.; Bepler, G.; Goldman, J.W.; Park, K.; et al. A Randomized, Open-Label Phase II Study Evaluating Emibetuzumab Plus Erlotinib and Emibetuzumab Monotherapy in MET Immunohistochemistry Positive NSCLC Patients with Acquired Resistance to Erlotinib. *Clin. Lung Cancer* **2022**, *23*, 300–310. [CrossRef]

99. Available online: https://oncologypro.esmo.org/meeting-resources/esmo-congress-2021/orchard-osimertinib-savolitinib-interim-analysis-a-biomarker-directed-phase-ii-platform-study-in-patients-pts-with-advanced-non-small-cell-lun (accessed on 16 September 2021).

100. Felip, E.; Minotti, V.; Soo, R.; Wolf, J.; Solomon, B.; Tan, D.; Ardizzoni, A.; Lee, D.; Sequist, L.; Barlesi, F.; et al. 1284P MET inhibitor capmatinib plus EGFR tyrosine kinase inhibitor nazartinib for EGFR-mutant non-small cell lung cancer. *Ann. Oncol.* **2020**, *31*, S829–S830. [CrossRef]

101. Available online: https://www.astrazeneca.com/media-centre/press-releases/2022/tagrisso-plus-savolitinib-demonstrated-49-objective-response-rate-in-lung-cancer-patients-in-savannah-phase-ii-trial.html (accessed on 8 August 2022).

Article

Rilotumumab Resistance Acquired by Intracrine Hepatocyte Growth Factor Signaling

Fabiola Cecchi [1], Karen Rex [2], Joanna Schmidt [2], Cathy D. Vocke [1], Young H. Lee [1], Sandra Burkett [3], Daniel Baker [2], Michael A. Damore [2], Angela Coxon [2], Teresa L. Burgess [2] and Donald P. Bottaro [1,*]

1   Urologic Oncology Branch, Center for Cancer Research, National Cancer Institute, National Institutes of Health, Bethesda, MD 20892, USA
2   Amgen, Inc., Thousand Oaks, CA 91320, USA
3   Molecular Cytogenetics Core Facility, Mouse Cancer Genetics Program, Center for Cancer Research, National Cancer Institute, National Institutes of Health, Frederick, MD 21702, USA
*   Correspondence: dbottaro@helix.nih.gov; Tel.: +1-301-793-7198

Simple Summary: Drug resistance is a long-standing impediment to effective systemic cancer therapy and acquired drug resistance is a growing problem for new therapeutics that otherwise have shown significant successes in disease control. Hepatocyte growth factor (HGF)/Met receptor pathway signaling is frequently involved in cancer and is widely targeted in drug development. We found that resistance to the HGF-neutralizing antibody drug candidate rilotumumab in glioblastoma cells was acquired through HGF overproduction and misfolding, which led to stress-response signaling and redirected transport inside cells that sequestered rilotumumab and misfolded HGF from native HGF and activated Met receptor. Resistant cells were more malignant but retained their sensitivity to Met kinase inhibition and gained sensitivity to inhibition of stress signaling and cholesterol biosynthesis. Defining this rapidly acquired, multisystem scheme improves our understanding of drug resistance and suggests strategies for early detection and intervention.

Abstract: Drug resistance is a long-standing impediment to effective systemic cancer therapy and acquired drug resistance is a growing problem for molecularly-targeted therapeutics that otherwise have shown unprecedented successes in disease control. The hepatocyte growth factor (HGF)/Met receptor pathway signaling is frequently involved in cancer and has been a subject of targeted drug development for nearly 30 years. To anticipate and study specific resistance mechanisms associated with targeting this pathway, we engineered resistance to the HGF-neutralizing antibody rilotumumab in glioblastoma cells harboring autocrine HGF/Met signaling, a frequent abnormality of this brain cancer in humans. We found that rilotumumab resistance was acquired through an unusual mechanism comprising dramatic HGF overproduction and misfolding, endoplasmic reticulum (ER) stress-response signaling and redirected vesicular trafficking that effectively sequestered rilotumumab and misfolded HGF from native HGF and activated Met. Amplification of *MET* and *HGF* genes, with evidence of rapidly acquired intron-less, reverse-transcribed copies in DNA, was also observed. These changes enabled persistent Met pathway activation and improved cell survival under stress conditions. Point mutations in the HGF pathway or other complementary or downstream growth regulatory cascades that are frequently associated with targeted drug resistance in other prevalent cancer types were not observed. Although resistant cells were significantly more malignant, they retained sensitivity to Met kinase inhibition and acquired sensitivity to inhibition of ER stress signaling and cholesterol biosynthesis. Defining this mechanism reveals details of a rapidly acquired yet highly-orchestrated multisystem route of resistance to a selective molecularly-targeted agent and suggests strategies for early detection and effective intervention.

Keywords: acquired drug resistance; glioblastoma; hepatocyte growth factor; Met; rilotumumab

Citation: Cecchi, F.; Rex, K.; Schmidt, J.; Vocke, C.D.; Lee, Y.H.; Burkett, S.; Baker, D.; Damore, M.A.; Coxon, A.; Burgess, T.L.; et al. Rilotumumab Resistance Acquired by Intracrine Hepatocyte Growth Factor Signaling. *Cancers* **2023**, *15*, 460. https://doi.org/10.3390/cancers15020460

Academic Editors: Jan Trøst Jørgensen and Jens Mollerup

Received: 5 December 2022
Revised: 4 January 2023
Accepted: 6 January 2023
Published: 11 January 2023

1. Introduction

Acquired drug resistance is a long-standing problem of cancer therapeutics. For example, chemoresistance to the DNA-alkylating agent temozolomide occurs in >90% of recurrent glioblastoma multiforme (GBM), often through increased expression of *MGMT* which encodes an alkyltransferase capable of repairing the DNA damage. The issue has become even more vexing with the development of highly selective agents such as those targeting the epidermal and hepatocyte growth factor (EGF and HGF, respectively) pathways. Acquired resistance to gefitinib or erlotinib in lung adenocarcinomas is a prevalent response to prolonged treatment and occurs through HGF pathway activation and other molecular mechanisms ([1–9], reviewed in [10,11]). Anticipating acquired resistance and understanding its basis should aid in the development of clinical strategies to prevent or circumvent its occurrence.

HGF, through its receptor tyrosine kinase Met, regulates mitogenesis, motogenesis, and morphogenesis in a range of cellular targets during development and homeostasis [12]. HGF/Met signaling also contributes to oncogenesis and tumor progression in many human malignancies, including GBM. The *HGF* and *MET* genes are expressed in human glioma and medulloblastoma, where their increased relative abundance frequently correlates with tumor grade, tumor blood vessel density and poor prognosis [12]. The *HGF* and *MET* genes are collectively altered (amplified, overexpressed and/or mutated) in published and provisional GBM datasets compiled by The Cancer Genome Atlas (TCGA) Research Network in 7.5% of 206 cases [13] and 18% of 291 [14] and 528 cases [15]; 2% of the latter 819 cases show concomitant overexpression or amplification of both genes [14,15]. Overexpression of *HGF* and/or *MET* in brain-tumor-derived cells enhances their tumorigenicity and growth, and inhibition of HGF or Met in experimental tumor xenografts suppresses tumor growth and angiogenesis [16–19]. Elevated levels of HGF protein in human cerebrospinal fluid are associated with mortality and GBM recurrence [20]. *MET* expression in GBM-derived cultures enriched for stem and progenitor cells was associated with mesenchymal and proneural gene signature GBM subtypes and with invasive and stem-like phenotypes [21]. Consistent with the suspected role of HGF in GBM progression, potent and highly selective antagonists of HGF–Met and Met–ATP binding interactions significantly inhibited subcutaneous and intracranial brain tumor growth in mice [22–24]. Although both paracrine and autocrine HGF are believed to enhance GBM growth, autocrine HGF/Met signaling in GBM cell lines reliably predicted sensitivity to pathway inhibition in vivo [25].

Rilotumumab (AMG102) is a fully human neutralizing monoclonal antibody against HGF that potently inhibited the growth of GBM-derived cell lines such as U87 MG, which has HGF/Met autocrine signaling, in culture and in tumor xenografts as a single agent. In addition, rilotumumab enhanced the efficacy of temozolomide or docetaxel in U87 MG tumor-bearing mice and enhanced the killing effect of ionizing radiation on U87 MG cells in vitro and in vivo [22,25–28]. Rilotumumab was found to be safe and well tolerated in phase I human clinical trials [29] and has been tested in multiple phase II clinical trials [30]. In anticipation of acquired resistance to rilotumumab in GBM, models were generated by growing U87 MG cells in maximally effective drug concentrations and by escalating dose treatment of mice implanted with U87 MG cells. Rilotumumab-resistant cell lines and tumors remained sensitive to a selective small molecule Met tyrosine kinase inhibitor, and therefore were dependent on HGF/Met signaling, which occurred through HGF overproduction and misfolding-induced ER stress-response signaling and rilotumumab uptake-driven subversion of vesicular trafficking that secluded Met activation from rilotumumab.

2. Materials and Methods

2.1. Reagents and Cell Culture

Rilotumumab was prepared by Amgen Inc. (Thousand Oaks, CA, USA) at a stock concentration of 30 mg/mL. An IgG2 antibody (Amgen Inc.) was used as an isotype control. Recombinant human HGF/NK1 (HGF variant 5) was prepared as described [31].

All three agents were stored at −80 °C until dilution. AMG517 (Ref. [32], identified therein as compound #22) was formulated in soybean oil at a concentration of 10mL/kg. Simvastatin (HMG-CoA reductase) and the PERK inhibitor GSK2656157 were purchased from SelleckChem.

The U87 MG-derived, rilotumumab-resistant cell line (designated U87 MG/HNR for <u>H</u>GF <u>N</u>eutralization <u>R</u>esistant) was generated by growing U87 MG cells (also known as HTB-14, obtained from ATCC, Manassas, VA, USA) continuously in normal growth media [2] plus rilotumumab at 100 nM for a period of 5 days followed by 115 days of rilotumumab at 600 nM for a total treatment period of 120 days. Aliquots of these cells were frozen for subsequent subculture and analysis. Subcultures of U87 MG/HNR grown in 1 μM rilotumumab showed no additional phenotypic changes. Cells grown for experiments were maintained in rilotumumab unless otherwise noted. For the in vivo experiments, U87 MG parental and U87 MG/HNR cell lines were cultured in DMEM High Glucose media (Gibco) with 10% fetal bovine serum (FBS Gibco) and 1X L-glutamine (Gibco). Other cell lines used were also obtained from ATCC (Manassas, VA, USA).

2.2. Quantitative Immunoassays

HGF, Met, and phosphoMet content in cell lysates, blood plasma, or conditioned media were determined using 2-site electrochemiluminescent immunoassays as described previously [31]. Met activation as indicated by phosphoMet content in cell lysates included parallel detection with anti-receptor antibodies and specific anti-phospho-receptor antibodies or the monoclonal anti-phosphotyrosine (anti-pY) antibody clone 4G10. Cultured cells were serum-deprived for 16–24 h in the presence or absence of various agents as noted in the text prior to stimulation for 20 min with HGF (1 nM) at 37 °C alone or in combination with rilotumumab or compound A at the indicated concentrations. The cells were extracted with ice cold buffer containing non-ionic detergent, protease and phosphatase inhibitors; cleared extracts were applied to plates containing immobilized HGF or Met capture antibody and detected with anti-HGF, anti-Met or anti-pY. For some of the experiments described in Figures 7 and 8, TX-100-insoluble cell extracts were solubilized in the same buffer containing 1% SDS and diluted 10-fold before measuring protein, HGF, rilotumumab, Met or pMet content.

All measurements were made on triplicate samples; the protein content of all samples was determined prior to immunoassay. All samples were adjusted to 0.5–0.7 mg/mL prior to immunoassay and all immunoassay values obtained were normalized to actual total protein values. HGF and Met content assays include purified recombinant reference standards for absolute quantitation. GraphPad Prism software version 5.0 was used for all statistical analyses. Protein content values were interpolated from standard curves by nonlinear regression analysis. All other statistical tests are described in the Results and Figure Legends.

2.3. Cell Proliferation and Anchorage Independent Growth Assays

For proliferation assays, U87 MG cell lines (5×10^4 cells per well) were seeded in 6-well culture plates in quadruplicate. The cell number per well was measured after 2, 4, 5 and 6 days of growth by removing cells with trypsin, collecting via centrifugation and counting suspended cells in triplicate in an automated cell counter. Assays for colony formation in soft agar were performed as described [31]; colonies were treated with 3,4,5-dimethyl thiazole-2-yl-2,5-diphenyltetrazolium bromide and the dye product was eluted with MeOH and quantitated by absorbance at 590 nm. Differences between mean values were determined by an unpaired *t*-test with Welch's correction using GraphPad Prism v5.0.

2.4. Tumorigenicity Assays

All experiments involving animals were performed in accordance with NIH Guidelines for Care and Use of Laboratory Animals, using institutionally reviewed and approved protocols at Amgen (Animal Protocol 2015-01243; Thousand Oaks, CA, USA) or the National

Cancer Institute (NIH Animal Study Protocol UOB-009; Bethesda, MD, USA). U87 MG and derived cell lines were injected subcutaneously into athymic nude mice (Jackson Laboratories and Harlan Laboratories; n = 10 per group) and tumor volumes were measured at regular intervals as described previously [26,31]. Mice were treated with rilotumumab by intraperitoneal injection every two days or with AMG517 by oral gavage every day at doses indicated in the text. Animals were sacrificed and tumors were removed for histopathology by conventional methods. Tumor growth curves were fitted by regression analysis ($R^2 > 0.95$) using GraphPad Prism software version 5.0. Measurement of plasma HGF protein levels was performed using the two-site electrochemiluminescent method [31], or by SDS-PAGE, electrophoretic transfer to PVDF membranes and immunoblotting using a polyclonal antibody directed against the human HGF amino-terminal sequence (Santa Cruz Biotechnology SC-1357). Levels of HGF in serum in U87 MG/HNR tumor-bearing mice were determined using either the Quantikine ELISA Human HGF Immunoassay (R&D Systems) or electrochemiluminescent immunoassay as described above. Data are expressed as mean $\pm$ SEM. A one-way ANOVA followed by Dunnett's post hoc testing was used to determine statistically significant differences. Other statistical analyses, curve fitting and IC_{50} determinations were performed using GraphPad Prism v5.0 as indicated in the Results and Figure Legends.

2.5. CGH and mRNA Profiling Arrays and Comparisons to TCGA Datasets

Genomic DNA was purified (DNeasy Blood and Tissue Kit, Qiagen, Inc., Valencia, CA) and mixed with either Cy3- or Cy5-labeled primer solution (0.5 OD; Cy-labeled primer, Trilink BioTechnologies, San Diego, CA) in 125 mM Tris HCl (pH 6.8) and 12.5 mM $MgCl_2$. After incubation at 99 °C for 10 min, reactions were brought to a total volume of 50 mL with 1 mM dNTPs and 50 U Exo-Klenow (Life Technologies, Carlsbad, CA). Reactions were incubated for 2 h at 37 °C prior to termination by heating to 65 °C for 10 min. Labeled product was purified using ethanol precipitation and quantified using a Nanodrop ND-1000 (Nanodrop Products, Wilmington, DE, USA).

Arrays were hybridized following the Oligonucleotide Array-Based comparative genomic hybridization (CGH) for Genomic DNA Analysis (Agilent Technologies, Santa Clara, CA, USA), with modifications. Briefly, 10 μg of Cy5-labeled experimental sample was mixed with 10 μg of Cy3-control DNA (pool of 40 normal male donor gDNAs) and incubated in a hybridization cocktail according to the manufacturer's instructions. Following incubation, samples were hybridized against the SurePrint G3 Human CGH Microarray 2 × 400K for 40 h at 65 °C, rotating at 20 rpm. Samples were washed according to the manufacturer's instructions and scanned at 2 μm using 100% PMT for Cy3 and 50% PMT for Cy5 on an Agilent DNA Microarray Scanner Model G2505C. Scanned data were feature extracted using Agilent Feature Extraction v10, and .txt files were imported to Agilent Genomic Workbench software to generate $\log_2$ ratio scores and aberration calls.

Total RNA was purified (RNeasy Mini Kit, Qiagen, Inc., Valencia, CA, USA) and profiled after the Agilent One-Color Microarray-Based Gene Expression Analysis Protocol, hybridizing to a customized Agilent Human Whole Genome V2 Microarray (Agilent Technologies, Santa Clara, CA, USA) with every reporter replicated at least four times (AMADID 026822). Array data were extracted using Agilent Feature Extraction Software (version 10.7) and imported into Rosetta Resolver software (version 7.2.2) for analysis.

Following background correction and normalization of microarray data, a difference set of significant mRNA expression changes between the parental and resistant cell lines was derived (defined as >1.17-fold and $p < 1.00 \times 10^{-4}$; 11,523 probe IDs) and a subset of 7688 genes that were significantly modulated 1.5-fold or greater (Table S1) were uploaded for Ingenuity Pathway Analysis (IPA) core analysis with either User Data Set or the Agilent Whole Genome Microarray 4 × 44k v2 probe set as reference set; both direct and indirect relationships were included. Default IPA settings were used for Networks (Interaction), Data Sources (all), Species (all), Tissues and Cell Lines (all) and Mutations (all). Probability (*p*) values for significant overlap ($p < 0.05$) with gene sets defining Networks, Biofunctions,

Canonical Pathways and Upstream Regulators were derived using the right-tailed Fisher's Exact test (single correlations) or the Benjamin–Hochberg Multiple Testing Correction (grouped correlations). All gene names and symbols listed conform to current HUGO Gene Nomenclature Committee convention.

A comparison of the expression profiling dataset (Table S1) with the TCGA GBM datasets [13–15] was performed using tools available through the cBioPortal [33,34].

2.6. Fluorescence In Situ Hybridization (FISH) Karyotyping

Multi-color FISH (M-FISH) karyotyping was performed in the Comparative Molecular Cytogenetic Core Facility at the National Cancer Institute. Briefly, chromosome preparations were obtained from established drug-resistant cell cultures, which were suspected to have generated double-minute chromosomes, by standard procedures. Slides were prepared and incubated overnight for use in FISH analysis of *MET* and *HGF* genes. For the detection of the *HGF* gene BAC Clone RP11-552M24 (Empire Genomics, Buffalo, NY, USA) was used with CEP 17 (Vysis, Abbott Molecular, Chicago, IL, USA) for verification of the location of the BAC clone. To analyze the *MET* gene, a DNA probe from Kreatech was used. Hybridization was carried out in a humidity chamber at 37 °C for 16 h according to standard protocols. The post-hybridization rapid-wash procedure was used with 0.4 × SSC at 72 °C for 4 min. Detection was carried out following the manufacturer's protocol. Spectral images of the hybridized metaphases were acquired using an SD301 SpectraCubeTM system (Applied Spectral Imaging Inc., Carlsbad, CA, USA) mounted on an epi-fluorescence Axioplan 2 microscope (Zeiss, Oberkochen, Germany) using Spectral Imaging 6.0 acquisition software (Applied Spectral Imaging Inc., Carlsbad, CA, USA).

2.7. Real-Time Quantitative PCR

RNA was isolated from cell pellets from subconfluent cell cultures using the RNeasy kit (Qiagen, Valencia, CA) following the manufacturer's instructions. Total RNA (20 ng) was reverse-transcribed with random primers to cDNA using the SuperScript III First-Strand Synthesis System (Invitrogen, Carlsbad, CA, USA). Variant-specific HGF primers were designed with the following DNA sequences: CV1: ACGAACACAGCTTTTTGCCTTC; CV1A: CCATGATACCACACGAACAC; CV2: CACACGAACACAGCTATCGG; CV3: CT-GAACACTGAGGAATGTCAC; CV4A: CCCACATGGCATTCAGGTT; CV5: CCATGGTGC-TATACTCTTGAC; CV6: GAAGTTCACCATCAGTTGAGAG; CV7: CTTGACCTTGGATG-CATTCAG. These primer pairs distinguish the HGF variant cDNAs as follows: CV1A/CV3 (all variants), CV1/CV3 (variants 1, 2 and 5), CV2/CV3 (variants 3 and 4), CV1/CV4A (variant 5), CV5/CV7 (variants 1 and 3), CV5/CV6 (variants 2 and 4). Real-time quantitative PCR was performed with SYBR-Green PCR Master Mix (Applied Biosystems, Foster City, CA, USA) using an ABI 7000 real-time PCR system (Applied Biosystems) following the manufacturer's protocols. All reactions were run in triplicate using *PPIA*, *GUSB* and *HPRT* as internal control genes. The relative level of gene expression was evaluated using the delta-delta CT method.

2.8. SDS-PAGE, Immunoblotting, HGF Affinity Chromatography and Ultrafiltration

SDS-PAGE was performed by conventional methods using precast gels (BioRad, Hercules, CA, USA). Electrophoretic transfer to PVDF membrane (Millipore, Burlington, MA, USA) and immunoblot analysis was performed using antibodies identified in the text and figure legends. Heparin-Sepharose CL-6B (GE Healthcare Life Sciences) affinity purification of HGF proteins was performed by batch loading conditioned media for 16 h at 4 °C, followed by stepwise elution with PBS containing increasing NaCl concentrations as listed in the text and figure legends. The 0.8 M NaCl fraction was further subjected to dialysis against PBS for 16 h at 4 °C, followed by centrifugal ultrafiltration using a 30 kDa Microcon concentrator (Millipore). The HGF content of both retentate and pass after filtration was determined by SDS-PAGE and immunoblotting using a standard curve made using purified recombinant HGF and NK1 proteins.

3. Results

3.1. HGF and Met Superabundance in Rilotumumab-Resistant U87 MG Cells

The HGF/Met dependent human GBM-derived cell line U87 MG was grown in continuous exposure to rilotumumab for 120 days to generate a cellular model of acquired resistance. For the first 5 days the rilotumumab concentration was 100 nM, which was then increased to 600 nM. Resistant cells (designated U87 MG/HNR for HGF Neutralization Resistant) displayed altered morphology in 2D culture relative to the parental cell line (Figure 1A,B). Resistant cells were modestly but consistently smaller, had fewer and shorter extended processes, and had fewer cell–cell interactions (visualized as cell clumping) than parental cells. Resistant cells were also noticeably less adherent to plastic or extracellular matrix substrata than the parental cell line. A relatively high number of floating cells also suggested an increased rate of cell death over the parental cell line. Indeed, U87 MG/HNR cells displayed substantially elevated activated caspase 3 relative to U87 MG, suggestive of increased apoptosis (Figure 1C; original immunoblot images for all figures are shown in Supplementary Materials Figure S1) despite a significantly higher rate of growth in culture (Figure 1D). Among the most distinguishing features of U87 MG/HNR was its >10,000-fold higher rate of HGF protein production compared to the parental U87 MG cells (Figure 1E). Met protein content and autophosphorylation level (phospho-Met) were also 8-fold and 80-fold higher than parental cell values, respectively (Figure 1F,G). The ratio of phospho-Met to total Met protein for U87 MG/HNR was similar to that of the normal mammary epithelial cell line 184B5 upon treatment with 1 nM exogenous HGF for 20 min at 37 °C (Figure 1G), indicating that Met maintained steady-state maximum kinase activity. Remarkably, U87 MG/HNR remained sensitive to the selective small-molecule Met kinase inhibitor AMG517 (Ref. [32], identified therein as compound #22); potent suppression of steady-state phospho-Met levels was observed in both U87 MG and U87 MG/HNR (Figure 1H), despite their significantly different total phospho-Met content (note the left vs. right y-axes scale difference).

3.2. U87 MG/HNR Tumorigenesis Is Rilotumumab-Resistant Yet Remains MET-Pathway Dependent

The tumor xenograft growth rate for the resistant cell line was significantly elevated over the parental line, even when one-sixth of the number of cells were implanted (Figure 2A). Plasma levels of human HGF correlated directly with U87 MG/HNR xenograft tumor mass, consistent with a prior report [22] for U87 MG (Figure 2B). In agreement with prior reports [22,26], tumor formation by U87 MG cells was potently suppressed by twice-weekly rilotumumab treatment; doses from 12 to 120 mg/kg caused complete regression of small, preformed xenograft tumors, as did a selective small-molecule Met kinase inhibitor AMG517 administered daily at 60 mg/kg (Figure 2C). As anticipated, rilotumumab treatment regimens that caused complete regression of U87 MG tumors had little impact on tumors arising from U87 MG/HNR cells, yet AMG517 was completely effective in blocking tumor growth by this cell line (Figure 2D). Similar to prior studies, rilotumumab sequestered human HGF in the blood of treated animals in a dose-dependent manner, indicating that target recognition by the antibody was not compromised (Figure 2E).

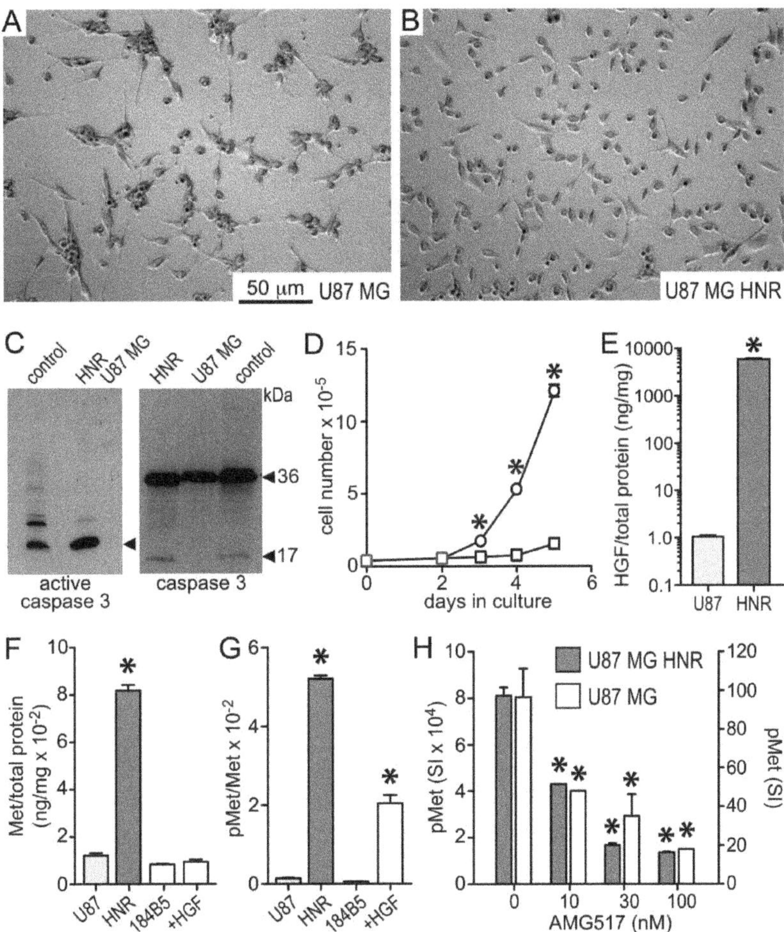

Figure 1. U87 MG and U87 MG/HNR cell morphology, proliferation, HGF, Met and phospho-Met content. Light micrographs of U87 MG (**A**) and U87 MG/HNR cells (**B**) in log growth phase. (**C**) Reduced SDS-PAGE and immunoblot analysis of U87 MG and U87 MG/HNR ("HNR") cell extracts for activated caspase 3 (17 kDa, left panel) and total caspase 3 protein (36 kDa, right panel). Lanes marked "control" contain positive control proteins for active or intact caspase 3 provided by the antibody manufacturer. (**D**) Growth rates of U87 MG (squares) and U87 MG/HNR cells (circles) in culture (mean cell number ± SD, n = 3). (**E**) Secreted HGF protein (mean ng/mg total protein ± SD, n = 3) present in 24 h media from cultured U87 MG (light gray bar, left) or U87 MG/HNR cells (dark gray bar, right). (**F**) Total Met protein content (mean ng Met/mg total protein ± SD, n = 3) of U87 MG (light gray bar), U87 MG/HNR (dark gray bar) and serum-deprived 184B5 normal human mammary epithelial cells (white bars) in the absence (left) or presence of 1 nM HGF (right). (**G**) Phospho-Met (pMet) content (mean phosphoMet/total Met signal intensity ratio per mg total protein ± SD, n = 3) of U87 MG (light gray bar), U87 MG/HNR (dark gray bar) and serum-deprived 184B5 cells (white bars) in the absence (left) or presence of 1 nM HGF (right). (**H**) Phospho-Met (pMet) content (mean phosphoMet/total Met signal intensity ratio per mg total protein ± SD, n = 3) of U87 MG/HNR (gray bars, left *Y*-axis) and U87 MG (white bars, right *Y*-axis) in the absence (0) or presence of indicated concentrations (nM) of AMG517, a selective small-molecule Met kinase inhibitor. Asterisks indicate a significant difference from control ($p < 0.05$). All the whole western blot figures can be found in the supplementary materials.

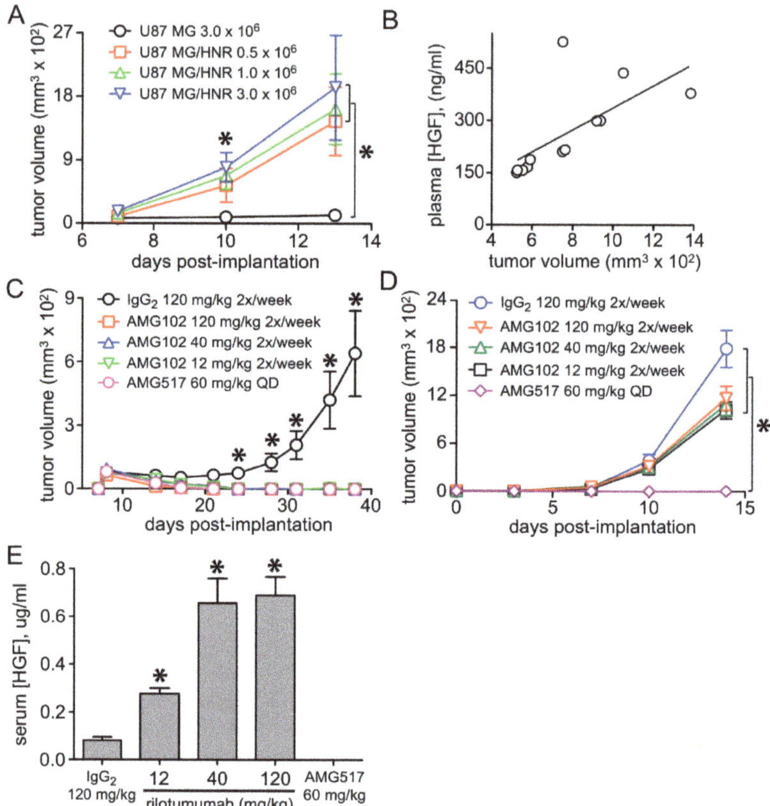

Figure 2. Rilotumumab-resistant U87 MG/HNR cell xenograft growth remains HGF-pathway dependent. (**A**) Tumor xenograft growth (mean tumor volume ± SEM, n = 10) in mice implanted with U87 MG (circles, 3×10^6 cells/animal) or U87 MG/HNR cells (squares, triangles and inverted triangles) at 0.5, 1 or 3×10^6 cells/animal, respectively. (**B**) Mean plasma HGF concentration (ng/mL) vs. tumor volume for mice implanted with U87 MG/HNR cells. (**A,C**) U87 MG tumor xenograft (5×10^6 cells/mouse) growth in mice (mean tumor volume ± SEM, n = 10) treated with control IgG (black circles) or rilotumumab (AMG102) at 12 (green inverted triangles), 40 (blue triangles) or 120 mg/kg (red squares) or compound A at 60 mg/kg (violet circles). (**D**) U87 MG HNR xenograft growth (0.5×10^6 cells/mouse, mean tumor volume ± SEM, n = 10) in mice treated with control IgG (circles) or rilotumumab (AMG102) at 12 (squares), 40 (triangles) or 120 mg/kg (inverted triangles) or the small-molecule kinase inhibitor AMG517 at 60 mg/kg (diamonds). (**E**) Mean (±SD) serum HGF concentration (μg/mL) at study termination in U87 MG/HNR tumor-bearing mice treated as in panel (**D**). Asterisks indicate a significant difference from control ($p < 0.05$).

Consistent with a highly stressed cell state (further detailed below), rilotumumab-resistant U87 MG/HNR xenograft tumors exhibited chronic inflammation, regional necrosis and scattered blood pools (Figure 3A vs. Figure 3B). Higher (20×) magnification views reveal that higher tumor growth rates were directly correlated with mitotic indices in the resistant xenografts (Figure 3C). This magnification also revealed serous accumulations, as well as increased but poorly-organized extracellular matrix production, features not observed in matched-size U87 MG parental tumors (Figure 3D–F).

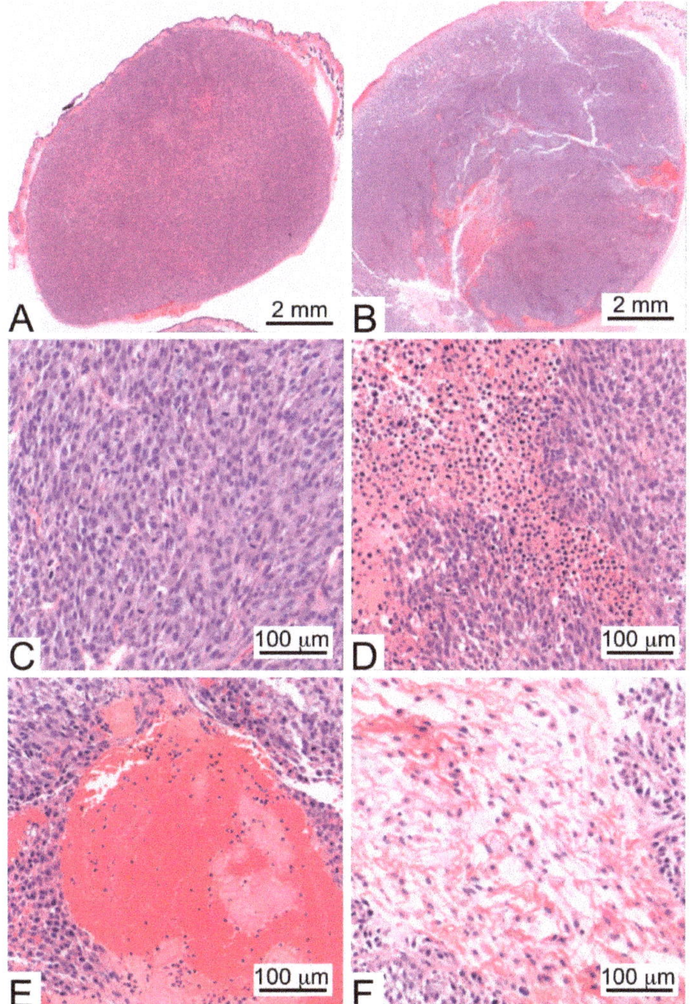

Figure 3. Representative histology of U87 MG and U87 MG/HNR xenograft tumors. Low magnification (1×) of representative tumors derived from (**A**) U87 MG parental and (**B**) U87 MG/HNR cells. (**C**) High magnification (20×) of a U87 MG/HNR tumor illustrating frequent mitotic figures. (**D**) High magnification (20×) of a U87 MG/HNR tumor showing regions of necrosis. (**E**) High magnification of a U87 MG/HNR tumor (20×) showing an example of a blood pool. (**F**) High magnification (20×) of a U87 MG/HNR tumor with an area of serous accumulation and poorly-organized extracellular matrix deposition.

In parallel with generating a rilotumumab-resistant cell line in culture, a group of mice were implanted with U87 MG cells and xenograft tumors were allowed to grow for 3 weeks; mice were then treated for 30 days with low-dose rilotumumab (Figure 4A, bracket at top "4 mg/kg"). All mice showed some degree of tumor regression early in this first treatment period, but later some mice showed renewed tumor growth (Figure 4A). The rilotumumab dose was then increased (Figure 4A, bracket at top "40 mg/kg") and two weeks later, persistent tumor growth in some mice suggested that they had acquired drug resistance (Figure 4A, brackets "R" vs. "S" at right). Eleven cell lines were established from these tumors, all of which showed significantly increased HGF production (Figure 4B; note *y*-axis "1" = 100-fold over the parental cell line) and increased phospho-Met content (Figure 4C) relative to the parental cell line. These cell lines also remained sensitive to Met inhibition by AMG517 with little deviation in IC_{50} concentration, suggesting the absence of *MET* mutation as a means of acquired resistance (Figure 4D).

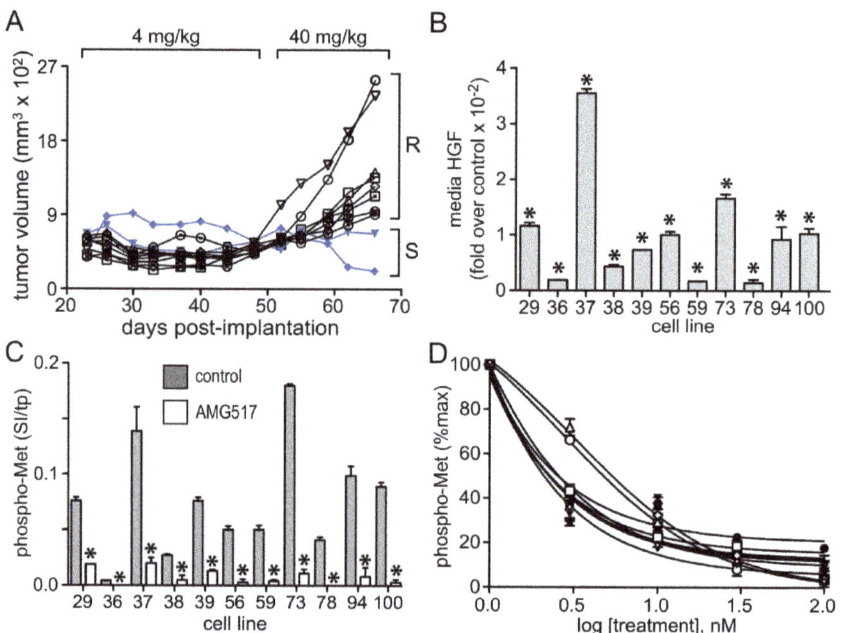

Figure 4. Acquired resistance to rilotumumab by U87 MG xenografts in vivo. (**A**) U87 MG tumor xenografts grown in mice (n = 10) treated with rilotumumab at 4 mg/kg until day 48 post-implantation and thereafter with 40 mg/kg displayed drug sensitivity ("S") or acquired drug resistance ("R") by day 68. (**B**) Mean HGF content (corrected for total cell protein; ±SD, n = 3), expressed as fold over the parental U87 MG cell line (control), in 24 h media conditioned by 11 cultured cell lines derived from rilotumumab-resistant U87 MG tumor xenografts that were generated as described in panel a. (**C**) Phospho-Met content (mean signal intensity/mg total protein ± SD, n = 3) in 11 cultured cell lines derived from rilotumumab-resistant U87 MG tumor xenografts in the absence (gray bars) or presence (white bars) of 100 nM AMG517. (**D**) Dose-dependent inhibition of phospho-Met (mean % maximum ± SD, n = 3) in 11 cultured cell lines derived from rilotumumab-resistant U87 MG tumor xenografts by the selective small-molecule Met kinase inhibitor AMG517. Asterisks indicate a significant difference from control ($p < 0.05$).

3.3. HGF and MET Gene Amplification in U87 MG/HNR

Comparative genomic hybridization (CGH) array analysis revealed focal amplification of both *HGF* and *MET* genes in U87 MG/HNR cells, but not U87 MG cells (Figure 5A, yellow and green arrows, respectively). Closer views of individual probe intensity values and moving averages for U87 MG (Figure 5B,C, blue line) or U87 MG/HNR cells (Figure 5B,C, tan line) indicate the presence of multiple extra copies of each gene in the resistant cell line. cDNA sequences derived from *HGF* and *MET* mRNA transcripts were normal in both the parental and resistant cell lines (i.e., 100% identity with UniProt P08581-1 (*MET*) and UniProt P14210-1 (*HGF*).

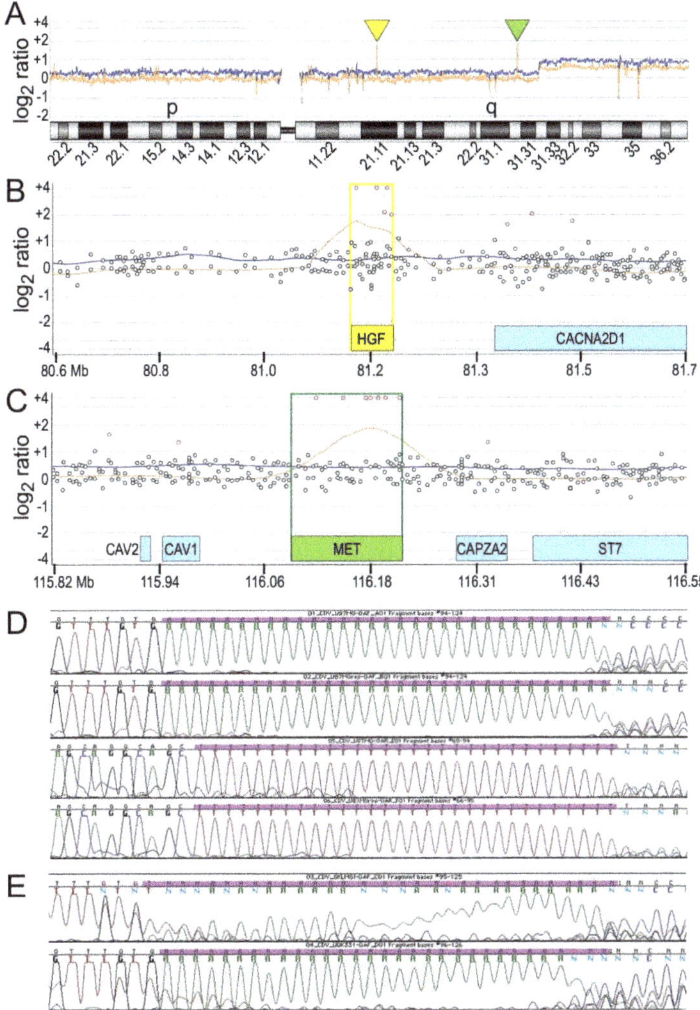

Figure 5. *HGF* and *MET* **gene amplification and** *HGF* **gene promoter DATE region analysis.** (**A**) View of human chromosome 7 CGH microarray analysis results showing moving averages of probe intensity for U87 MG (blue line) or U87 MG/HNR cells (tan line). Inverted triangles indicate the positions of *HGF* (yellow) and *MET* (green) genes. Horizontal lines above and below the center indicate probe intensities corresponding to whole copy number changes. (**B,C**) Zoomed view of chromosome 7 regions encoding the genes for *HGF* ((**B**), yellow box) and *MET* ((**C**), green box). Other nearby gene loci are indicated by blue boxes. As in (**A**), moving averages of probe intensities for U87 MG (blue) or U87 MG/HNR cells (tan) are shown. Circles indicate individual probes with intensity values corresponding to gain or loss of less than one gene copy (black) or greater than one gene copy (red). (**D**) DNA sequencing chromatogram encompassing the DATE region in the *HGF* gene promoter. Coding strand sequence (green) for U87 MG DNA (top panel) and U87 MG/HNR DNA (second panel from the top) and non-coding strand sequences for each cell line (third and fourth panels from the top, respectively) shown normal DATE region length in both U87 MG and U87 MG/HNR cells. (**E**) Control samples for DATE sequence truncation (coding strands only) obtained from the leiomyosarcoma cell line SK-LMS-1 (**top**) and for normal DATE region length obtained from the clear cell renal cell carcinoma cell line UOK331 (**bottom**).

Although the observed increases in gene copy number are consistent with the increased relative abundance of HGF and Met proteins in U87 MG/HNR, the extraordinary level of HGF protein production suggested that additional mechanisms of upregulation might also contribute. Disruption of an *HGF* gene promoter region termed DATE (for deoxyadenosine tract element) that acts as a transcriptional repressor and consists of 30 tandem deoxyadenosines has been reported to increase *HGF* expression in breast cancer [35]. DNA sequencing the DATE region of the *HGF* promoter in U87 MG and U87 MG/HNR revealed that it was of normal length in both cell lines (Figure 5D). The leiomyosarcoma cell line SK-LMS1, which also has autocrine HGF/Met signaling, and the clear cell renal cell carcinoma cell line UOK331 provided unambiguous positive and negative controls for DATE region truncation, respectively (Figure 5E).

Interestingly, not all probes in the CGH array within either the *HGF* or *MET* gene regions had (log_2 ratio) intensity values indicative of gene amplification. In fact, probes that fell within introns had normal copy number values, probes within exons had uniformly elevated values and probes that fell on intron/exon borders had intermediate values such that a significant linear relationship was observed between relative CGH probe intensity and the proportion (%) of probe/exon overlap (Figure 6A,B, $r^2 = 0.8695$ and 0.9920 for *HGF* and *MET*, respectively). This suggested the possibility that extra gene copies had been acquired by retrotransposition, a process with potentially diverse impacts on cancer development [36]. To our knowledge, this is the first evidence of *HGF* or *MET* somatic reverse-transcribed gene amplification linked to acquired drug resistance. If functional, acquired *HGF* and *MET* pseudogenes could increase the rate of encoded protein production by eliminating the need for RNA splicing, in addition to doing so by increasing template abundance. Moreover, apart from the functionality of their encoded proteins, the RNA transcripts of these pseudogenes could also act as decoys that undermine miRNA regulation [37,38]. The additional possibility that multiple copies of *HGF* and *MET* genes might amplify through a chronology of autonomous circular DNA replication, double-minute chromosome formation and chromosomal integration, as described for other cancer-associated gene amplification events [39–41], prompted us to perform fluorescence in situ hybridization (FISH) to visualize *HGF* and *MET* genes in the parental and resistant cell lines (Figure 6C,D). Consistent with CGH array results, U87 MG had normal *HGF* and *MET* chromosomal localization and copy number (Figure 6C), whereas U87 MG/HNR displayed homogenously staining regions suggestive of amplification within chromosome 7 and trisomy of chromosome 7 (Figure 6D). Uniquely present in U87 MG/HNR were numerous double-minute chromosomes that were positive for *HGF* or *MET* FISH probes (Figure 6E,F); these DNA fragments were visible within 3 weeks of rilotumumab treatment and increased with longer periods of drug treatment. To independently confirm the relationship between CGH probe intensities and the proportion of exon overlap, PCR primers were generated corresponding to the CGH probe regions of genomic *HGF* and *MET* DNA sequences. qRT-PCR was performed using these primers, and comparing relative PCR product abundance with the Agilent CGH probe intensity values produced a significant linear relationship (Figure 6G; $p = 0.0109$ for regression non-linearity). DNA sequencing of several of these PCR products further revealed complete continuity between adjacent exons for both genes (data not shown). Further evidence that a portion of HGF protein superproduction by U87 MG/HNR cells resulted from *HGF* gene amplification by reverse transcription was obtained by growing the resistant and parental cell lines in the presence of rilotumumab and in the presence or absence of the reverse transcriptase inhibitor azidothymidine (AZT) and measuring HGF protein production over time. As shown in Figure 6H, AZT treatment significantly reduced HGF production by U87 MG/HNR but not by the parental cells.

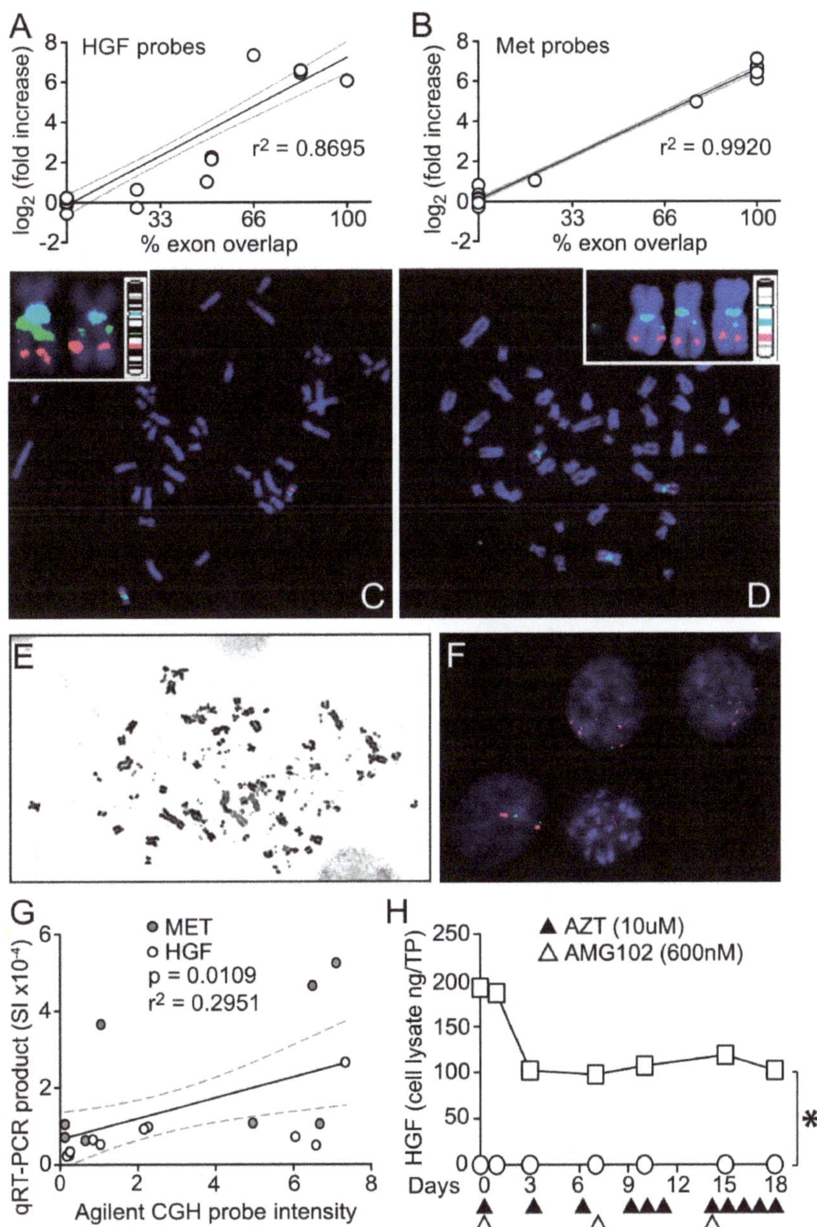

Figure 6. *HGF* and *MET* gene CGH array probe analysis, FISH, qRT-PCR and HGF protein reduction by AZT indicate that amplification involved reverse transcription. Scatter plot of intensity scores (*y*-axis, $\log_2$ (fold increase)) from individual CGH array probes for *HGF* (**A**) or *MET* (**B**) vs. percent overlap between probe sequence and matching complementary gene exon sequence (*x*-axis). Linear regression analysis lines (black), 95% confidence intervals (dashed gray lines) and r^2 values are shown. FISH analysis of U87 MG (**C**) and U87 MG/HNR (**D**); *MET* probes are red, *HGF* probes are green and centromeric probes are cyan. Insets in each panel show copies of chromosome 7 adjacent to ideograms of probe locations. Inset for U87 MG/HNR (**D**) shows double-minute chromosomes

positive for *HGF* only at left. (**E**) U87 HNR DNA analyzed for metaphases after being cultured for 4 weeks in the presence of rilotumumab; numerous double minute (DM) chromosomes and fragments are visible. (**F**) FISH probes for *HGF* (green) and *MET* (red) hybridized with interphase chromatin show numerous copies of each gene. (**G**) Scatter plot of Agilent CGH array probe intensities for *HGF* and *MET* genes in U87 MG/HNR (*x*-axis) vs. relative content of qRT-PCR products generated using primers corresponding to CGH probe sequences (*y*-axis). Linear regression analysis (black line), 95% confidence limits (gray dashed lines), r^2 and *p* values of non-linearity are shown. (**H**) HGF protein content (ng/mg total protein, *y*-axis) in samples (n = 3) of cultured U87 MG (circles) or U87 MG/HNR cells (squares) grown for 18 days (*x*-axis) in the presence of added rilotumumab (open triangles below *x*-axis) and AZT (closed triangles below *x*-axis). Error bars (SD) at all time points are not visible because they are smaller than the symbol size. Asterisks indicate a significant difference from control ($p < 0.05$).

3.4. HGF Transcript Variant 5: A Possible but Unobserved Route to Resistance

Rilotumumab binds to a region encompassing the nascent amino terminus of the light (or beta) chain in the mature, 2-chain HGF protein, which undergoes conformational rearrangement and exposure upon proteolytic activation of the single-chain pro-HGF precursor [42–44]. Because the primary Met-binding epitope in HGF resides in the amino (N)-terminal and kringle 1 (K1) domains and because the NK1 protein (also known as HGF variant 5) is a naturally occurring and biologically active truncated HGF isoform [12 and references therein], NK1 production offers a theoretical route to rilotumumab resistance while sustaining Met signaling. This was confirmed by analyzing the rilotumumab dose–response of phospho-Met levels in 184B5 cells stimulated with purified recombinant full-length HGF or NK1 proteins. As shown in Figure 7A, NK1-induced Met activation remained high in the presence of rilotumumab at concentrations that completely suppressed activation by full-length HGF.

Steady-state levels of the five *HGF* transcript variants present in the parental and rilotumumab-resistant cell lines were compared by quantitative PCR. The left side of Figure 7B schematically depicts the domain structures of the *HGF* transcript variants (v): v1 and v3 encode full-length HGF protein forms that differ by the absence in v3 of five amino acid residues in the kringle 1 domain (v3 is thus known as delta 5 or d5); v2 is identical to v1 but truncated after the region encoding kringle 2 (also known as NK2); v4 is identical to v3 but truncated after kringle 2 (NK2 d5); and v5 encodes NK1, identical to the v1 product, but truncated after kringle 1 [12 and references therein]. Quantitative PCR analysis showed that the mRNA transcript levels for *HGF* v1 and v3, the full-length variants, were more than 600- and 100-fold more abundant in the resistant cell line relative to the parental, respectively (Figure 7B, right); v5 (encoding NK1) abundance was unchanged, suggesting that it was unlikely to have provided a route to resistance. Immunoblot analysis of 24-h-conditioned media from U87 MG/HNR confirmed that full-length HGF production was significantly higher than the parental cell line (Figure 7C, left side, "starting material"; note that U87 MG/HNR media was diluted >1000-fold relative to U87 MG media), but also revealed the presence of an HGF protein with similar molecular mass to a purified recombinant NK1 protein standard ("stds") and that was recognized by an HGF antibody directed against the amino terminus, indicating that it contained the primary Met binding epitope. The precise amino-acid sequence and how this NK1-like protein was produced by U87 MG/HNR were unclear, but an assessment of its biological activity was undertaken.

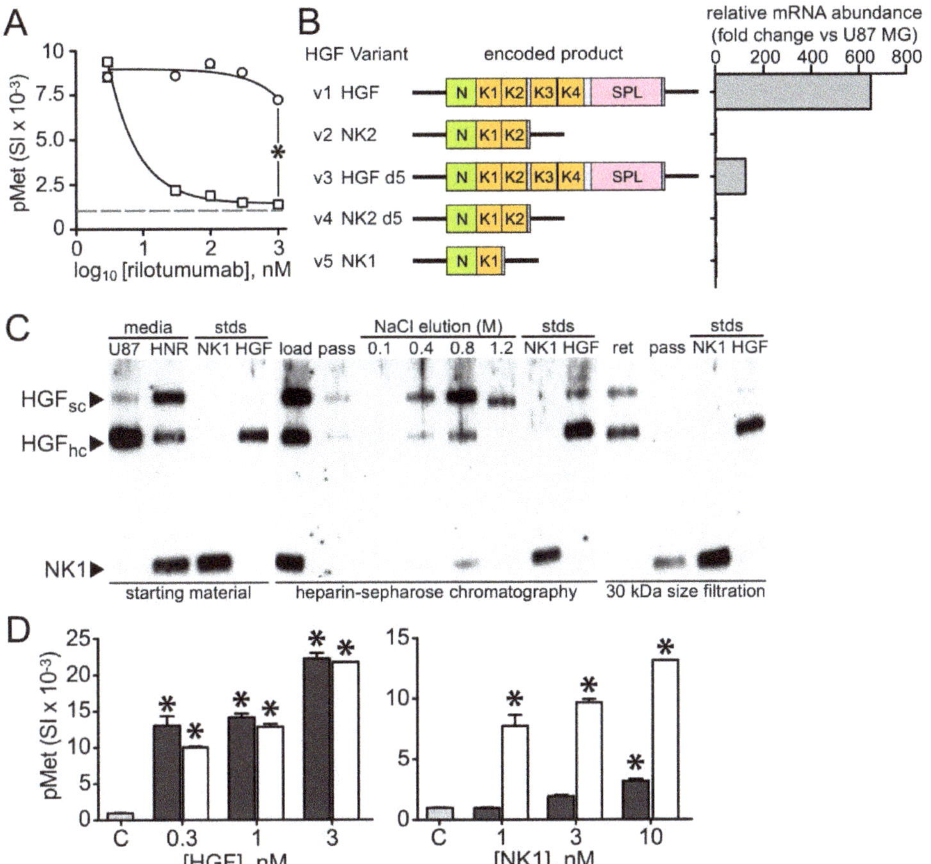

Figure 7. Potential route of rilotumumab resistance by production of a non-neutralizable truncated HGF protein. (**A**) Rilotumumab potently neutralizes Met autophosphorylation (pMet; mean signal intensity, SI ± SD, n = 3) in intact 184B5 cells induced by purified recombinant full-length HGF (squares) but not by the purified recombinant truncated HGF variant NK1 (circles). (**B**) Relative abundance (fold change vs. U87 MG; right) of all *HGF* mRNA transcript variants (listed and represented schematically at left) in U87 MG HNR as determined by quantitative PCR. (**C**) Reduced SDS-PAGE and immunoblot analysis of partial purification and size separation of HGF protein species present in U87 MG/HNR-conditioned media. Starting material (left panel) shows media ("media") from U87 MG ("U87") and U87 MG/HNR ("HNR") contain single-chain ("HGF$_{sc}$") and mature HGF heavy-chain ("HGF$_{hc}$"); an NK1-like protein ("NK1") similar in size to purified recombinant NK1 (21 kDa, "stds") is present only in the latter. HGF proteins from U87 MG/HNR media were partially purified using heparin-sepharose (center panel). Starting material before ("load") and after ("pass") application to the column are at left; the fraction ("NaCl elution") eluting at 0.8 M NaCl was then dialyzed and sizing filtration (right panel) separated full-length (retained, "ret") and truncated HGF ("pass") forms. (**D**) Met activation (pMet) by partially purified full-length HGF (left panel, dark gray bars) and truncated HGF (NK1; right panel, dark gray bars) at indicated concentrations (nM) were compared with control media ("C", light gray bars) and to purified recombinant HGF and NK1 protein standards (white bars). Asterisks indicate a significant difference from control ($p < 0.05$).

Media conditioned by the resistant cell line was partially purified by heparin Sepharose chromatography. SDS-PAGE and immunoblot analysis (Figure 7C, center) showed that the majority of both full-length and truncated HGF proteins eluted with 0.8 M NaCl, consistent the presence of the amino-terminal domain containing the primary heparin binding site in the truncated form [12 and references therein]. Protein eluted in this fraction was dialyzed and subjected to ultrafiltration using a membrane with a 30 kDa cutoff; full-length HGF was retained, whereas the truncated, NK1-like HGF was not (Figure 7C, right side). The bioactivities of these HGF proteins were compared to each other and to purified, recombinant HGF and NK1 proteins by their abilities to stimulate Met activation in 184B5 mammary epithelial cells. As shown in the left panel of Figure 7D, Met activation (pMet) by partially-purified full-length HGF was indistinguishable from fully-purified recombinant HGF protein. In contrast, the truncated 21 kDa HGF form purified from U87 MG/HNR-conditioned media (Figure 7D, right side) had very little activity when compared with purified recombinant NK1 protein. This NK1-like protein was therefore unlikely to provide rilotumumab resistance.

3.5. Mechanistic Insights from Gene Expression Profiling and Pathway Analysis

A subset of 7688 genes that were significantly modulated 1.5-fold or greater up or down in the rilotumumab-resistant cell line relative to the parental cells (listed in Supplementary Table S1) were analyzed using Ingenuity Pathway Core Analysis software (IPA). Not surprisingly, the most significant IPA "Disease and Disorder" identified (IPA Summary, Supplementary Table S2, pp. 2–3) was "Cancer" (p value range: 8.35×10^{-4}–1.20×10^{-13}; number of modulated molecules in overlap: 4515) and the most significant IPA "Molecular and Cellular Function" identified was "Cellular Growth and Proliferation" (p value range: 7.12×10^{-4}–2.79×10^{-25}; number of modulated molecules in overlap: 2146). Twenty-six IPA "Canonical Pathways" were identified as having significant overlap with the dataset by right-tailed Fisher's Exact test and the Benjamini–Hochberg (B–H) multiple test correction (Supplementary Table S3). These pathways could be logically grouped into 6 functional categories to provide a broader phenotypic portrait of acquired changes in U87 MG/HNR: (1) stress signaling; (2) immune-related; (3) cell proliferation and survival; (4) established cancer networks; (5) second messengers and nuclear receptors; and (6) nervous-system-related (Supplementary Table S3). The relative abundance of stress and immune-related inflammatory signaling pathways, and those related to proliferation, survival and established cancer networks (34% and 39% of pathways, respectively) is consistent with an aggressive GBM phenotype [45].

Signaling pathways identified using IPA suggested potential intracellular cellular processes and signaling pathways underlying acquired rilotumumab resistance. Significant overlap with pathways in the phenotypic portrait groups 3, 4 and 5 as defined above might be expected for many particularly aggressive tumor types (including GBM), and group 6 pathways are consistent with a cellular origin of GBM. Phenotypic groups 1 and 2 however—stress signaling and immune-related pathways—suggested processes and pathways by which autocrine HGF/Met activation could evade contact with rilotumumab.

3.6. A Coordinated, Multiplex Route to Rilotumumab Resistance

Transcriptional reprogramming downstream of endoplasmic reticulum (ER) stress signaling provides tumor cells with the needed proteostasis machinery and contributes to adaptation and cell survival even in the face of increased cell death [46]. Biochemical evidence of increased ER stress signaling included substantially increased levels of active caspase 3 (Figure 1C), PERK and calreticulin proteins (Figure 8A) in U87 MG/HNR relative to the parental cells. Increased total eIF2a protein and phospho-(ser51)-eIF2a were also observed in the resistant cell line (Figure 8B, left side), and phosphorylation of eIF2a at ser51 was inhibited by treatment with the selective PERK inhibitor GSK2656157 [47] (Figure 8B, right side). One potential connection between HGF superproduction and increased ER stress signaling was very predictable: full-length HGF is structurally complex, being comprised of 6 subdomains with 20 intrachain disulfide bonds. HGF cannot be expressed recombinantly in bacterial systems without denaturation and refolding, and even yeast systems (e.g., *P. pastoris*) provide low yields of active protein unless they are specifically engineered to overexpress protein disulfide isomerase(s) to facilitate the correct disulfide bond pairing required for proper folding [48]. To further investigate HGF misfolding, parental and resistant cells were extracted in buffer containing 1% Triton X-100 (TX100) non-ionic detergent, and insoluble cell material pelleted after centrifugation was extracted with the same buffer containing 1% SDS detergent. All samples were then dissolved in Laemmli (SDS) sample buffer for resolving by SDS-PAGE in the absence of a disulfide-bond-reducing agent such as dithiothreitol or beta-mercapto-ethanol (i.e., "non-reducing SDS-PAGE"). As shown in Figure 8C, subsequent immunoblotting for HGF revealed monomeric HGF (~90 kDa) in both cell lines, as well as an abundance of misfolded, disulfide-bonded HGF dimers (~180 kDa), trimers (~270 kDa) and lower levels of higher-order multimers in U87 MG/HNR (left-most lane) but not U87 MG (right lanes). The HGF multimers in U87 MG/HNR were present only in the cell material pelleted after TX100 extraction (Figure 8C, "pel"), an ER-rich fraction. In fact, low disulfide isomerase activity (and consequent HGF misfolding) is very likely to have been a positive selection factor for acquired resistance to rilotumumab; 9 out of 10 genes encoding thioredoxin-domain-containing ER-resident protein disulfide isomerases (*PDIA3, PDIA5, PDIA6, TMX1, TMX3, TMX4, TXNDC5, P4HB* and *ERP44*) showed reduced expression in U87 MG/HNR (−1.5 to −3.3-fold; Table S1) relative to the parental line. Consistent with an ER-stress-induced unfolded protein response, the superproduction of misfolded HGF protein was associated with its retention inside cells; the overall ratio of intracellular to secreted HGF in U87 MG/HNR was completely reversed relative to the parental cells (Figure 8D).

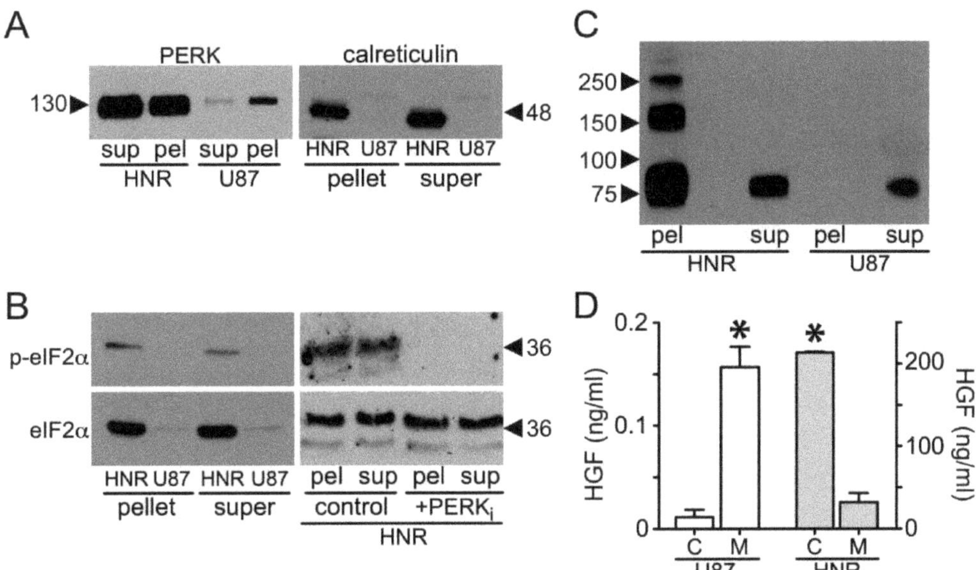

Figure 8. Acquisition of ER stress through HGF protein misfolding in U87 MG/HNR. (**A**) Reduced SDS-PAGE and immunoblot analysis of U87 MG ("U87") and U87 MG/HNR ("HNR") cell extracts for PERK (left panel) and calreticulin (right panel). Cells were extracted with ice-cold buffer containing TX-100 detergent, protease and phosphatase inhibitors before centrifugation to separate soluble ("sup", left, or "super" right) and insoluble ("pel", left, or "pellet", right; these abbreviations also apply to panels (**B,C**)) cell fractions; insoluble fractions were subsequently solubilized in buffer containing 1% SDS prior to analysis. (**B**) Left, reduced SDS-PAGE and immunoblot analysis of U87 MG ("U87") and U87 MG/HNR ("HNR") TX-100 cell extracts for phospho-EIF2a ("p-eIF2a";

upper panels) and total EIF2α (lower panels); as in panel (**A**), TX-100 extracts were separated into soluble and insoluble fractions prior to analysis. Right, reduced SDS-PAGE and immunoblot analysis for phospho-EIF2a (upper panel) and total EIF2α (lower panel) for U87 MG/HNR ("HNR") in the absence ("control") or presence ("+PERK$_i$") of the selective PERK antagonist GSK2656157. As in panel (**A**), TX-100 extracts were separated into soluble and insoluble fractions prior to analysis. (**C**) Non-reduced (i.e., not treated with dithiothreitol or beta-mercaptoethanol so that disulfide bonds remain intact) SDS-PAGE and immunoblot analysis of U87 MG ("U87") and U87 MG/HNR ("HNR") cell TX100 extracts for HGF; as in panel (**A**), extracts were separated into soluble and insoluble fractions prior to analysis. (**D**) HGF protein concentration (ng/mL) in U87 MG cell ("U87", white bars, left *y*-axis) lysates ("C") and 24-h-conditioned media ("M") and U87 MG/HNR cell lysate and media ("HNR", gray bars, "C" and "M", respectively, right *y*-axis). Mean values (ng/mL $\pm$ SD) from triplicate samples were normalized to total cell protein. Asterisks indicate a significant difference from control (*p* < 0.05). All the whole western blot figures can be found in the supplementary materials.

Most significant among the immune-related IPA canonical pathway changes acquired with rilotumumab resistance was "Antigen Presentation" (75% pathway overlap, B–H multiple test $p = 1.74 \times 10^{-3}$, Table S3). We suspected that substantial alterations in vesicular trafficking might effectively partition rilotumumab away from HGF and Met. Because lipid rafts are critical for many vesicular traffic systems [49,50] and these cholesterol-rich lipid regions are usually insoluble in TX100 detergent, we performed a series of experiments analyzing TX100 detergent soluble and insoluble cell lysate fractions of the resistant and parental cell lines for HGF, Met, pMet and rilotumumab content. Consistent with a role in acquired resistance, the rate of rilotumumab uptake by U87 MG/HNR was over 28-fold higher than that of the parental cells under identical conditions (Figure 9A). Moreover, whereas nearly all (88%) of the antibody taken up by the parental cell line was found in the TX100-soluble cell fraction (Figure 9A, left, gray bar), most rilotumumab uptake (~60%) in resistant cells was found in the TX100-soluble cell fraction (Figure 9A, right, white bar), consistent with a shift in rilotumumab uptake to a lipid-raft-dependent system. The cholesterol dependence of rilotumumab uptake was tested using the HMG CoA reductase inhibitor simvastatin. Although dose-dependent rilotumumab uptake by the parental cell line was unaltered by simvastatin treatment (Figure 9B, orange/yellow vs. gray/white bars), uptake by the resistant cell line was inhibited >98% (Figure 9C, orange/yellow vs. gray/white bars). Together these findings show that substantial quantitative and qualitative changes in rilotumumab uptake accompanied resistance.

Remarkably, increased rilotumumab uptake by the resistant cell line was also linked to Met activation. Met protein levels in the TX100-soluble and -insoluble cell fractions of U87 MG/HNR were similar in the absence of rilotumumab (Figure 9D, 0 nM rilotumumab). Rilotumumab addition did not change Met content in the TX100-soluble cell fraction (Figure 9D, gray bars), but was associated with a significant dose-dependent reductions of Met (Figure 9D, white bars) and phosphoMet (Figure 9E, white bars) in the TX100-insoluble cell fraction. Rilotumumab-driven Met and pMet loss from the TX100-insoluble cell fraction was blocked by simvastatin (Figure 9D,E, white vs. yellow bars), suggesting that rilotumumab uptake triggered a proteolytic process that was dependent on lipid-raft-mediated transport, potentially related to antigen presentation [51,52]. Although Met protein abundance was similar in the TX100-soluble and -insoluble fractions of untreated cells (Figure 9D, 0 nM rilotumumab), Met activation (pMet content) was three-fold higher in the TX100-soluble fraction (Figure 9E, 0 nM rilotumumab). Note that HGF protein in this fraction was not multimeric (Figure 8C) and thus was likely to be properly folded and capable of activating Met. Remarkably, rilotumumab treatment was associated with a significant and dose-dependent increase in Met activation in the TX100-soluble cell fraction and overall (Figure 9E, gray bars). This rilotumumab-driven increase in Met kinase activation was also blocked completely by simvastatin treatment (Figure 9E, orange vs. gray bars). These findings link U87 MG/HNR's adaptive changes in the route of rilotumumab internalization to maintaining—and enhancing—HGF/Met pathway activation.

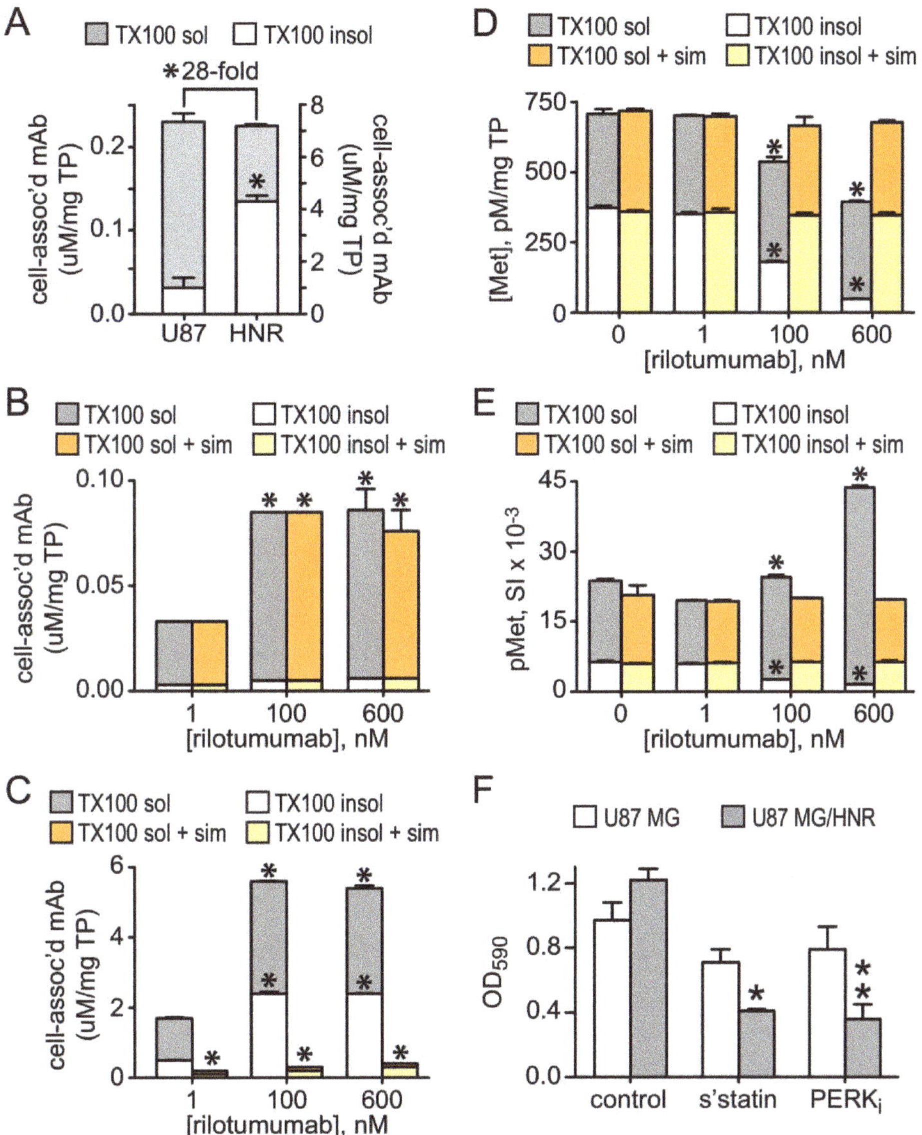

Figure 9. Qualitative and quantitative changes in vesicular trafficking accompany rilotumumab resistance. As in Figure 8, cells were extracted with ice-cold buffer containing TX-100 detergent, protease and phosphatase inhibitors before centrifugation to separate soluble ("sup", left, or "super" right) and insoluble ("pel", left, or "pellet", right) cell fractions; insoluble fractions were subsequently solubilized in buffer containing 1% SDS prior immunoassay analysis (see Methods). (**A**) Cell-associated mAb (mean uM rilotumumab/mg total protein $\pm$ SD, n = 3) measured in U87 MG ("U87", left *y*-axis) or U87 MG/HNR ("HNR", right *y*-axis) cell Triton X-100 extracts separated into soluble (gray bars) and insoluble (white bars) fractions prepared after 24 h rilotumumab exposure. (**B,C**): Cell-associated mAb (mean uM rilotumumab/mg total protein $\pm$ SD, n = 3) in Triton X-100 soluble (gray or orange bars) and insoluble (white or yellow bars) fractions prepared from U87 MG cells (**B**) or U87 MG/HNR cells (**C**) exposed for 16 h to rilotumumab at the indicated concentrations

Cancers **2023**, *15*, 460

in the absence (gray/white bars) or presence (yellow/orange bars) of simvastatin (2 mM). (**D,E**): Met protein content ((**D**), mean pM/mg total protein ± SD, n = 3) and phosphoMet content ((**E**), mean signal intensity/mg total protein ± SD, n = 3) in U87 MG/HNR cell-derived Triton X-100 extracts separated into soluble (gray or orange bars) and insoluble (white or yellow bars) fractions exposed for 16 h to rilotumumab at the indicated concentrations in the absence (gray/white bars) or presence (yellow/orange bars) of simvastatin (2 mM). (**F**) Soft agar colony formation by U87 MG (clear bars) or U87 MG/HNR cells (gray bars) left untreated ("control") or treated with simvastatin ("s'statin") or the selective PERK inhibitor GSK2656157 ("PERK$_i$"). Values are mean ± SD from triplicate samples, * (p = 0.0024) and ** (p = 0.001) indicate statistical significance as determined by unpaired t-test with Welch's correction. Asterisks indicate a significant difference from control ($p < 0.05$).

These new features underlying rilotumumab resistance—ER stress signaling and altered vesicular transport—also rendered U87 MG/HNR cells susceptible to inhibition of anchorage-independent growth by the PERK antagonist GSK2656157 and simvastatin, relative to the parental cell line (Figure 9F). Neither simvastatin nor GSK2656157 significantly inhibited soft agar colony formation by U87 MG, but these agents suppressed colony formation by U87 MG/HNR by 66% and 70%, respectively (p = 0.0024 and 0.0010), indicating an acquired dependence on these pathways and processes for rilotumumab resistance and enhanced HGF/Met-driven oncogenicity (Figure 9F).

4. Discussion

Acquired drug resistance to targeted single-agent therapies has become a serious obstacle to their long-term efficacy and use. For several malignancies such as non-small-cell lung cancer (NSCLC), breast, ovarian, and gastric cancers and melanoma, acquired resistance frequently involves activation of the HGF/Met pathway [1–9,53–57]. Aberrant HGF/Met pathway activation can result in increased tumor invasiveness, angiogenesis and metastasis, and is correlated with poor prognosis in many tumor types [58]. It is critical to understand the mechanisms by which acquired resistance develops, and preclinical studies have begun to focus on methods for detecting acquired resistance involving HGF/Met pathway activation as well as strategies to circumvent or remedy its manifestations [8,59–65].

Standard-of-care treatment for GBM is surgical resection followed by concurrent radiotherapy and chemotherapy with the DNA-alkylating agent temozolomide; this combination represents an improvement over prior approaches, but ultimately fails in almost all patients. Acquired resistance to radio- and chemotherapy for GBM is frequently related to the loss of DNA damage checkpoint mechanisms and/or enhanced expression of DNA repair machinery [66]. Together with several independent studies over the last decade, the genomic characterization of GBM undertaken by The Cancer Genome Atlas Research Network has helped to identify clinically-relevant subgroups of GBM that may benefit from therapeutics targeting growth factor pathways such as those of platelet-derived growth factor, EGF, HGF, and critical downstream effectors such as *PIK3CA* [13–15,45].

In anticipation that targeting HGF with rilotumumab in GBM may lead to acquired resistance, we developed the U87 MG-derived models described here. In U87 MG cells grown in the continuous presence of a maximally effective concentration of rilotumumab, amplification of both *HGF* and *MET* genes was associated with significantly increased HGF and Met abundance. CGH array data suggested that many of the extra gene copies lack introns and thus may have been acquired by retrotransposition. The roles of Alu and LINE-1 retrotransposons in cancer is an area of growing interest [36]; evidence of involvement in gene-amplification-associated acquired drug resistance and the remarkably short time span from deployment to impact expands current theory and warrants further investigation. Dramatically increased HGF secretion was also observed in 11 cell lines derived from rilotumumab-resistant U87 MG-derived tumor xenografts in mice, suggesting that HGF superproduction is an important route to resistance and may serve as a biomarker for its emergence in patients treated with HGF inhibitors. Although at first glance this appears to be a relatively simple mechanism relative to the spectrum and complexity of those reported for trastuzumab, cetuximab and bevacizumab, closer scrutiny indicated that the route to rilotumumab resistance was also multifaceted. Although overproduction of HGF variant 5 could have led to rilotumumab resistance, quantitative PCR analysis ruled against it and a truncated HGF protein in U87 MG/HNR-conditioned media showed little ability to activate Met.

Expression profiling, pathway analysis and confirmatory experiments pointed to a distinct and more complex mechanism. Of all 26 significant Ingenuity "Canonical Pathway" changes acquired by the resistant cell line, the "Antigen Presentation Pathway" had greatest overlap ratio (75%), followed closely by the "Endoplasmic Reticulum Stress Pathway" (71%). Confirmatory experiments showed that HGF superproduction specifically increased intracellular full-length HGF protein content, where a large portion of this full-length HGF protein was misfolded, multimeric and partitioned away from the monomeric soluble intracellular HGF protein. The misfolding of full-length HGF is very likely to have initiated ER stress signaling, enabling downstream events that promote cell survival despite increased apoptosis. These changes were also linked to significant quantitative and qualitative changes in rilotumumab internalization comprising a major shift to a lipid-raft-dependent transport system. This shift partitioned rilotumumab, misfolded HGF and a fraction of Met content destined for degradation to the same cell fraction, while sequestering soluble HGF and a distinct fraction of Met to a separate compartment that facilitated Met kinase activation.

The cellular origins of glioblastoma have been traced to neural stem cells, neural-stem-cell-derived astrocytes and oligodendrocyte precursor cells [67]. In the brain, all of these cell types have roles in antigen presentation; this characteristic may predispose GBM cells to adapt its underlying processes for drug resistance to mAbs. The autocrine HGF-Met dependence of U87 MG is consistent with Met as a functional marker of a glioblastoma stem cell phenotype [21] and the well-documented frequency of functional HGF-Met signaling in GBM [14–25]. ER stress signaling and lipid-raft-mediated rilotumumab uptake were essential components for the enhanced aggressive phenotype of the resistant cells and also presented new points of vulnerability; both simvastatin and PERK inhibitors became effective antagonists of anchorage-independent cell growth.

Remarkably, rilotumumab resistance in our cultured cell model was not associated with mutations in either *HGF* or *MET*, nor did it depend on alterations in other signaling pathways or mediators, since a selective small molecule Met TK inhibitor remained effective in blocking cell growth in vitro and tumorigenesis in vivo. The absence of target mutation and retention of target pathway dependence in our model of acquired rilotumumab resistance is thus unlike many other mechanisms of acquired drug resistance, where mutation of the gene encoding the drug target (e.g., *EGFR*), loss of prominent negative regulators downstream of the drug target (e.g., *PTEN*), or activation of alternative mitogenic pathways parallel to the target (e.g., *MET* in response to EGFR inhibitors) are prevalent means of restoring tumor cell proliferative and invasive activities. Our model is generally similar to others in that restoration of signaling via the PI3K and MAPK pathways was achieved, whether resistance was acquired to neutralizing antibodies [reviewed in [68]] or small TK inhibitors [54–57,69–71]. Despite the apparent diversity of the subcellular systems involved in the mechanism of acquired resistance to rilotumumab, its rapid development suggests that GBM cells in which HGF is an important oncogenic driver are predisposed to its adaptation as a route to survival. Measuring HGF/Met pathway activity in GBM patients is a logical basis for selecting those most likely to benefit from HGF/Met-targeted therapeutics; the results presented here further suggest that monitoring Met pathway activity and HGF production in those patients could provide early indications of acquired resistance, that Met kinase inhibitors may still be efficacious when resistance occurs and that targeting the other critical mediators of resistance identified here may provide effective alternative or combinatorial treatments.

5. Conclusions

Acquired drug resistance obstructs the effective treatment of many cancers and occurs through various mechanisms, often involving the elimination of drug from target cells or new defects in proto-oncogenes or tumor suppressors that revitalize essential growth and survival signaling pathways. Here we report a novel route to acquired resistance. In glioblastoma cells that require autocrine hepatocyte growth factor (HGF)/Met signaling for proliferation and survival, resistance to the HGF neutralizing monoclonal antibody rilotumumab was acquired through a complex interplay of several intracellular systems: (1) *HGF* and *MET* gene amplification and HGF protein super-production; (2) downregulation of ER-resident disulfide isomerases contributing to significant HGF protein misfolding; (3) induction of ER stress-response signaling and intracellular HGF and Met protein retention; and (4) dramatically increased rilotumumab uptake and degradation through a shift to caveolar endocytosis and activation of antigen presentation pathways. Together, these changes provided intracellular seclusion of properly folded HGF and Met proteins from rilotumumab and maintained HGF/Met signaling dependence for cell growth and survival.

Unlike other acquired drug-resistance mechanisms, mutation of the gene encoding the drug target, loss of critical negative regulators downstream of the drug target, and/or activation of alternative mitogenic pathways parallel to the target were not observed. Despite the number and diversity of the subcellular systems involved, resistance developed rapidly in GBM cells in which HGF is an important oncogenic driver. Defining this mechanism also revealed targetable co-acquired dependencies for survival in resistance cells: cholesterol synthesis needed for caveolar uptake and ER-stress signaling as well as continued sensitivity to small-molecule Met tyrosine kinase inhibitors. These findings suggest strategies for the early detection of this form of resistance and for effective intervention.

Supplementary Materials: The following supporting information can be downloaded at: https://www.mdpi.com/article/10.3390/cancers15020460/s1, Figure S1: The original Western blot images; Table S1: IPA Core Analysis Molecules; Table S2: Ingenuity IPA Analysis Summary; Table S3: Ingenuity IPA Analysis Canonical Pathways.

Author Contributions: F.C., K.R., J.S., C.D.V., Y.H.L., S.B., D.B., M.A.D. and D.P.B. performed experiments; F.C., K.R., J.S., C.D.V., Y.H.L., S.B., D.B., M.A.D., A.C., T.L.B. and D.P.B. participated in experimental design and results interpretation; F.C., C.D.V., A.C., S.B., T.L.B. and D.P.B. participated in manuscript production. All authors have read and agreed to the published version of the manuscript.

Funding: This research was funded by National Institutes of Health grant number ZIA BC011124 awarded to D.P.B. and performed under the Collaborative Research and Development Agreement No. 01943 between Amgen, Inc. and the National Cancer Institute, National Institutes of Health.

Institutional Review Board Statement: Animal studies were conducted under approval by the Animal Care and Use Committees of the National Cancer Institute, Center for Cancer Research (Protocol UOB-009) and Amgen (Animal Protocol 2015-01243).

Informed Consent Statement: Not applicable.

Data Availability Statement: Data presented in this study that are not contained in the article are available on request from the corresponding author. The data are not publicly available due to agreements set forth in NIH CRADA No. 01943 between the National Cancer Institute and Amgen, Inc.

Acknowledgments: This work was supported by the Intramural Research Program of the NIH, National Cancer Institute, Center for Cancer Research. We would like to thank members of the Pathology Department, Amgen Thousand Oaks, for the histological processing and interpretation of the U87 MG- and U87 MG/HNR-derived tumors.

Conflicts of Interest: Karen Rex, Joanna Schmidt, Daniel Baker, Michael A. Damore, Angela Coxon, and Teresa L. Burgess are current or former employees of Amgen Inc. Authors declare that there are no conflicts of interest.

References

1. Engelman, J.A.; Zejnullahu, K.; Mitsudomi, T.; Song, Y.; Hyland, C.; Park, J.O.; Lindeman, N.; Gale, C.M.; Zhao, X.; Christensen, J.; et al. MET amplification leads to gefitinib resistance in lung cancer by activating ERBB3 signaling. *Science* **2007**, *316*, 1039–1043. [CrossRef] [PubMed]
2. Bean, J.; Brennan, C.; Shih, J.Y.; Riely, G.; Viale, A.; Wang, L.; Chitale, D.; Motoi, N.; Szoke, J.; Broderick, S.; et al. MET amplification occurs with or without T790M mutations in EGFR mutant lung tumors with acquired resistance to gefitinib or erlotinib. *Proc. Natl. Acad. Sci. USA* **2007**, *104*, 20932–20937. [CrossRef]
3. Suda, K.; Murakami, I.; Katayama, T.; Tomizawa, K.; Osada, H.; Sekido, Y.; Maehara, Y.; Yatabe, Y.; Mitsudomi, T. Reciprocal and complementary role of MET amplification and EGFR T790M mutation in acquired resistance to kinase inhibitors in lung cancer. *Clin. Cancer Res.* **2010**, *16*, 5489–5498. [CrossRef]
4. Yano, S.; Wang, W.; Li, Q.; Matsumoto, K.; Sakurama, H.; Nakamura, T.; Ogino, H.; Kakiuchi, S.; Hanibuchi, M.; Nishioka, Y.; et al. Hepatocyte growth factor induces gefitinib resistance of lung adenocarcinoma with epidermal growth factor receptor-activating mutations. *Cancer Res.* **2008**, *68*, 9479–9487. [CrossRef]
5. Yano, S.; Yamada, T.; Takeuchi, S.; Tachibana, K.; Minami, Y.; Yatabe, Y.; Mitsudomi, T.; Tanaka, H.; Kimura, T.; Kudoh, S.; et al. Hepatocyte growth factor expression in EGFR mutant lung cancer with intrinsic and acquired resistance to tyrosine kinase inhibitors in a Japanese cohort. *J. Thorac. Oncol.* **2011**, *6*, 2011–2017. [CrossRef]
6. Cappuzzo, F.; Jänne, P.A.; Skokan, M.; Finocchiaro, G.; Rossi, E.; Ligorio, C.; Zucali, P.A.; Terracciano, L.; Toschi, L.; Roncalli, M.; et al. MET increased gene copy number and primary resistance to gefitinib therapy in non-small-cell lung cancer patients. *Ann. Oncol.* **2009**, *20*, 298–304. [CrossRef]
7. Turke, A.B.; Zejnullahu, K.; Wu, Y.L.; Song, Y.; Dias-Santagata, D.; Lifshits, E.; Toschi, L.; Rogers, A.; Mok, T.; Sequist, L.; et al. Preexistence and clonal selection of MET amplification in EGFR mutant NSCLC. *Cancer Cell* **2010**, *17*, 77–88. [CrossRef]
8. Garofalo, M.; Romano, G.; Di Leva, G.; Nuovo, G.; Jeon, Y.J.; Ngankeu, A.; Sun, J.; Lovat, F.; Alder, H.; Condorelli, G.; et al. EGFR and MET receptor tyrosine kinase-altered microRNA expression induces tumorigenesis and gefitinib resistance in lung cancers. *Nat. Med.* **2011**, *18*, 74–82. [CrossRef]
9. Yamada, T.; Takeuchi, S.; Kita, K.; Bando, H.; Nakamura, T.; Matsumoto, K.; Yano, S. Hepatocyte growth factor induces resistance to anti-epidermal growth factor receptor antibody in lung cancer. *J. Thorac. Oncol.* **2012**, *7*, 272–280. [CrossRef]
10. Johnson, M.; Garassino, M.C.; Mok, T.; Mitsudomi, T. Treatment strategies and outcomes for patients with EGFR-mutant non-small cell lung cancer resistant to EGFR tyrosine kinase inhibitors: Focus on novel therapies. *Lung Cancer* **2022**, *21*, 41–51. [CrossRef]
11. Shah, M.P.; Neal, J.W. Targeting Acquired and Intrinsic Resistance Mechanisms in Epidermal Growth Factor Receptor Mutant Non-Small-Cell Lung Cancer. *Drugs* **2022**, *82*, 649–662. [CrossRef]
12. Matsumoto, K.; Umitsu, M.; De Silva, D.M.; Roy, A.; Bottaro, D.P. Hepatocyte growth factor/MET in cancer progression and biomarker discovery. *Cancer Sci.* **2017**, *108*, 296–307. [CrossRef]
13. The Cancer Genome Atlas Research Network. Comprehensive genomic characterization defines human glioblastoma genes and core pathways. *Nature* **2008**, *455*, 1061–1068. [CrossRef] [PubMed]
14. Brennan, C.W.; Verhaak, R.G.; McKenna, A.; Campos, B.; Noushmehr, H.; Salama, S.R.; Zheng, S.; Chakravarty, D.; Sanborn, J.Z.; Berman, S.H.; et al. TCGA Research Network. The somatic genomic landscape of glioblastoma. *Cell* **2013**, *155*, 462–477. [CrossRef] [PubMed]
15. The Cancer Genome Atlas Research Network. National Cancer Institute GDC Data Portal: TCGA-GBM. 2022. Available online: https://portal.gdc.cancer.gov/projects/TCGA-GBM (accessed on 9 January 2023).
16. Moriyama, T.; Kataoka, H.; Koono, M.; Wakisaka, S. Expression of hepatocyte growth factor/scatter factor and its receptor c-Met in brain tumors: Evidence for a role in progression of astrocytic tumors. *Int. J. Mol. Med.* **1999**, *3*, 531–536. [CrossRef]
17. Li, Y.; Lal, B.; Kwon, S.; Fan, X.; Saldanha, U.; Reznik, T.E.; Kuchner, E.B.; Eberhart, C.; Laterra, J.; Abounader, R. The scatter factor/hepatocyte growth factor: C-met pathway in human embryonal central nervous system tumor malignancy. *Cancer Res.* **2005**, *65*, 9355–9362. [CrossRef]
18. Kim, K.J.; Wang, L.; Su, Y.C.; Gillespie, G.Y.; Salhotra, A.; Lal, B.; Laterra, J. Systemic anti-hepatocyte growth factor monoclonal antibody therapy induces the regression of intracranial glioma xenografts. *Clin. Cancer Res.* **2006**, *12*, 1292–1298. [CrossRef] [PubMed]
19. Abounader, R.; Laterra, J. Scatter factor/hepatocyte growth factor in brain tumor growth and angiogenesis. *Neuro Oncol.* **2005**, *7*, 436–451. [CrossRef]
20. Garcia-Navarrete, R.; Garcia, E.; Arrieta, O.; Sotelo, J. Hepatocyte growth factor in cerebrospinal fluid is associated with mortality and recurrence of glioblastoma, and could be of prognostic value. *J. Neurooncol.* **2010**, *97*, 347–351. [CrossRef]
21. De Bacco, F.; Casanova, E.; Medico, E.; Pellegatta, S.; Orzan, F.; Albano, R.; Luraghi, P.; Reato, G.; D'Ambrosio, A.; Porrati, P.; et al. The MET oncogene is a functional marker of a glioblastoma stem cell subtype. *Cancer Res.* **2012**, *72*, 4537–4550. [CrossRef]
22. Burgess, T.; Coxon, A.; Meyer, S.; Sun, J.; Rex, K.; Tsuruda, T.; Chen, Q.; Ho, S.Y.; Li, L.; Kaufman, S.; et al. Fully human monoclonal antibodies to hepatocyte growth factor with therapeutic potential against hepatocyte growth factor/c-Met-dependent human tumors. *Cancer Res.* **2006**, *66*, 1721–1729. [CrossRef] [PubMed]
23. Martens, T.; Schmidt, N.O.; Eckerich, C.; Fillbrandt, R.; Merchant, M.; Schwall, R.; Westphal, M.; Lamszus, K. A novel one-armed anti-c-Met antibody inhibits glioblastoma growth in vivo. *Clin. Cancer Res.* **2006**, *12*, 6144–6152. [CrossRef] [PubMed]

24. Guessous, F.; Zhang, Y.; diPierro, C.; Marcinkiewicz, L.; Sarkaria, J.; Schiff, D.; Buchanan, S.; Abounader, R. An orally bioavailable c-Met kinase inhibitor potently inhibits brain tumor malignancy and growth. *Anticancer Agents Med. Chem.* **2010**, *10*, 28–35. [CrossRef] [PubMed]
25. Xie, Q.; Bradley, R.; Kang, L.; Koeman, J.; Ascierto, M.L.; Worschech, A.; De Giorgi, V.; Wang, E.; Kefene, L.; Su, Y.; et al. Hepatocyte growth factor (HGF) autocrine activation predicts sensitivity to MET inhibition in glioblastoma. *Proc. Natl. Acad. Sci. USA* **2012**, *109*, 570–575. [CrossRef]
26. Jun, H.T.; Sun, J.; Rex, K.; Radinsky, R.; Kendall, R.; Coxon, A.; Burgess, T.L. AMG 102, a fully human anti-hepatocyte growth factor/scatter factor neutralizing antibody, enhances the efficacy of temozolomide or docetaxel in U-87 MG cells and xenografts. *Clin. Cancer Res.* **2007**, *13*, 6735–6742. [CrossRef]
27. Gao, C.F.; Xie, Q.; Zhang, Y.W.; Su, Y.; Zhao, P.; Cao, B.; Furge, K.; Sun, J.; Rex, K.; Osgood, T.; et al. Therapeutic potential of hepatocyte growth factor/scatter factor neutralizing antibodies: Inhibition of tumor growth in both autocrine and paracrine hepatocyte growth factor/scatter factor: c-Met-driven models of leiomyosarcoma. *Mol. Cancer Ther.* **2009**, *8*, 2803–2810. [CrossRef]
28. Buchanan, I.M.; Scott, T.; Tandle, A.T.; Burgan, W.E.; Burgess, T.L.; Tofilon, P.J.; Camphausen, K. Radiosensitization of glioma cells by modulation of Met signalling with the hepatocyte growth factor neutralizing antibody, AMG102. *J. Cell. Mol. Med.* **2011**, *15*, 1999–2006. [CrossRef]
29. Gordon, M.S.; Sweeney, C.S.; Mendelson, D.S.; Eckhardt, S.G.; Anderson, A.; Beaupre, D.M.; Branstetter, D.; Burgess, T.L.; Coxon, A.; Deng, H.; et al. Safety, pharmacokinetics, and pharmacodynamics of AMG 102, a fully human hepatocyte growth factor-neutralizing monoclonal antibody, in a first-in-human study of patients with advanced solid tumors. *Clin. Cancer Res.* **2010**, *16*, 699–710. [CrossRef]
30. Cecchi, F.; Wright, C.; Bottaro, D.P. Experimental Cancer Therapeutics Targeting the Hepatocyte Growth Factor/Met Signaling Pathway. 2018. Available online: https://ccrod.cancer.gov/confluence/display/CCRHGF/Home (accessed on 9 January 2023).
31. Cecchi, F.; Pajalunga, D.; Fowler, C.A.; Uren, A.; Rabe, D.C.; Peruzzi, B.; Macdonald, N.J.; Blackman, D.K.; Stahl, S.J.; Byrd, R.A.; et al. Targeted disruption of heparan sulfate interaction with hepatocyte and vascular endothelial growth factors blocks normal and oncogenic signaling. *Cancer Cell* **2012**, *22*, 250–262. [CrossRef]
32. Boezio, A.A.; Berry, L.; Albrecht, B.K.; Bauer, D.; Bellon, S.F.; Bode, C.; Chen, A.; Choquette, D.; Dussault, I.; Fang, M.; et al. Discovery and optimization of potent and selective triazolopyridazine series of c-Met inhibitors. *Bioorg. Med. Chem. Lett.* **2009**, *19*, 6307–6312. [CrossRef]
33. Cerami, E.; Gao, J.; Dogrusoz, U.; Gross, B.E.; Sumer, S.O.; Aksoy, B.A.; Jacobsen, A.; Byrne, C.J.; Heuer, M.L.; Larsson, E.; et al. The cBio cancer genomics portal: An open platform for exploring multidimensional cancer genomics data. *Cancer Discov.* **2012**, *2*, 401–404. [CrossRef]
34. Gao, J.; Aksoy, B.A.; Dogrusoz, U.; Dresdner, G.; Gross, B.; Sumer, S.O.; Sun, Y.; Jacobsen, A.; Sinha, R.; Larsson, E.; et al. Integrative analysis of complex cancer genomics and clinical profiles using the cBioPortal. *Sci. Signal.* **2013**, *6*, pl1. [CrossRef]
35. Ma, J.; DeFrances, M.C.; Zou, C.; Johnson, C.; Ferrell, R.; Zarnegar, R. Somatic mutation and functional polymorphism of a novel regulatory element in the HGF gene promoter causes its aberrant expression in human breast cancer. *J. Clin. Investig.* **2009**, *119*, 478–491. [CrossRef] [PubMed]
36. Cooke, S.L.; Shlien, A.; Marshall, J.; Pipinikas, C.P.; Martincorena, I.; Tubio, J.M.; Li, Y.; Menzies, A.; Mudie, L.; Ramakrishna, M.; et al. Processed pseudogenes acquired somatically during cancer development. *Nat. Commun.* **2014**, *5*, 3644. [CrossRef]
37. Brighenti, M. MicroRNA and MET in lung cancer. *Ann. Transl. Med.* **2015**, *3*, 68. [PubMed]
38. Poliseno, L.; Salmena, L.; Zhang, J.; Carver, B.; Haveman, W.J.; Pandolfi, P.P. A coding-independent function of gene and pseudogene mRNAs regulates tumour biology. *Nature* **2010**, *465*, 1033–1038. [CrossRef] [PubMed]
39. Wahl, G.M. The importance of circular DNA in mammalian gene amplification. *Cancer Res.* **1989**, *49*, 1333–1340. [PubMed]
40. Nielsen, J.L.; Walsh, J.T.; Degen, D.R.; Drabek, S.M.; McGill, J.R.; von Hoff, D.D. Evidence of gene amplification in the form of double minute chromosomes is frequently observed in lung cancer. *Cancer Genet. Cytogenet.* **1993**, *65*, 120–124. [CrossRef]
41. Shimizu, N. Molecular mechanisms of the origin of micronuclei from extrachromosomal elements. *Mutagenesis* **2011**, *26*, 119–123. [CrossRef]
42. Burgess, T.L.; Sun, J.; Meyer, S.; Tsuruda, T.S.; Sun, J.; Elliott, G.; Chen, Q.; Haniu, M.; Barron, W.F.; Juan, T.; et al. Biochemical characterization of AMG 102: A neutralizing, fully human monoclonal antibody to human and nonhuman primate hepatocyte growth factor. *Mol. Cancer Ther.* **2010**, *9*, 400–409. [CrossRef] [PubMed]
43. Kirchhofer, D.; Yao, X.; Peek, M.; Eigenbrot, C.; Lipari, M.T.; Billeci, K.L.; Maun, H.R.; Moran, P.; Santell, L.; Wiesmann, C.; et al. Structural and functional basis of the serine protease-like hepatocyte growth factor beta-chain in Met binding and signaling. *J. Biol. Chem.* **2004**, *279*, 39915–39924. [CrossRef]
44. Kirchhofer, D.; Lipari, M.T.; Santell, L.; Billeci, K.L.; Maun, H.R.; Sandoval, W.N.; Moran, P.; Ridgway, J.; Eigenbrot, C.; Lazarus, R.A. Utilizing the activation mechanism of serine proteases to engineer hepatocyte growth factor into a Met antagonist. *Proc. Natl. Acad. Sci. USA* **2007**, *104*, 5306–5311. [CrossRef] [PubMed]
45. Verhaak, R.G.; Hoadley, K.A.; Purdom, E.; Wang, V.; Qi, Y.; Wilkerson, M.D.; Miller, C.R.; Ding, L.; Golub, T.; Mesirov, J.P.; et al. Cancer Genome Atlas Research Network. Integrated genomic analysis identifies clinically relevant subtypes of glioblastoma characterized by abnormalities in PDGFRA, IDH1, EGFR, and NF1. *Cancer Cell* **2010**, *17*, 98–110. [CrossRef] [PubMed]
46. Chevet, E.; Hetz, C.; Samali, A. Endoplasmic reticulum stress-activated cell reprogramming in oncogenesis. *Cancer Discov.* **2015**, *5*, 586–597. [CrossRef]

47. Atkins, C.; Liu, Q.; Minthorn, E.; Zhang, S.Y.; Figueroa, D.J.; Moss, K.; Stanley, T.B.; Sanders, B.; Goetz, A.; Gaul, N.; et al. Characterization of a novel PERK kinase inhibitor with antitumor and antiangiogenic activity. *Cancer Res.* **2013**, *73*, 1993–2002. [CrossRef] [PubMed]

48. Karbalaei, M.; Rezaee, S.A.; Farsiani, H. Pichia pastoris: A highly successful expression system for optimal synthesis of heterologous proteins. *J. Cell. Physiol.* **2020**, *235*, 5867–5881. [CrossRef] [PubMed]

49. Low, J.Y.; Laiho, M. Caveolae-associated molecules, tumor stroma, and cancer drug resistance: Current findings and future perspectives. *Cancers* **2022**, *25*, 589. [CrossRef] [PubMed]

50. Surma, M.A.; Klose, C.; Simons, K. Lipid-dependent protein sorting at the trans-Golgi network. *Biochim. Biophys. Acta* **2012**, *1821*, 1059–1067. [CrossRef]

51. Zuidscherwoude, M.; de Winde, C.M.; Cambi, A.; van Spriel, A.B. Microdomains in the membrane landscape shape antigen-presenting cell function. *J. Leukoc. Biol.* **2014**, *95*, 251–263. [CrossRef]

52. Grao-Cruces, E.; Lopez-Enriquez, S.; Martin, M.E.; Montserrat-de la Paz, S. High-density lipoproteins and immune response: A review. *Int. J. Biol. Macromol.* **2022**, *195*, 117–123. [CrossRef]

53. Shattuck, D.L.; Miller, J.K.; Carraway, K.L., 3rd; Sweeney, C. Met receptor contributes to trastuzumab resistance of Her2-overexpressing breast cancer cells. *Cancer Res.* **2008**, *68*, 1471–1477. [CrossRef] [PubMed]

54. Vergani, E.; Vallacchi, V.; Frigerio, S.; Deho, P.; Mondellini, P.; Perego, P.; Cassinelli, G.; Lanzi, C.; Testi, M.A.; Rivoltini, L.; et al. Identification of MET and SRC activation in melanoma cell lines showing primary resistance to PLX4032. *Neoplasia* **2011**, *13*, 1132–1142. [CrossRef] [PubMed]

55. Chen, C.T.; Kim, H.; Liska, D.; Gao, S.; Christensen, J.G.; Weiser, M.R. MET activation mediates resistance to lapatinib inhibition of HER2-amplified gastric cancer cells. *Mol. Cancer Ther.* **2012**, *11*, 660–669. [CrossRef] [PubMed]

56. Wilson, T.R.; Fridlyand, J.; Yan, Y.; Penuel, E.; Burton, L.; Chan, E.; Peng, J.; Lin, E.; Wang, Y.; Sosman, J.; et al. Widespread potential for growth-factor-driven resistance to anticancer kinase inhibitors. *Nature* **2012**, *487*, 505–509. [CrossRef] [PubMed]

57. Straussman, R.; Morikawa, T.; Shee, K.; Barzily-Rokni, M.; Qian, Z.R.; Du, J.; Davis, A.; Mongare, M.M.; Gould, J.; Frederick, D.T.; et al. Tumour micro-environment elicits innate resistance to RAF inhibitors through HGF secretion. *Nature* **2012**, *487*, 500–504. [CrossRef]

58. Malik, R.; Mambetsariev, I.; Fricke, J.; Chawla, N.; Nam, A.; Pharaon, R.; Salgia, R. MET receptor in oncology: From biomarker to therapeutic target. *Adv. Cancer Res.* **2020**, *147*, 259–301.

59. Dussault, I.; Bellon, S.F. c-Met inhibitors with different binding modes: Two is better than one. *Cell Cycle* **2008**, *7*, 1157–1160. [CrossRef]

60. Puccini, A.; Marín-Ramos, N.I.; Bergamo, F.; Schirripa, M.; Lonardi, S.; Lenz, H.J.; Loupakis, F.; Battaglin, F. Safety and tolerability of c-MET inhibitors in cancer. *Drug Saf.* **2019**, *42*, 211–233. [CrossRef] [PubMed]

61. Passiglia, F.; Van Der Steen, N.; Raez, L.; Pauwels, P.; Gil-Bazo, I.; Santos, E.; Santini, D.; Tesoriere, G.; Russo, A.; Bronte, G.; et al. The role of cMet in non-small cell lung cancer resistant to EGFR-inhibitors: Did we really find the target? *Curr. Drug Targets* **2014**, *15*, 1284–1292. [CrossRef]

62. Wang, C.; Cao, Y.; Yang, C.; Bernards, R.; Qin, W. Exploring liver cancer biology through functional genetic screens. *Nat. Rev. Gastroenterol. Hepatol.* **2021**, *18*, 690–704. [CrossRef]

63. Elshazly, A.M.; Gewirtz, D.A. An overview of resistance to Human epidermal growth factor receptor 2 (Her2) targeted therapies in breast cancer. *Cancer Drug Resist.* **2022**, *5*, 472–486. [CrossRef]

64. Parchment, R.E.; Doroshow, J.H. Pharmacodynamic endpoints as clinical trial objectives to answer important questions in oncology drug development. *Semin. Oncol.* **2016**, *43*, 514–525. [CrossRef]

65. Moosavi, F.; Giovannetti, E.; Peters, G.J.; Firuzi, O. Combination of HGF/MET-targeting agents and other therapeutic strategies in cancer. *Crit. Rev. Oncol. Hematol.* **2021**, *160*, 103234. [CrossRef]

66. Rodríguez-Camacho, A.; Flores-Vázquez, J.G.; Moscardini-Martelli, J.; Torres-Ríos, J.A.; Olmos-Guzmán, A.; Ortiz-Arce, C.S.; Cid-Sánchez, D.R.; Pérez, S.R.; Macías-González, M.D.S.; Hernández-Sánchez, L.C.; et al. Glioblastoma treatment: State-of-the-art and future perspectives. *Int. J. Mol. Sci.* **2022**, *23*, 7207. [CrossRef] [PubMed]

67. Yao, M.; Li, S.; Wu, X.; Diao, S.; Zhang, G.; He, H.; Bian, L.; Lu, Y. Cellular origin of glioblastoma and its implication in precision therapy. *Cell. Mol. Immunol.* **2018**, *15*, 737–739. [CrossRef]

68. Arteaga, C.L.; Sliwkowski, M.X.; Osborne, C.K.; Perez, E.A.; Puglisi, F.; Gianni, L. Treatment of HER2-positive breast cancer: Current status and future perspectives. *Nat. Rev. Clin. Oncol.* **2011**, *9*, 16–32. [CrossRef]

69. Metcalfe, C.; de Sauvage, F.J. Hedgehog fights back: Mechanisms of acquired resistance against Smoothened antagonists. *Cancer Res.* **2011**, *71*, 5057–5061. [CrossRef]

70. Bedard, P.L.; Azambuja, E.D.; Cardoso, F. Beyond Trastuzumab: Overcoming resistance to targeted HER-2 therapy in breast cancer. *Curr. Cancer Drug Targ.* **2009**, *9*, 148–162. [CrossRef] [PubMed]

71. Fedorenko, I.V.; Paraiso, K.H.T.; Smalley, K.S.M. Acquired and intrinsic BRAF inhibitor resistance in BRAF V600E mutant melanoma. *Biochem. Pharmacol.* **2011**, *82*, 201–209. [CrossRef] [PubMed]

Review

The Development and Role of Capmatinib in the Treatment of MET-Dysregulated Non-Small Cell Lung Cancer—A Narrative Review

Robert Hsu [1], David J. Benjamin [2] and Misako Nagasaka [3,*]

[1] Division of Medical Oncology, Department of Internal Medicine, Norris Comprehensive Cancer Center, University of Southern California, Los Angeles, CA 90033, USA; robert.hsu@med.usc.edu
[2] Hoag Family Cancer Institute, Newport Beach, CA 92663, USA; david.benjamin@hoag.org
[3] Division of Hematology and Oncology, Department of Medicine, Chao Family Comprehensive Cancer Center, University of California Irvine School of Medicine, Orange, CA 92868, USA
* Correspondence: nagasakm@hs.uci.edu

Simple Summary: In this narrative review, we discuss the development of capmatinib, a reversible *MET* tyrosine kinase inhibitor that received approval for advanced non-small cell lung cancer (NSCLC) harboring *MET* exon 14 skipping mutation. Capmatinib was first discovered in 2011 and has been shown to have promising antitumor activity. Early-phase trials identified a recommended dose of 400 mg twice daily in tablet formulation. The GEOMETRY mono-1 trial showed efficacy in *MET* exon 14 skipping mutation, leading to FDA approval for capmatinib. Currently, ongoing clinical trials evaluating combination therapy with capmatinib, including amivantamab, trametinib, and immunotherapy, are being conducted to improve efficacy and broaden indications of capmatinib with new drug agents such as antibody–drug conjugates being developed to treat *MET* dysregulated NSCLC.

Abstract: Non-small cell lung cancer (NSCLC) is a leading cause of death, but over the past decade, there has been tremendous progress in the field with new targeted therapies. The mesenchymal–epithelial transition factor (*MET*) proto-oncogene has been implicated in multiple solid tumors, including NSCLC, and dysregulation in NSCLC from *MET* can present most notably as *MET* exon 14 skipping mutation and amplification. From this, *MET* tyrosine kinase inhibitors (TKIs) have been developed to treat this dysregulation despite challenges with efficacy and reliable biomarkers. Capmatinib is a Type Ib *MET* TKI first discovered in 2011 and was FDA approved in August 2022 for advanced NSCLC with *MET* exon 14 skipping mutation. In this narrative review, we discuss preclinical and early-phase studies that led to the GEOMETRY mono-1 study, which showed beneficial efficacy in *MET* exon 14 skipping mutations, leading to FDA approval of capmatinib along with Foundation One CDx assay as its companion diagnostic assay. Current and future directions of capmatinib are focused on improving the efficacy, overcoming the resistance of capmatinib, and finding approaches for new indications of capmatinib such as acquired *MET* amplification from epidermal growth factor receptor (*EGFR*) TKI resistance. Clinical trials now involve combination therapy with capmatinib, including amivantamab, trametinib, and immunotherapy. Furthermore, new drug agents, particularly antibody–drug conjugates, are being developed to help treat patients with acquired resistance from capmatinib and other TKIs.

Keywords: NSCLC; MET dysregulation; capmatinib; tyrosine kinase inhibitor; detection

Citation: Hsu, R.; Benjamin, D.J.; Nagasaka, M. The Development and Role of Capmatinib in the Treatment of MET-Dysregulated Non-Small Cell Lung Cancer—A Narrative Review. *Cancers* **2023**, *15*, 3561. https://doi.org/10.3390/cancers15143561

Academic Editors: Jan Trøst Jørgensen and Jens Mollerup

Received: 24 May 2023
Revised: 4 July 2023
Accepted: 6 July 2023
Published: 10 July 2023

1. Introduction

Non-small cell lung cancer (NSCLC) is a leading cause of death, accounting for an estimated 1.8 million deaths according to GLOBOCAN in 2020 [1]. Over the past decade, there has been tremendous progress in the discovery and development of targeted therapies for *EGFR*; *KRAS* G12C; *BRAF* V600E mutations; *ALK, ROS1*; *RET* gene rearrangements; *MET* alterations, including *MET* exon 14 skipping mutations, *ERBB2* (HER2) mutations, and *NTRK* 1/2/3 gene mutations [2–10]. This has led to the personalization of medicine in NSCLC.

The mesenchymal–epithelial transition factor (*MET*) gene is located in human chromosome 7 (7q21–q31), comprising 21 exons and 21 introns, and encodes a protein that is approximately 120 kDa in size. The ligand for MET is hepatocyte growth factor (HGF), which is a soluble cytokine and is synthesized by mesenchymal cells, fibroblasts, and smooth muscle cells [11]. HGF will bind to MET, and this will trigger the autophosphorylation of Tyr-1234 and Tyr-1235 in the intracellular tyrosine kinase domain, which then undergoes further autophosphorylation of Tyr-1340 and Tyr-1356 in the C-terminal docking site [11,12]. This then facilitates the recruitment of intracellular effector molecules such as GRB2, SRC, PIK3, and GAB1, leading to the activation of downstream pathways. Normally, MET/HGF signaling pathway mediates embryogenesis, tissue regeneration, wound healing, and the formation of nerves and muscles [11–13].

In cancer, the *MET* proto-oncogene is abnormally activated and stimulates other signaling pathways in tumor cells, notably PI3K/AKT, JAK/STAT, Ras/MAPK, SRC, and Wnt/beta-catenin [11] (Figure 1). *MET* overexpression can be found in inflammation and hypoxia, leading to proliferation and migration, and is seen in a large variety of cancer types, including epithelial, mesenchymal, and hematological malignancies [14]. In NSCLC, it has been shown to be overexpressed in 35–72% of cases [14]. High levels of *MET* expression have been found to correlate with early disease recurrence [15]. *MET* dysregulation in NSCLC can present in a variety of ways—gene overexpression; HGF expression that can cause ligand-induced activation, leading to sustained or altered signaling; gene amplification, which can lead to overexpression and reduce the requirement for ligand activation, leading to sustained or altered signaling of the *MET* receptor; gene rearrangement, which may reduce or remove the requirement for ligand activation, leading to sustained altered signaling properties of the *MET* receptor; and downstream MET signaling alterations [11,12,15]. Notably, cigarette smoking can upregulate c-MET and the downstream Akt pathway [16]. It also affects the sensitivity of *EGFR* TKIs as cigarette smoke attenuates the AMP-activated protein-kinase (AMPK)-dependent inhibition of mTOR which then decreases the sensitivity of NSCLC cells with wild-type *EGFR* to TKI and thereby represses the expression of liver kinase B1 (LKB1) [17]. Finally, *MET* dysregulation can occur via gene mutation, most notably the *MET* exon 14 skipping mutation seen in about 3–4% of adenocarcinoma and 2% of squamous cell carcinoma but in higher frequencies in adenosquamous carcinoma (6%) and pulmonary sarcomatoid carcinoma (9–22%) [15,18].

MET exon 14 skipping mutations are processes in which the 47-amino-acid juxtamembrane domain is deleted, altered, or disrupted by intronic regions surrounding exon 14, leading to fusion in mature mRNA between exon 13 and exon 15 [19,20]. *MET* exon 14 skipping mutations have been shown to be exclusive from other driver mutations but coexist with other *MET* amplification or copy number gains [21]. Meanwhile, the amplification of the *MET* gene, which is defined as a gain in copy number (GCN), has been seen both de novo and as an acquired resistance mechanism [22]. *MET* amplification is seen in *EGFR*-acquired resistance and can occur with or without the loss of T790M [23]. In the analysis of resistance mechanisms in the AURA 3 study (*n* = 78), *MET* amplification was seen in (14/78,18%) of samples, *EGFR* C797S (14/78,18%) of cases, and 15 patients having >1 resistance-related genomic alteration [23,24]. *MET* amplification is also considered an acquired resistance mechanism of *ALK* inhibitors, as *MET* amplification has been observed in about 15% of next-generation *ALK* inhibitor resistance [25]. Both *MET* exon 14 skipping mutations and *MET* high-level amplification have been shown to portend poor prognosis [21]. Without the use of *MET* inhibitors, a retrospective study by Awad et al. showed that the median OS was 8.1 months [26]. *MET* exon 14 skipping mutations are seen more frequently in females than in males, and the median age of *MET* exon 14 skip mutation patients ranged from 71.4 to 76.7 years [6,18].

Compared with other driver mutations, *MET* exon 14 skip mutation patients tend to be smokers, with only about 36% being never smokers in a previous retrospective analysis [27].

MET Signaling and Blockade by MET Inhibitors

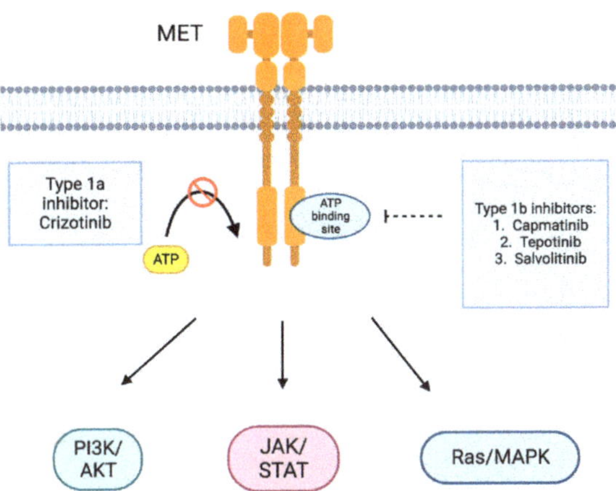

Figure 1. MET signaling pathway and blockade by MET inhibitors. In cancer, the *MET* proto-oncogene is abnormally activated and stimulates other signaling pathways in tumor cells, notably PI3K/AKT, JAK/STAT, Ras/MAPK, SRC, and Wnt/beta-catenin [11]. Type 1a inhibitor crizotinib blocks ATP binding to prevent the phosphorylation of the receptor, whereas type 1b inhibitors such as capmatinib are more specific and bind to a pocket adjacent to the ATP binding site. This figure was generated by BioRender.

MET tyrosine kinase inhibitors have been developed to treat *MET*-dysregulated NSCLC, classified as Type I, Type II, and Type III inhibitors. Type I inhibitors compete with ATP for the binding of the ATP-binding pocket of the active conformation of *MET*. Specifically, Type Ia inhibitors such as crizotinib interact with the Y1230 residue in the hinge region and are dependent on binding with the G1163 residue [28,29]. Type Ib inhibitors such as capmatinib, tepotinib, and savolitinib also connect with the Y1230 residue but are not dependent on G1163 binding [28,30–32]. Meanwhile, Type II inhibitors, which include cabozantinib, meresitinib, and gleasatanib, bind the ATP pocket in an inactive state [32–35]. Type III inhibitors bind to allosteric sites different from the ATP site and are not competitive; tivantinib has been studied in NSCLC but was not found to show any benefit in interim analysis and therefore was discontinued [32,34,36].

This review specifically focuses on capmatinib (INC280), which received U.S. Food and Drug Administration (FDA) approval for *MET* exon 14 skip mutations in metastatic NSCLC on 10 August 2022 and by the European Medicines Agency (EMA) on 20 June 2022 specifically for those patients who have received immunotherapy or platinum-based chemotherapy or both [37,38]. Herein, we review clinical development trials involving capmatinib, notably the GEOMETRY mono-1 study, which led to FDA approval and the companion diagnostic assay for the detection of MET exon 14 skipping mutations.

2. Crizotinib

Prior to capmatinib, crizotinib was the first *MET* TKI to show efficacy in *MET* exon 14 skipping mutation in advanced NSCLC. The PROFILE 1001 trial showed an overall response rate (ORR) of 32% (95% CI 21–45) among 65 response-evaluable patients, with a median duration of response (DOR) of 9.1 months (95% CI 6.4–12.7) and a progression-free survival (PFS) rate of 7.3 months (95% CI 5.4–9.1), with two additional Phase II crizotinib trials showing ORR of around 30% [39–41]. However, crizotinib confers resistance to G1163R mutation not seen in *MET* Type Ib TKIs such as capmatinib, and thus treatment for *MET* dysregulation has shifted towards *MET* Type Ib TKIs [42]. Currently, crizotinib is approved for *ALK*- and *ROS1*-positive advanced NSCLC by the FDA and EMA [43,44].

3. Preclinical Studies

Capmatinib was first reported in 2011 by Liu et al., who showed that in both in vivo and in vitro mice studies using human cell lines, capmatinib had a 10,000-fold selectivity for c-met over a large panel of human kinase [45]. They showed that capmatinib can block the c-MET phosphorylation and activation of downstream targets, including HGF. They further showed that activated c-met upregulates cancer-promoting EGFR and HER-3 pathways [45]. Baltschukat et al. further investigated capmatinib in NSCLC [46]. They investigated the affinity of capmatinib in a set of 442 kinases and demonstrated a selectivity in *MET* of over 1000 fold [46]. Furthermore, they demonstrated that capmatinib is highly selective to Y1230 and D1228 and observed resistance when using cell lines bearing mutations to Y1230 and D1228 [46]. *MET* amplification and HGF expression in vitro were also associated with capmatinib sensitivity in vitro [46].

4. Pharmacodynamics/Pharmacokinetics

Capmatinib is a selective Type Ib ATP-competitive tyrosine kinase inhibitor targeting *MET*. Capmatinib has an average IC_{50} value of 0.13 nM and a cell-based IC_{50} of 0.3–0.7 nM in lung cancer cell lines [28,46] (Figure 2). Capmatinib has linear pharmacokinetics, with exposure increasing approximately dose-proportionally over a dose range of 200–400 mg. It is rapidly absorbed, with peak plasma concentration (C_{max}) obtained about 1–2 h after a 400 mg dose is given. There is similar absorption when taken with and without food. The effective elimination half-life is 6.5 h. The plasma protein binding is 96% [38,47].

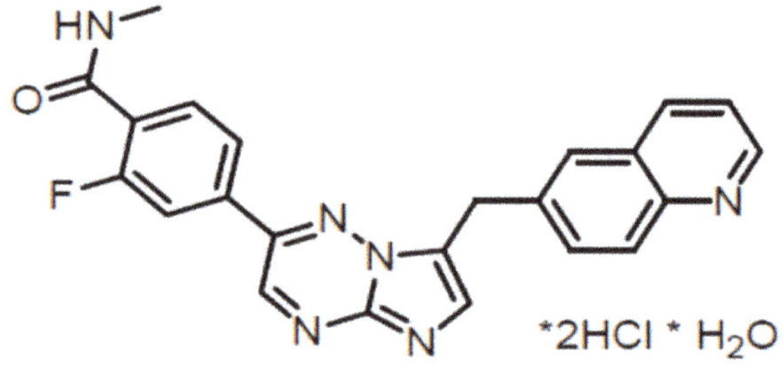

Figure 2. Chemical structure of capmatinib; the asterisk (*) represents the chiral carbons that are part of the chemical structure.. The chemical name for capmatinib is 2-Fluoro-N-methyl-4-[7-(quinolin-6-ylmethyl)imidazo[1,2 b][1,2,4]triazin-2-yl]benzamide—hydrogen chloride—water (1/2/1). The molecular formula for capmatinib hydrochloride is $C_{23}H_{21}Cl_2FN_6O_2$ [38].

Capmatinib is metabolized by CYP3A4 and aldehyde oxidase. In a single oral dose, 78% of total radioactivity was recovered in feces with 42% as unchanged and 22% recovered in urine. There are no specific significant effects on the pharmacokinetic parameters of capmatinib identified in the following covariates assessed: age, sex, race, mild-to-moderate renal impairment, and hepatic impairment [38,47].

In drug interaction studies, coadministration with itraconazole, a strong CYP3A inhibitor, increased capmatinib's area under the curve ($AUC_{0\text{-}INF}$) by 42%, with no change in C_{max}. Coadministration with rifampicin, a strong CYP3A inducer, decreased capmatinib $AUC_{0\text{-}INF}$ by 67% and decreased C_{max} by 56%. Coadministration with protein pump inhibitors (rabeprazole) decreased capmatinib by $AUC_{0\text{-}INF}$ 25% and decreased C_{max} by 38%. Coadministration with rosuvastatin, a BRCP substrate, increased rosuvastatin $AUC_{0\text{-}INF}$ by 108% and increased C_{max} by 204% [38,47].

5. Phase I Clinical Trials

Multiple open-label, multicenter, Phase I studies in advanced solid tumors have evaluated capmatinib. A Phase I study comprising 44 adult Japanese patients, including 15 NSCLC patients, found that the highest studied dose determined to be safe was 400 mg administered orally (po) twice a day (b.i.d.) as a tablet. The median duration of treatment exposure was 7 weeks (range 0.4–32.3 weeks), with disease progression being the primary reason for the discontinuation occurring in 38 patients (86.4%). There were two drug-limiting toxicities (DLTs), which consisted of Grade 2 suicidal ideation in a patient taking 600 mg po b.i.d. and Grade 3 depression in a patient taking 400 mg po b.i.d. [48]. Another global Phase I study, comprising 38 patients primarily with gastrointestinal cancers, had a recommended Phase II dose (R2PD) of 600 mg po b.i.d. in a capsule formulation and 400 mg po b.i.d. in a tablet formulation. The most frequent Grade 3 or 4 adverse events were an increase in levels of blood bilirubin (11%), fatigue (8%), and AST increase (8%) [49].

Schuler et al. investigated 55 patients with advanced *MET*-dysregulated NSCLC, which included 40 patients with prior systemic therapies. All patients discontinued treatment, mostly due to disease progression (69.1%), with a median duration of 10.4 weeks. While the overall response rate (ORR) by RECIST for the entire cohort was 20%, *MET* with a gene copy number $\geq$6 had an ORR of 47%, with median progression-free survival (PFS) of 9.3 months, and all patients with *MET* exon 14 skip mutations had a response. The most common toxicities were nausea (42%), peripheral edema (33%), and vomiting (31%) [50]. Another Phase Ib/II study involving capmatinib investigated *EGFR*-mutated, *MET*-dysregulated NSCLC in combination with gefitinib, an *EGFR* TKI, in patients with acquired *EGFR* TKI resistance. The ORR across the cohort was 27%, with a 47% ORR in patients with a *MET* copy number $\geq$6. The drug was relatively well tolerated, with the most common Grade 3–4 adverse event being increased amylase and lipase levels (6% in both). The R2PD was capmatinib 400 mg po b.i.d. plus gefitinib 250 mg po daily [51] (Table 1).

Table 1. Early-stage studies on capmatinib.

Publication	*n*	Indication	R2PD	ORR
Esaki et al. [48]	44 (15 NSCLC)	Advanced solid tumors	400 mg po bid	
Bang et al. [49]	38 (1 NSCLC)	Advanced solid tumors	600 mg po bid (capsule)/ 400 mg po bid (tablet)	
Schuler et al. [50]	55	Advanced NSCLC	600 mg po bid (capsule)/ 400 mg po bid (tablet)	47%
Wu et al. [51]	61 Phase Ib/100 Phase II	Advanced NSCLC in patients with acquired *EGFR* TKI resistance	400 mg po b.i.d. plus gefitinib 250 mg po daily	27% (47% in patients with MET GCN $\geq$ 6)

6. GEOMETRY Mono-1 Trial

The GEOMETRY mono-1 trial was a multicohort Phase II study in patients with *MET*-dysregulated advanced NSCLC. The patients were either in Stage IIIB or IV NSCLC, had no *EGFR* mutation, and were negative for *ALK* rearrangement. All subjects took capmatinib 400 mg po b.i.d. A total of 364 patients were enrolled, with 97 having a *MET* exon 14 skipping mutation and 210 having *MET* amplification. There were seven cohorts to the study: In previously treated patients (1–2 lines of therapy), Cohort 1 consisted of

MET amplification with (a) GCN $\geq$ 10 (n = 69) or (b) GCN 6–9 (n = 42); Cohort 2 consisted of *MET* amplification with GCN 4–5 (n = 54); Cohort 3 consisted of *MET* amplification with GCN < 4 (n = 30); Cohort 4 consisted of *MET* exon 14 skipping mutation with any GCN (n = 69); and Cohort 6 consisted of *MET* amplification with GCN > 10 (n = 3) or *MET* exon 14 skipping mutation with any GCN (n = 31) who had received one line of therapy (n = 34). In the untreated group, Cohort 5a consisted of *MET* amplification with GCN $\geq$ 10 (n = 15); Cohort 5b consisted of *MET* exon 14 skipping mutation with any GCN (n = 28); and Cohort 7 consisted of treatment-naïve *MET* exon 14 skipping mutation with any GCN (n = 23). *MET* exon 14 skipping mutation patients had a slightly higher median age (71 years) than patients with *MET* amplification (60–70 years) on diagnosis. Patients with *MET* exon 14 skipping mutation were more likely to be women and to have never smoked [6].

Among patients with *MET* exon 14 skip mutations, ORR was seen in 41% (95% CI 29–53) of 69 previously treated patients and 68% (95% CI 48–84) of 28 previously untreated patients. The median duration of response (DOR) was 9.7 months (95% CI 5.6–13.0) among the treated patients and 12.6 months (95% CI 5.6—not reached) in previously untreated patients. Most patients (82% in treated and 68% in untreated) had a response at the first tumor evaluation following the start of capmatinib therapy. The median PFS was 5.4 months (95% CI 4.2–7.0) in previously treated patients and 12.4 months (95% CI 8.2—not reached) in previously untreated patients. Notably, 12 of 13 patients with exon 14 skipping mutations who had brain metastasis had intracranial disease control. The primary reason for discontinuation was progressive disease (58% in previously treated patients and 46% in untreated patients) [6].

In patients with GCN < 10, the cohorts were closed due to futility, as PFS for GCN 6–9 and 4 or 5 was only 2.7 months. In GCN $\geq$ 10, there was activity; the ORR was 29% (95% CI 19–41) in previously treated patients and 40% (95% CI 16–68) in previously untreated patients, but this fell below the predefined clinical efficacy. The median DOR was 8.3 months (95% CI 4.2–15.4) in treated patients and 7.5 months (95% CI 2.6–14.3) in untreated patients. The median PFS was 4.1 months (95% CI 2.9–4.8) in treated patients and 4.2 months (95% CI 1.4–6.9) in untreated patients [6] (Table 2).

Across all cohorts, the most reported adverse events were peripheral edema, nausea, and vomiting. Overall, 67% of patients had adverse events of Grade 3 or 4; the most frequent of these were peripheral edema, nausea, vomiting, and increased blood creatinine level. Treatment-related adverse events led to the discontinuation of treatment in 39 patients (11%), with treatment-related peripheral edema leading to discontinuation in 6 patients (2%) [6].

The post hoc analysis involving 69 *MET* exon 14 skipping mutation patients that focused on 19 patients in the cohort who had previously received immunotherapy (IO) showed ORR 57.9% (n = 11/19; 95% CI 33.5–79.5%), with a median DOR of 11.2 months (95% CI 3.35—not reached). Safety findings were similar, according to which capmatinib showed efficacy irrespective of prior treatment with IO and was also well tolerated in post-IO patients [52]. Moreover, capmatinib was associated with clinically meaningful improvements in cough and preserved the quality of life in patient-reported surveys [53]. There was also a subgroup analysis on 45 Japanese patients, which showed an ORR of 36% (95% 10.9–69.2) and good tolerability [54].

Table 2. Responses to capmatinib treatment relative to the cohort in GEOMETRY mono-1 trial (6).

Response	NSCLC with *MET* Exon 14 Skipping Mutation		NSCLC with *MET* Amplification				
Best Response—No (%)	**Cohort 4 *n* = 69, any GCN with 1–2 Lines of Therapy**	**Cohort 5b *n* = 28, any GCN with No Previous Therapy**	**Cohort 1a *n* = 69, GCN ≥ 10 with 1–2 Lines of Therapy**	**Cohort 5a *n* = 15, GCN ≥ 10 with No Previous Therapy**	**Cohort 1b *n* = 42, GCN 6–9 with 1–2 Lines of Therapy**	**Cohort 2 *n* = 54, GCN 4 or 5 with 1–2 Lines of Therapy**	**Cohort 3 *n* = 30, GCN < 4 with 1–2 Lines of Therapy**
Complete response	0	1 (4)	1 (1)	0	0	0	0
Partial Response	28 (41)	18 (64)	19 (28)	6 (40)	5 (12)	5 (9)	2 (7)
Stable disease	25 (36)	7 (25)	28 (41)	4 (27)	17 (40)	20 (37)	14 (47)
Incomplete response or nonprogressive disease	1 (1)	1 (4)	1 (1)	0	1 (2)	0	0
Unknown or could not be evaluated	9 (13)	0	8 (12)	1 (7)	4 (10)	8 (15)	8 (27)
Overall response							
No. of patients with overall response	28	19	20	6	5	5	2
Percent of patients (95% CI)	41 (29–53)	68 (48–84)	29 (19–41)	40 (16–68)	12 (4–26)	9 (3–20)	7 (1–22)
Disease control							
No. of patients with disease control	54	27	49	10	23	25	16
Percent of patients (95% CI)	78 (67–87)	96 (82–100)	71 (59–81)	67 (38–88)	55 (39–70)	46 (33–60)	53 (34–72)
Duration of Response							
No. of events/No. of patients with response	23/28	11/19	15/20	6/6	3/5	4/5	2/2
Median duration of response (95% CI)—mo	9.7 (5.6–13.0)	12.6 (5.6–NE)	8.3 (4.2–15.4)	7.5 (2.6–14.3)	24.9 (2.7–24.9)	9.7 (4.2–NE)	4.2 (4.2–4.2)
Progression-free survival							
Progression or death—No. of patients	60	17	58	15	34	50	22
Median progression-free survival (95% CI)—mo	5.4 (4.2–7.0)	12.4 (8.2–NE)	4.1 (2.9–4.8)	4.2 (1.4–6.9)	2.7 (1.4–3.1)	2.7 (1.4–4.1)	3.6 (2.2–4.2)

A recent real-world analysis was carried out that investigated *MET* exon 14 skipping mutation and brain metastasis patients; of the 68 patients that fit the criteria, the real-world response rate was 90.9%, with 87.3% intracranial response along with a median PFS rate of 14.1 months [55]. Another real-world retrospective study examined 81 cases of NSCLC with advanced NSCLC and *MET* exon 14 skipping mutation who were treated with capmatinib from March 2019 to December 2021 [56]. The ORR to capmatinib was 58% (95% CI 47–69), including 68% (95% CI 50–82) for treatment-naïve and 50% (95% CI 35–65) for pretreated patients. The median PFS was 9.5 months (95% CI 4.7–14.3), and the median OS was 18.2 months (95% CI 13.2) for the entire cohort, including a median PFS of 10.6 months (95% CI 5.5–15.7) for untreated patients [56].

Thus, the GEOMETRY mono-1 trial evaluated *MET*-dysregulated, advanced NSCLC, with promising ORR and PFS seen in *MET* exon 14 skip mutations, though the results showed a lack of effect in *MET* GCN < 10, leading to FDA and EMA approval for capmatinib only in advanced NSCLC with *MET* exon 14 skipping mutations. Subsequent real-world data have shown response to capmatinib among patients with *MET* exon 14 skipping mutations, with IO exposure and brain metastasis [52,56].

7. Tepotinib and Savolitinib

Two other MET selective Type Ib inhibitors have been investigated in *MET* alterations, namely tepotinib and savolitinib [30,31]. Tepotinib received accelerated approval from the FDA for *MET* exon 14 skipping mutations in advanced NSCLC after the open-label Phase II VISION study [31]. It also received approval from the EMA for those with advanced NSCLC *MET* exon 14 skipping mutations who require systemic therapy following immunotherapy and/or platinum-based therapy [57]. In this study, 152 patients with *MET* exon 14 skipping mutations were followed, and the ORR was 46% (95% CI 36–57), including 44.2% (95% CI 29.1–60.1) in untreated patients and 48.2 (95% CI 34.7–62.0) in previously treated patients. The median DOR was 11.1 months (95% CI 7.2—not reached), and the PFS was 8.5 months (95% CI 5.1–11.0) [31]. There were 11 patients with brain metastasis in the study, with a median PFS of 10.9 months (95% CI 8.0—not reached) [31].

Meanwhile, a Phase II, single-arm, open-label study in China involved 84 patients with *MET* exon 14 skipping mutations who had positive pulmonary sarcomatoid carcinoma or other NSCLC subtypes and received savolitinib [30]. The ORR was 42.9% (95% CI 31.1–55.3) [20] (Table 3). Savolitinib received conditional approval in China in 2021 for the treatment of metastatic NSCLC with *MET* exon 14 skipping mutations in patients who have progressed after or who are unable to tolerate platinum-based chemotherapy [58].

Table 3. Key trials involving *MET* selective Type 1b inhibitors.

	Capmatinib [6]	Tepotinib [31]	Savolitinib [30]
N (with MET exon 14 skipping mutation)	97	152 (99 evaluable)	84 (70 evaluable)
Overall response rate (%) (95% CI)	68% (48–84) in untreated patients (*n* = 28) and 41 (29–53) in previously treated patients (*n* = 69)	46 (36–57); 44.2% (29.1–60.1) in untreated patients (*n* = 43) and 48.2 (34.7–62.0) in previously treated patients (*n* = 56)	42.9 (31.1–53.3); 46.4 (27.5–66.1) in untreated patients (*n* = 28) and 40.5 (25.6–56.7) in previously treated patients (*n* = 42)
Duration of response mo (95% CI)	12.6 (5.6—NE) in untreated patients and 9.7 (5.6–13.0) in previously treated patients	11.1 (7.2—NE)	8.3 (5.3–16.6); 5.6 (4.1–9.6) in untreated patients and 9.7 (4.9—NE) In previously treated patients
Progression-free survival mo (95% CI)	12.4 (8.2—NE) in untreated patients and 5.4 (4.2–7.0) in previously treated patients	8.5 (5.1–11.0)	6.8 (4.2–9.6); 5.6 (4.1–9.6) in untreated patients and 6.9 (4.1–9.3) in previously treated patients

8. Companion Diagnostic Assay

One of the challenges in the success of finding successful *MET*-targeted therapies has been finding a reliable biomarker. For example, in previous studies where *MET* GCN $\geq$ 6, the ORR outcomes ranged from 16% to 67%, while for immunohistochemistry (IHC) 2+ and 3+, the ORR outcomes ranged from 14% to 68% [6,39–41,50,51,59]. Another way to assess *MET* overexpression has been the *MET*/chromosome 7 centromere (CEP7) ratio, in which ORR outcomes range from 33% to 67% [51,59] (Table 4).

Table 4. Predictive biomarkers and methods for FDA-approved, MET-targeted drugs in NSCLC [59].

Publication	Drug	Method	Biomarker	N	ORR%
Moro-Sibilot et al. [39]	Crizotinib	FISH	*MET* GCN $\geq$ 6	25	16
		NGS	*MET* exon 14 skip	25	12
Landi et al. [40]	Crizotinib	FISH	*MET*/CEP7 > 2.2	16	31
		NGS	*MET* exon 14 skip	10	20
Drilon et al. [41]	Crizotinib	NGS	*MET* exon 14 skip	65	32
Schuler et al. [50]	Capmatinib	FISH	*MET* GCN < 4	17	6
			MET GCN 4–6	12	25
			MET GCN $\geq$ 6	15	47
			MET/CEP7 > 2.0	9	44
			MET/CEP7 < 2.0	32	22
		IHC	MET IHC 2+	14	14
			MET IHC 3_	37	27
Wu et al. [51]	Capmatinib with gefitinib	FISH	*MET* GCN < 4	41	12
			MET GCN 4–6	18	22
			MET GCN $\geq$ 6	36	47
		IHC	MET IHC 2+	16	19
			MET IHC 3+_	37	27
Wolf et al. [6]	Capmatinib	NGS	*MET* exon 14 skip (Previously treated)	69	41
			MET exon 14 skip (Untreated)	28	64
		NGS	*MET* GCN < 4 (Previously treated)	30	7
			MET GCN 4–5 (Previously treated)	54	9
			MET GCN > 6–9 (Previously treated)	42	12
			MET GCN $\geq$ 10 (Previously treated)	69	28
			MET GCN $\geq$ 10 (Untreated)	15	40
Paik et al. [31]	Tepotinib	NGS	*MET* exon 14 skip	99	46

Some thoughts as to the lack of reliability in *MET* amplification have been that gene copy number gains can occur through both polysomy and amplification and thus the gene copy number could be a result of polysomy, not true amplification [59–61]. Another possible problem has been the use of NGS-based assays with a control group using CEP7 [59,61]. A previous study has shown that a *MET*/CEP7 ratio >5 is reliable for *MET* inhibitor response, but the issue is that many below this ratio have other oncogenes and may not be truly *MET*-addicted cases [61]. Guo et al. demonstrated that MET expression via mass spectrometry, IHC, and H-score $\geq$ 200 had significantly improved PFS but saw no association based on copy number [62].

Another challenging aspect of finding a reliable companion diagnostic assay has been the discrepancy between circulating tumor DNA (ctDNA) and tumor next-generation sequencing (NGS) testing. Ikeda et al. studied the ctDNA of 438 patients, and among the 31 patients with *MET* alterations, only 2 of the 18 patients who also received tissue testing were found to have *MET* alterations in the tissue [63]. Another study involving paired plasma and tissue samples in advanced NSCLC patients showed 77.6% concordance between tissue and plasma NGS; 26% of the cohort who received both ctDNA and tissue testing had *MET* alterations on ctDNA testing, but only 17.8% of the 26% total also had *MET* alterations on tissue testing [64]. Overall, when compared to tumor NGS testing, ctDNA had 67.7% sensitivity and 88.8% specificity in pretreated patients, whereas in treated patients, it revealed a sensitivity of 68.4% but only a specificity of 16.7% [64]. Yet, *MET* alterations have been found in both circulating-free DNA (cfDNA) and circulating tumor cells (CTCs) both at diagnosis and at resistance to *EGFR* TKIs [65]. Moreover, Peng et al. examined 48 paired samples and showed a 92.4% concordance between the absolute copy number variant > 6 and the NGS detection of *MET* amplification in tumor tissue [66]. This all has significant ramifications clinically when it comes to making sure *MET* dysregulation is captured on diagnosis but then also on acquired resistance because sometimes patients may not have adequate tissue for testing, which limits them only to liquid biopsy testing, or clinicians may choose to only perform liquid biopsy testing upon the progression of the disease. Thus, finding a trustable biomarker, whether it is a specific *MET* GCN or *MET*/CEP7 ratio threshold that can be used in both tissue testing and ctDNA testing, will go a long way towards determining which *MET* amplification patients would benefit from capmatinib and other *MET*-targeted agents and to ensure that as many *MET* exon 14 skipping mutations are detected as possible.

In *MET* exon 14 skipping mutations, there is also some variability in the ORR, with ranges from 32% to 64%, though these studies do originate from patients on different lines of therapy and different *MET* TKI inhibitors [30,31,41]. However, in the GEOMETRY mono-1 trial, a clinical bridging study was carried out to show analytical and clinical agreement between the enrollment assay and the Foundation One CDx assay [59,67]. The Foundation One CDx assay, developed by Foundation Medicine in collaboration with Novartis, is performed at Foundation Medicine Inc. using DNA isolated from fresh-frozen paraffin-embedded (FFPE) tumor tissue specimens. In previously treated patients, the positive percent agreement (PPA) was 96.8%, the negative percent agreement (NPA) was 100%, and the overall agreement (OA) was 100%. In untreated patients, the PPA, NPA, and OA were all 100%. This led to the FDA approval of the Foundation One CDx assay as the only assay associated with a *MET* inhibitor [59,67].

9. Toxicities

In the GEOMETRY mono-1 trial, across all cohorts, the most reported adverse events were peripheral edema, nausea, and vomiting. Notably, 67% of patients had adverse events of Grade 3 or 4; the most frequent of these were peripheral edema, nausea, vomiting, and increased blood creatinine level. Treatment-related adverse events led to the discontinuation of treatment in 39 patients (11%), with treatment-related peripheral edema leading to discontinuation in 6 patients (2%) [6].

In the VISION study, 28% of patients had Grade 3–4 adverse events, with peripheral edema (7%) being the greatest [31]. Other Grade 3–4 adverse events with greater than 1% incidence included increased amylase (3%), increased lipase (3%), pleural effusion (3%), increased ALT (3%), increased AST (2%), and general edema (3%) [31]. Meanwhile, in the study involving savolitinib, treatment-related adverse events occurred in 46% of the patients, with increased aspartate aminotransferase (n = 9), alanine aminotransferase

(n = 7), and peripheral edema (n = 6) being the most common serious adverse side effect. There was one death in the study due to tumor lysis syndrome, likely treatment-related [30] (Table 5).

Table 5. Adverse events in all cohorts (n = 364) in the GEOMETRY mono-1 trial [6].

Adverse Event	Total	Grade 3 or 4
Any event—No. (%)	355 (98)	244 (67)
Most common events—No. (%)		
Peripheral edema	186 (51)	33 (9)
Nausea	163 (45)	9 (2)
Vomiting	102 (28)	9 (2)
Blood creatinine increased	89 (24)	0
Dyspnea	84 (23)	24 (7)
Fatigue	80 (22)	16 (4)
Decreased appetite	76 (21)	3 (1)
Constipation	66 (18)	3 (1)
Diarrhea	64 (18)	2 (1)
Cough	58 (16)	2 (1)
Back Pain	54 (15)	3 (1)
Pyrexia	50 (14)	3 (1)
ALT increased	48 (13)	23 (6)
Asthenia	42 (12)	13 (4)
Pneumonia	39 (11)	17 (5)
Weight loss	36 (10)	2 (1)
Noncardiac chest pain	35 (10)	4 (1)
Serious adverse event—No. (%)	184 (51)	152 (42)
Event leading to discontinuation—No. (%)	56 (15)	35 (10)

10. Discussion and Future Directions

Although capmatinib has been approved by both the FDA and EMA, there has not been a Phase III trial comparing capmatinib versus chemotherapy and immunotherapy in the first-line setting for *MET* exon 14 skipping mutations despite the National Comprehensive Cancer Network (NCCN) recommending capmatinib as first-line therapy in advanced NSCLC with *MET* exon 14 skipping mutations [68]. In pretreated populations, the GEOMETRY-III (NCT04427072) trial is a study that involves approximately 90 previously treated advanced NSCLC patients harboring *MET* exon 14 skipping mutation and compares the efficacy of capmatinib with docetaxel [69]. Furthermore, capmatinib has been studied in 20 patients previously treated with a *MET* inhibitor, including 15 with *MET* exon 14 skipping mutation. The DCR was 80%. Notably, circulating tumor DNA analysis was carried out on these patients, and a secondary *MET* mutation was detected in four patients with *MET* D1228H and Y1230H, along with three patients having MAPK signaling alterations [70]. Furthermore, capmatinib and other Type Ib *MET* inhibitors have not been directly compared with Type Ia *MET* inhibitors.

Meanwhile, the challenge remains in finding reliable combinations to both improve the efficacy of capmatinib and broaden the indications of capmatinib use beyond MET exon 14 skipping mutations (Table 6).

Table 6. Current key ongoing studies involving capmatinib.

Clinical Trial Number	Phase	Purpose
NCT04427072	Phase III	Previously treated advanced NSCLC patients with *MET* exon 14 skipping mutation treated with capmatinib versus docetaxel
NCT04926831	Phase II	Efficacy and safety of neoadjuvant and adjuvant capmatinib
NCT05435846	Phase I/Ib	Capmatinib plus trametinib in patients with *MET* exon 14 skipping mutation
NCT04677595	Phase II	Chinese patients who are *EGFR* wt and *ALK* rearrangement negative with *MET* exon 14 skipping mutation
NCT05110196	Phase IV	Indian patients with *MET* exon 14 skipping mutation
NCT05488314	Phase I/II	Combination therapy of capmatinib and amivantamab in unresectable Stage IV NSCLC in patients with *MET* exon 14 skipping mutations or *MET* amplification
NCT05642572	Phase II	Combination therapy of capmatinib with osimertinib +/− ramucirumab in *EGFR* mutant, *MET*-amplified, Stage IV or recurrent NSCLC

Within population subgroups, there are ongoing studies on capmatinib in Asia, which may give insight into its efficacy within specific Asian subgroup populations, including one in China (GEOMETRY-C study, NCT04677595) and one in India (NCT05110196). For early-stage NSCLC, the GEOMETRY-N (NCT04926831) study is a Phase II, two-cohort, two-stage study evaluating the efficacy and safety of neoadjuvant and adjuvant capmatinib therapy in improving the major pathological rate (MPR) and outcomes in patients with MET exon 14 skipping or high-level MET amplification NSCLC [71]. As there has been success with EGFR mutations and the use of osimertinib in an adjuvant setting with the ADUARA trial, it will be interesting to note the results of the major pathological response rate in this study [72].

Currently, there is a Phase I/Ib trial underway that investigates capmatinib and trametinib, a MEK inhibitor (NCT05435846), which may be of benefit to patients with progression on crizotinib. Meanwhile, there has not been much success with capmatinib in combination with immunotherapy due to limited activity and tolerability. A retrospective study at two academic institutions showed an ORR of 17% (95% CI 6–36) in MET exon 14 skip mutations receiving PD-L1 blockade [73]. A Phase II study (NCT04323436) looking at the efficacy and safety of capmatinib plus spartalizumab, a PD-1 monoclonal antibody, did not demonstrate significant antitumor benefit, with a high dose reduction/interruption (80.6%) and discontinuation rate (35.5%) [74]. Another Phase II randomized, open-label study (NCT04139317) evaluated the efficacy and safety of combination therapy with capmatinib and pembrolizumab versus pembrolizumab alone in first-line therapy among advanced NSCLC patients with PD-L1 tumor proportion score (TPS) $\geq$ 50% and no *EGFR* mutation or *ALK* rearrangements. However, the trial closed due to concerns from the drug sponsor of tolerability in patients [75]. Finally, there was another study that investigated the efficacy of capmatinib plus nivolumab or nazartinib (EGF816) plus nivolumab in previously treated NSCLC patients (NCT02323126). This study was also terminated due to low accrual, but in its primary endpoint of PFS at 6 months, capmatinib plus nivolumab showed a 68.9% (95% CI 48.85–85.7) PFS at 6 months in high cMet and 50.9% (95% CI 35.6–66.4) in low cMet (NCT02323126). However, there continue to be clinical trials, particularly with cabozantinib and atezolizumab (NCT03170960 and NCT04471428) targeting the *MET* pathway, as *MET* expression has been found to be implicated through its pathway with MET/HGF and is involved in the regulation of the inflamed tumor microenvironment, leading towards the upregulation of inhibitory molecules such as PD-L1 and the downregulation of immune stimulators such as CD137, CD252, and CD70 [76].

Another important role of capmatinib in the future is in patients with acquired *MET* amplification, as observed in about 15% of patients who received first-line osimertinib and in 12–22% of patients receiving second-line osimertinib [60]. As mentioned earlier, Wu et al. saw efficacy using capmatinib and gefitinib, and the TATTON trial, which incorporated osmertinib and savolitinib, showed ORR of 23–66% between the two arms of treatment [51,77]. The GEOMETRY-E study (NCT04816214) was a Phase III study involving osimertinib with capmatinib but recently closed due to a business decision, but a recent Phase II LUNG-MAP trial with SWOG (NCT05642572) recently opened that investigates capmatinib with osimertinib $+/-$ ramucirumab in *EGFR* mutant, *MET*-amplified Stage IV or recurrent NSCLC. Meanwhile, NCT03040973 is a rollover study currently accruing in patients who were part of a Novartis-sponsored clinical trial to continue receiving capmatinib as a single agent or in combination with other treatments.

As with all TKIs, it will be important to note the recurring resistance mechanisms with capmatinib to aid with future directions. Previous studies have shown that in Type I MET TKIs, secondary mutations at residue Y1230 may cause resistance, as Type I MET TKIs do interact with Y1230, specifically Y1230C [42,78,79]. However, notably, D1228 mutations have also been seen in capmatinib and other Type I TKIs [42,46,79,80].

While switching to Type II MET TKIs has been believed to help overcome resistance to capmatinib, novel drugs that can bypass the *MET* signaling pathway may provide the answer for treatment in the post-capmatinib treatment setting [79]. Amivantamab, a bispecific, monoclonal antibody targeting *EGFR* and *MET* is a promising combination that can be considered in conjunction with capmatinib. In the CHYRSALIS study specifically involving patients with *MET* exon 14 skipping mutation whose disease had progressed or had declined standard-of-care therapy, the ORR was 21% (4/19) in patients with prior *MET* inhibitor therapy and 46% (5/11) in patients with no prior *MET* inhibitor therapy. The median DOR was not reached, and 67% (8/13) had DOR $\geq$ 6 months [81]. Meanwhile, in another cohort of patients with *EGFR* exon 19 deletion or L858R NSCLC who had progressed on an *EGFR* TKI, ORR with amivantamab and lazertinib, an *EGFR* inhibitor, was 36% (95% CI 23–51), and 39% had a DOR $\geq$ 6 months [82]. An ongoing clinical trial (NCT05488314) is currently underway that investigates the combination of amivantamab and capmatinib in advanced NSCLC with *MET* exon 14 skipping mutation or MET amplification and may provide a promising new combination. Another promising class of novel drugs includes antibody–drug conjugates (ADCs) in which the monoclonal antibody binds to a specific protein and can deliver a cytotoxic drug to its intended target [83]. telisotuzumab vedotin (Teliso-V) is an antibody–drug conjugate composed of a c-Met antibody (ABT-700) and a microtubule inhibitor (monomethyl auristatin E); the ongoing Phase II M14-239 LUMINOSITY trial (NCT03539536) showed a 52% ORR in patients with previously treated c-MET overexpressors with nonsquamous pathology and *EGFR* wild-type [84]. ABBV-400 is another ADC, which targets c-Met and topoisomerase-1, with an ongoing Phase I study (NCT05029882) involving c-Met overexpression in advanced solid tumors. In addition, a biparatopic MET x MET ADC REGN 5093-M114 has shown promising preclinical activity in both MET-overexpressed, TKI-naïve, EGFR-mutant NSCLC cells regardless of MET gene copy number as well as cell lines of EGFR-mutant NSCLC with PTEN loss or MET Y1230C mutation after the progression of prior osimertinib and savolitinib treatment [85]. A Phase I study (NCT04982224) is ongoing that involves the study of REGN5093-M114 in MET overexpression in advanced solid tumors.

Finally, it is worth noting the tolerability of capmatinib, as 67% of patients in the GEOMETRY mono-1 trial had a Grade 3 or 4 toxicity, and 42% of patients had serious adverse events [6]. The most frequent etiologies for Grade 3–4 toxicity include peripheral edema (9%), dyspnea (7%), fatigue (4%), and asthenia (4%), which all can severely impact the quality of life in patients [6]. While some of these side effects like peripheral edema can be controlled with supportive care, the toxicity profile of capmatinib merits further comparison with other standard-of-care options in a Phase III study and real-world prospective studies that evaluate side effects of capmatinib in clinical practice [86].

Thus, future directions in capmatinib and other combinations and novel agents in MET-dysregulated NSCLC will focus on the efficacy of these drugs, tolerability, and given the multiple new drugs, the sequence of these agents.

11. Conclusions

The dysregulation of *MET* in NSCLC has proven challenging when it comes to finding therapeutic options given the lack of activity and reliability of biomarkers. Capmatinib, a Type Ib *MET* TKI that is not dependent on G1163, as crizotinib is, has proven to have efficacy, as shown in the GEOMETRY mono-1 study. Subsequent post hoc analyses have shown similar efficacy regardless of the prior treatment used and patient-reported improvement in quality of life. In addition, real-world analysis has shown similar efficacy with a promising intracranial response. The Foundation One CDx assay has been shown to be a reliable companion assay and remains the only FDA-approved assay for *MET*-targeted therapies. However, there have been no completed Phase III studies comparing capmatinib to first-line chemotherapy and immunotherapy or second-line chemotherapy. Furthermore, there was a notable percentage of Grade 3–4 toxicities. Future studies include investigations of capmatinib with *MEK* inhibition, combination therapy with amivantamab, and new classes of drugs, particularly ADCs. Capmatinib's role in a perioperative setting in early-stage NSCLC may provide further treatment options for early stage patients with MET exon 14 skipping NSCLC, but the sequencing of these drugs and tolerability will be key factors, along with finding a more reliable biomarker.

Author Contributions: Conceptualization, M.N.; formal analysis, R.H., D.J.B. and M.N.; investigation, R.H., D.J.B. and M.N.; writing—original draft preparation, R.H., D.J.B. and M.N.; writing—review and editing, R.H., D.J.B. and M.N.; visualization, R.H. and D.J.B.; supervision, M.N. All authors have read and agreed to the published version of the manuscript.

Funding: This research received no external funding.

Conflicts of Interest: RH is a consultant for Targeted Oncology and received honoraria from DAVA Oncology and The Dedham Group. D.J.B. has received consulting fees from Seagen. MN has received consulting fees from AstraZeneca, Caris Life Sciences, Daiichi Sankyo, Novartis, EMD Serono, Pfizer, Lilly and Genentech, has received travel support from AnHeart Therapeutics and is a speaker for Takeda, Janssen, Mirati and Blueprint Medicines.

Abbreviations

NSCLC	Non-small cell lung cancer
EGFR	Epidermal growth factor receptor
KRAS	Kirsten rat sarcoma virus
BRAF	v-raf murine sarcoma viral oncogene homolog B1
ALK	Anaplastic lymphoma kinase
ROS1	Proto-oncogene tyrosine–protein kinase ROS
RET	Rearranged during transfection proto-oncogene
MET	Mesenchymal–epithelial transition
ERBB2	erb-b2 receptor tyrosine kinase 2
NTRK	Neurotrophic tyrosine receptor kinase
HGF	Hepatocyte growth factor
AMPK	AMP-activated protein kinase
LKB1	Liver kinase B1
GCN	Gain of copy number
FDA	U.S. Food and Drug Administration
EMA	European Medicines Agency
po	Oral
DLT	Drug limiting toxicity
b.i.d.	Twice a day
R2PD	Recommended Phase II dose
ORR	Overall response rate
PFS	Progression-free survival
DOR	Duration of response
IO	Immunotherapy

IHC	Immunohistochemistry
CEP7	Chromosome 7 centromere
ctDNA	Circulating tumor DNA
cfDNA	Circulating-free DNA
CTCs	Circulating tumor cells
FFPE	Fresh-frozen paraffin-embedded
PPA	Positive percent agreement
NPA	Negative percent agreement
OA	Overall agreement
NCCN	National Comprehensive Cancer Network
Teliso-V	Telisotuzumab vedotin

References

1. Sung, H.; Ferlay, J.; Siegel, R.L.; Laversanne, M.; Soerjomataram, I.; Jemal, A.; Bray, F. Global Cancer Statistics 2020: GLOBOCAN Estimates of Incidence and Mortality Worldwide for 36 Cancers in 185 Countries. *CA Cancer J. Clin.* **2021**, *71*, 209–249. [CrossRef] [PubMed]
2. Planchard, D.; Smit, E.F.; Groen, H.J.M.; Mazieres, J.; Besse, B.; Helland, Å.; Giannone, V.; D'Amelio, A.M.; Zhang, P.; Mookerjee, B.; et al. Dabrafenib plus trametinib in patients with previously untreated BRAFV600E-mutant metastatic non-small-cell lung cancer: An open-label, phase 2 trial. *Lancet Oncol.* **2017**, *18*, 1307–1316. [CrossRef] [PubMed]
3. Shaw, A.T.; Riely, G.J.; Bang, Y.-J.; Kim, D.-W.; Camidge, D.R.; Solomon, B.J.; Varella-Garcia, M.; Iafrate, A.J.; Shapiro, G.I.; Usari, T.; et al. Crizotinib in ROS1-rearranged advanced non-small-cell lung cancer (NSCLC): Updated results, including overall survival, from PROFILE 1001. *Ann Oncol.* **2019**, *30*, 1121–1126. [CrossRef] [PubMed]
4. Drilon, A.; Siena, S.; Dziadziuszko, R.; Barlesi, F.; Krebs, M.G.; Shaw, A.T.; de Braud, F.; Rolfo, C.; Ahn, M.-J.; Wolf, J.; et al. Entrectinib in ROS1 fusion-positive non-small-cell lung cancer: Integrated analysis of three phase 1–2 trials. *Lancet Oncol.* **2020**, *21*, 261–270. [CrossRef] [PubMed]
5. Drilon, A.; Oxnard, G.R.; Tan, D.S.W.; Loong, H.H.F.; Johnson, M.; Gainor, J.; McCoach, C.E.; Gautschi, O.; Besse, B.; Cho, B.C.; et al. Efficacy of Selpercatinib in RET Fusion-Positive Non-Small-Cell Lung Cancer. *N. Engl. J. Med.* **2020**, *383*, 813–824. [CrossRef]
6. Wolf, J.; Seto, T.; Han, J.-Y.; Reguart, N.; Garon, E.B.; Groen, H.J.M.; Tan, D.S.W.; Hida, T.; de Jonge, M.; Orlov, S.V.; et al. Capmatinib in *MET* Exon 14–Mutated or *MET*-Amplified Non–Small-Cell Lung Cancer. *N. Engl. J. Med.* **2020**, *383*, 944–957. [CrossRef]
7. Ramalingam, S.S.; Yang, J.C.-H.; Lee, C.K.; Kurata, T.; Kim, D.-W.; John, T.; Nogami, N.; Ohe, Y.; Mann, H.; Rukazenov, Y.; et al. Osimertinib As First-Line Treatment of EGFR Mutation-Positive Advanced Non-Small-Cell Lung Cancer. *J. Clin. Oncol.* **2018**, *36*, 841–849. [CrossRef]
8. Skoulidis, F.; Li, B.T.; Dy, G.K.; Price, T.J.; Falchook, G.S.; Wolf, J.; Italiano, A.; Schuler, M.; Borghaei, H.; Barlesi, F.; et al. Sotorasib for Lung Cancers with *KRAS* p.G12C Mutation. *N. Engl. J. Med.* **2021**, *384*, 2371–2381. [CrossRef]
9. Doebele, R.C.; Drilon, A.; Paz-Ares, L.; Siena, S.; Shaw, A.T.; Farago, A.F.; Blakeley, C.M.; Seto, T.; Cho, B.C.; Tosi, D.; et al. Entrectinib in patients with advanced or metastatic NTRK fusion-positive solid tumours: Integrated analysis of three phase 1-2 trials. *Lancet Oncol.* **2020**, *21*, 271–282. [CrossRef]
10. Li, B.T.; Smit, E.F.; Goto, Y.; Nakagawa, K.; Udagawa, H.; Mazières, J.; Nagasaka, M.; Bazhenova, L.; Saltos, A.N.; Felip, E.; et al. Trastuzumab Deruxtecan in *HER2*-Mutant Non–Small-Cell Lung Cancer. *N. Engl. J. Med.* **2022**, *386*, 241–251. [CrossRef]
11. Zhang, Y.; Xia, M.; Jin, K.; Wang, S.; Wei, H.; Fan, C.; Wu, Y.; Li, X.; Li, X.; Li, G.; et al. Function of the c-Met receptor tyrosine kinase in carcinogenesis and associated therapeutic opportunities. *Mol. Cancer* **2018**, *17*, 45. [CrossRef] [PubMed]
12. Cecchi, F.; Rabe, D.C.; Bottaro, D.P. Targeting the HGF/Met signalling pathway in cancer. *Eur. J. Cancer* **2010**, *46*, 1260–1270. [CrossRef]
13. Pilotto, S.; Carbognin, L.; Karachaliou, N.; Ma, P.C.; Rosell, R.; Tortora, G.; Bria, E. Tracking MET de-addiction in lung cancer: A road towards the oncogenic target. *Cancer Treat. Rev.* **2017**, *60*, 1–11. [CrossRef] [PubMed]
14. Lee, M.; Jain, P.; Wang, F.; Ma, P.C.; Borczuk, A.; Halmos, B. MET alterations and their impact on the future of non-small cell lung cancer (NSCLC) targeted therapies. *Expert Opin. Ther. Targets* **2021**, *25*, 249–268. [CrossRef] [PubMed]
15. Salgia, R. MET in Lung Cancer: Biomarker Selection Based on Scientific Rationale. *Mol. Cancer Ther.* **2017**, *16*, 555–565. [CrossRef] [PubMed]
16. Tu, C.-Y.; Cheng, F.-J.; Chen, C.-M.; Wang, S.-L.; Hsiao, Y.-C.; Chen, C.-H.; Tsia, T.-C.; He, Y.-H.; Wang, B.-W.; Hsieh, I.-S.; et al. Cigarette smoke enhances oncogene addiction to c-MET and desensitizes EGFR-expressing non-small cell lung cancer to EGFR TKIs. *Mol. Oncol.* **2018**, *12*, 705–723. [CrossRef]
17. Cheng, F.-J.; Chen, C.-H.; Tsai, W.-C.; Wang, B.-W.; Yu, M.-C.; Hsia, T.-C.; Wei, Y.-L.; Hsiao, Y.-C.; Hu, D.-W.; Ho, C.-Y.; et al. Cigarette smoke-induced LKB1/AMPK pathway deficiency reduces EGFR TKI sensitivity in NSCLC. *Oncogene* **2021**, *40*, 1162–1175. [CrossRef]

18. Hong, L.; Zhang, J.; Heymach, J.V.; Le, X. Current and future treatment options for MET exon 14 skipping alterations in non-small cell lung cancer. *Ther. Adv. Med. Oncol.* **2021**, *13*, 1758835921992976. [CrossRef]

19. Cortot, A.B.; Kherrouche, Z.; Descarpentries, C.; Wislez, M.; Baldacci, S.; Furlan, A.; Tulasne, D. Exon 14 Deleted MET Receptor as a New Biomarker and Target in Cancers. *JNCI J. Natl. Cancer Inst.* **2017**, *109*, djw262. [CrossRef]

20. Socinski, M.A.; Pennell, N.A.; Davies, K.D. *MET* Exon 14 Skipping Mutations in Non–Small-Cell Lung Cancer: An Overview of Biology, Clinical Outcomes, and Testing Considerations. *JCO Precis. Oncol.* **2021**, *5*, 653–663. [CrossRef]

21. Tong, J.H.; Yeung, S.F.; Chan, A.W.H.; Chung, L.Y.; Chau, S.L.; Lung, R.W.M.; Tong, C.Y.; Chow, C.; Tin, E.K.Y.; Yu, Y.H.; et al. MET Amplification and Exon 14 Splice Site Mutation Define Unique Molecular Subgroups of Non-Small Cell Lung Carcinoma with Poor Prognosis. *Clin. Cancer Res.* **2016**, *22*, 3048–3056. [CrossRef] [PubMed]

22. Schubart, C.; Stöhr, R.; Tögel, L.; Fuchs, F.; Sirbu, H.; Seitz, G.; Seggewiss-Bernhardt, R.; Leistner, R.; Sterlacci, W.; Vieth, M.; et al. MET Amplification in Non-Small Cell Lung Cancer (NSCLC)—A Consecutive Evaluation Using Next-Generation Sequencing (NGS) in a Real-World Setting. *Cancers* **2021**, *13*, 5023. [CrossRef] [PubMed]

23. Leonetti, A.; Sharma, S.; Minari, R.; Perego, P.; Giovannetti, E.; Tiseo, M. Resistance mechanisms to osimertinib in EGFR-mutated non-small cell lung cancer. *Br. J. Cancer* **2019**, *121*, 725–737. [CrossRef]

24. Chmielecki, J.; Mok, T.; Wu, Y.-L.; Han, J.-Y.; Ahn, M.; Ramalingam, S.S.; John, T.; Okamoto, I.; Yang, J.C.-H.; Shepherd, F.A.; et al. Analysis of acquired resistance mechanisms to osimertinib in patients with EGFR-mutated advanced non-small cell lung cancer from the AURA3 trial. *Nat Commun.* **2023**, *14*, 1071. [CrossRef] [PubMed]

25. Dagogo-Jack, I.; Yoda, S.; Lennerz, J.K.; Langenbucher, A.; Lin, J.J.; Rooney, M.M.; Prutisto-Chang, K.; Oh, A.; Adams, N.A.; Yeap, B.Y.; et al. MET Alterations Are a Recurring and Actionable Resistance Mechanism in ALK-Positive Lung Cancer. *Clin. Cancer Res.* **2020**, *26*, 2535–2545. [CrossRef]

26. Awad, M.M.; Leonardi, G.C.; Kravets, S.; Dahlberg, S.E.; Drilon, A.; Noonan, S.A.; Camidge, D.R.; Ou, S. H-I; Costa, D.B.; Gadgeel, S.M.; et al. Impact of MET inhibitors on survival among patients with non-small cell lung cancer harboring MET exon 14 mutations: A retrospective analysis. *Lung Cancer* **2019**, *133*, 96–102. [CrossRef]

27. Awad, M.M.; Oxnard, G.R.; Jackman, D.M.; Savukoski, D.O.; Hall, D.; Shivdasani, P.; Heng, J.C.; Dahlberg, S.E.; Janne, P.A.; Verma, S.; et al. MET Exon 14 Mutations in Non-Small-Cell Lung Cancer Are Associated With Advanced Age and Stage-Dependent MET Genomic Amplification and c-Met Overexpression. *J. Clin. Oncol.* **2016**, *34*, 721–730. [CrossRef]

28. Bahcall, M.; Paweletz, C.P.; Kuang, Y.; Taus, L.J.; Sim, T.; Kim, N.D.; Dholakia, K.H.; Lau, C.J.; Gokhale, P.C.; Chopade, P.R.; et al. Combination of Type I and Type II MET Tyrosine Kinase Inhibitors as Therapeutic Approach to Prevent Resistance. *Mol. Cancer Ther.* **2022**, *21*, 322–335. [CrossRef]

29. Cui, J.J. Targeting receptor tyrosine kinase MET in cancer: Small molecule inhibitors and clinical progress. *J. Med. Chem.* **2014**, *57*, 4427–4453. [CrossRef]

30. Lu, S.; Fang, J.; Li, X.; Cao, L.; Zhou, J.; Guo, Q.; Liang, Z.; Cheng, Y.; Jiang, L.; Yang, N.; et al. Once-daily savolitinib in Chinese patients with pulmonary sarcomatoid carcinomas and other non-small-cell lung cancers harbouring MET exon 14 skipping alterations: A multicentre, single-arm, open-label, phase 2 study. *Lancet Respir. Med.* **2021**, *9*, 1154–1164. [CrossRef]

31. Paik, P.K.; Felip, E.; Veillon, R.; Sakai, H.; Cortot, A.B.; Garassino, M.C.; Mazieres, J.; Viteri, S.; Senellart, H.; Van Meerbeeck, J.; et al. Tepotinib in Non–Small-Cell Lung Cancer with *MET* Exon 14 Skipping Mutations. *N. Engl. J. Med.* **2020**, *383*, 931–943. [CrossRef] [PubMed]

32. Santarpia, M.; Massafra, M.; Gebbia, V.; D'Aquino, A.; Garipoli, C.; Altavilla, G.; Rosell, R. A narrative review of MET inhibitors in non-small cell lung cancer with MET exon 14 skipping mutations. *Transl. Lung Cancer Res.* **2021**, *10*, 1536–1556. [CrossRef] [PubMed]

33. Yan, S.B.; Um, S.L.; Peek, V.L.; Stephens, J.R.; Zeng, W.; Konicek, B.W.; Ling, L.; Manro, J.R.; Wacheck, W.; Walgren, R.A. MET-targeting antibody (emibetuzumab) and kinase inhibitor (merestinib) as single agent or in combination in a cancer model bearing MET exon 14 skipping. *Investig. New Drugs* **2018**, *36*, 536–544. [CrossRef] [PubMed]

34. Calles, A.; Kwiatkowski, N.; Cammarata, B.K.; Ercan, D.; Gray, N.S.; Jänne, P.A. Tivantinib (ARQ 197) efficacy is independent of MET inhibition in non-small-cell lung cancer cell lines. *Mol. Oncol.* **2015**, *9*, 260–269. [CrossRef]

35. Klempner, S.J.; Borghei, A.; Hakimian, B.; Ali, S.M.; Ou, S.-H.I. Intracranial Activity of Cabozantinib in MET Exon 14-Positive NSCLC with Brain Metastases. *J. Thorac. Oncol.* **2017**, *12*, 152–156. [CrossRef]

36. Scagliotti, G.; von Pawel, J.; Novello, S.; Ramlau, R.; Favaretto, A.; Barlesi, F.; Akerley, W.; Orlov, S.; Santoro, A.; Spigel, D.; et al. Phase III Multinational, Randomized, Double-Blind, Placebo-Controlled Study of Tivantinib (ARQ 197) Plus Erlotinib Versus Erlotinib Alone in Previously Treated Patients With Locally Advanced or Metastatic Nonsquamous Non–Small-Cell Lung Cancer. *J. Clin. Oncol.* **2015**, *33*, 2667–2674. [CrossRef]

37. Tabrecta. Available online: https://www.ema.europa.eu/en/medicines/human/EPAR/tabrecta (accessed on 1 July 2023).

38. Novartis. TABRECTA (Capmatinib): US Prescribing Information. 2020. Available online: https://www.accessdata.fda.gov/drugsatfda_docs/label/2020/213591s000lbl.pdf (accessed on 8 March 2023).

39. Moro-Sibilot, D.; Cozic, N.; Pérol, M.; Mazières, J.; Otto, J.; Souquet, P.J.; Bahleda, R.; Wislez, M.; Zalcman, G.; Guibert, S.D. Crizotinib in c-MET- or ROS1-positive NSCLC: Results of the AcSé phase II trial. *Ann. Oncol.* **2019**, *30*, 1985–1991. [CrossRef]

40. Landi, L.; Chiari, R.; Tiseo, M.; D'Incà, F.; Dazzi, C.; Chella, A.; Delmonte, A.; Bonanno, L.; Giannarell, D.; Cortinovis, D.L.; et al. Crizotinib in *MET* -Deregulated or *ROS1* -Rearranged Pretreated Non–Small Cell Lung Cancer (METROS): A Phase II, Prospective, Multicenter, Two-Arms Trial. *Clin. Cancer Res.* **2019**, *25*, 7312–7319. [CrossRef]

41. Drilon, A.; Clark, J.W.; Weiss, J.; Ou, S.-H.I.; Camidge, D.R.; Solomon, B.J.; Otterson, G.A.; Villaruz, L.C.; Riely, G.J.; Heist, R.S.; et al. Antitumor activity of crizotinib in lung cancers harboring a MET exon 14 alteration. *Nat. Med.* **2020**, *26*, 47–51. [CrossRef]

42. Recondo, G.; Bahcall, M.; Spurr, L.F.; Che, J.; Ricciuti, B.; Leonardi, G.C.; Lo, Y.-C.; Li, Y.Y.; Lamberti, G.; Nguyen, T.; et al. Molecular Mechanisms of Acquired Resistance to MET Tyrosine Kinase Inhibitors in Patients with MET Exon 14-Mutant NSCLC. *Clin. Cancer Res.* **2020**, *26*, 2615–2625. [CrossRef]

43. Xalkori (Crizotinib). Available online: https://www.accessdata.fda.gov/drugsatfda_docs/label/2017/202570s021lbl.pdf (accessed on 1 July 2023).

44. Xalkori, INN-Crizotinib. Available online: https://www.ema.europa.eu/en/documents/product-information/xalkori-epar-product-information_en.pdf (accessed on 1 July 2023).

45. Liu, X.; Wang, Q.; Yang, G.; Marando, C.; Koblish, H.K.; Hall, L.M.; Fridman, J.S.; Behshad, E.; Wynn, R.; Li, Y.; et al. A Novel Kinase Inhibitor, INCB28060, Blocks c-MET–Dependent Signaling, Neoplastic Activities, and Cross-Talk with EGFR and HER-3. *Clin. Cancer Res.* **2011**, *17*, 7127–7138. [CrossRef] [PubMed]

46. Baltschukat, S.; Engstler, B.S.; Huang, A.; Hao, H.-X.; Tam, A.; Wang, H.Q.; Liang, J.; DiMare, M.T.; Bhang, H.-E.C.; Wang, Y.; et al. Capmatinib (INC280) Is Active Against Models of Non–Small Cell Lung Cancer and Other Cancer Types with Defined Mechanisms of MET Activation. *Clin. Cancer Res.* **2019**, *25*, 3164–3175. [CrossRef] [PubMed]

47. Dhillon, S. Capmatinib: First Approval. *Drugs* **2020**, *80*, 1125–1131. [CrossRef] [PubMed]

48. Esaki, T.; Hirai, F.; Makiyama, A.; Seto, T.; Bando, H.; Naito, Y.; Yoh, K.; Ishihara, K.; Kakizume, T.; Natsume, K.; et al. Phase I dose-escalation study of capmatinib (INC280) in Japanese patients with advanced solid tumors. *Cancer Sci.* **2019**, *110*, 1340–1351. [CrossRef]

49. Bang, Y.-J.; Su, W.-C.; Schuler, M.; Nam, D.-H.; Lim, W.T.; Bauer, T.M.; Azaro, A.; Poon, R.T.P.; Hong, D.; Lin, C.-C.; et al. Phase 1 study of capmatinib in MET-positive solid tumor patients: Dose escalation and expansion of selected cohorts. *Cancer Sci.* **2020**, *111*, 536–547. [CrossRef]

50. Schuler, M.; Berardi, R.; Lim, W.-T.; de Jonge, M.; Bauer, T.M.; Azaro, A.; Gottfried, M.; Han, J.-Y.; Lee, D.-H.; Wollner, M.; et al. Molecular correlates of response to capmatinib in advanced non-small-cell lung cancer: Clinical and biomarker results from a phase I trial. *Ann. Oncol.* **2020**, *31*, 789–797. [CrossRef]

51. Wu, Y.-L.; Zhang, L.; Kim, D.-W.; Liu, X.; Lee, D.H.; Yang, J.C.-H.; Ahn, M.-J.; Vansteenkiste, J.F.; Su, W.-C.; Felip, E.; et al. Phase Ib/II Study of Capmatinib (INC280) Plus Gefitinib After Failure of Epidermal Growth Factor Receptor (EGFR) Inhibitor Therapy in Patients With EGFR-Mutated, MET Factor-Dysregulated Non-Small-Cell Lung Cancer. *J. Clin. Oncol.* **2018**, *36*, 3101–3109. [CrossRef]

52. Vansteenkiste, J.F.; Smit, E.F.; Groen, H.J.M.; Garon, E.B.; Heist, R.S.; Hida, T.; Nishio, M.; Kokowski, K.; Grohe, C.; Reguart, N.; et al. 1285P Capmatinib in patients with METex14-mutated advanced non-small cell lung cancer who received prior immunotherapy: The phase II GEOMETRY mono-1 study. *Ann. Oncol.* **2020**, *31*, S830. [CrossRef]

53. Wolf, J.; Garon, E.B.; Groen, H.J.M.; Tan, D.S.W.; Gilloteau, I.; Le Mouhaer, S.; Hampe, M.; Cai, C.; Chassot-Agostinho, A.; Reynolds, M.; et al. Patient-reported outcomes in capmatinib-treated patients with METex14-mutated advanced NSCLC: Results from the GEOMETRY mono-1 study. *Eur. J. Cancer* **2023**, *183*, 98–108. [CrossRef]

54. Seto, T.; Ohashi, K.; Sugawara, S.; Nishio, M.; Takeda, M.; Aoe, K.; Moizumi, S.; Nomura, S.; Tajima, T.; Hida, T. Capmatinib in Japanese patients with MET exon 14 skipping-mutated or MET-amplified advanced NSCLC: GEOMETRY mono-1 study. *Cancer Sci.* **2021**, *112*, 1556–1566. [CrossRef]

55. Paik, P.K.; Goyal, R.K.; Cai, B.; Price, M.A.; Davis, K.L.; Ansquer, V.D.; Caro, N.; Saliba, T.R. Real-world outcomes in non-small-cell lung cancer patients with MET Exon 14 skipping mutation and brain metastases treated with capmatinib. *Future Oncol.* **2023**, *19*, 217–228. [CrossRef]

56. Illini, O.; Fabikan, H.; Swalduz, A.; Vikström, A.; Krenbek, D.; Schumacher, M.; Dudnik, E.; Studnicka, M.; Ohman, R.; Wurm, R.; et al. Real-world experience with capmatinib in MET exon 14-mutated non-small cell lung cancer (RECAP): A retrospective analysis from an early access program. *Ther. Adv. Med. Oncol.* **2022**, *14*, 17588359221103206. [CrossRef]

57. TEPMETKO, INN. Available online: https://www.ema.europa.eu/en/documents/product-information/tepmetko-epar-product-information_en.pdf (accessed on 1 July 2023).

58. Markham, A. Savolitinib: First Approval. *Drugs* **2021**, *81*, 1665–1670. [CrossRef] [PubMed]

59. Jørgensen, J.T.; Mollerup, J. Companion Diagnostics and Predictive Biomarkers for MET-Targeted Therapy in NSCLC. *Cancers* **2022**, *14*, 2150. [CrossRef] [PubMed]

60. Coleman, N.; Hong, L.; Zhang, J.; Heymach, J.; Hong, D.; Le, X. Beyond epidermal growth factor receptor: MET amplification as a general resistance driver to targeted therapy in oncogene-driven non-small-cell lung cancer. *ESMO Open* **2021**, *6*, 100319. [CrossRef] [PubMed]

61. Noonan, S.A.; Berry, L.; Lu, X.; Gao, D.; Barón, A.E.; Chesnut, P.; Sheren, J.; Aisner, D.L.; Merrick, D.; Doebele, R.C.; et al. Identifying the Appropriate FISH Criteria for Defining MET Copy Number-Driven Lung Adenocarcinoma through Oncogene Overlap Analysis. *J. Thorac. Oncol.* **2016**, *11*, 1293–1304. [CrossRef]

62. Guo, R.; Offin, M.; Brannon, A.R.; Chang, J.; Chow, A.; Delasos, L.; Girshman, J.; Wilkins, O.; McCarthy, C.G.; Makhnin, A. MET Exon 14-altered Lung Cancers and MET Inhibitor Resistance. *Clin. Cancer Res.* **2021**, *27*, 799–806. [CrossRef]

63. Ikeda, S.; Schwaederle, M.; Mohindra, M.; Fontes Jardim, D.L.; Kurzrock, R. MET alterations detected in blood-derived circulating tumor DNA correlate with bone metastases and poor prognosis. *J. Hematol. Oncol.* **2018**, *11*, 76. [CrossRef]

64. Park, S.; Olsen, S.; Ku, B.M.; Lee, M.-S.; Jung, H.-A.; Sun, J.-M.; Lee, S.-H.; Ahn, J.S.; Park, K.; Choi, Y.-L.; et al. High concordance of actionable genomic alterations identified between circulating tumor DNA–based and tissue-based next-generation sequencing testing in advanced non–small cell lung cancer: The Korean Lung Liquid Versus Invasive Biopsy Program. *Cancer* **2021**, *127*, 3019–3028. [CrossRef]

65. Mondelo-Macía, P.; Rodríguez-López, C.; Valiña, L.; Aguín, S.; León-Mateos, L.; García-González, J.; Abalo, A.; Rapado-Gonzalez, O.; Suarez-Cunqueiro, M.; Díaz-Lagares, A.; et al. Detection of MET Alterations Using Cell Free DNA and Circulating Tumor Cells from Cancer Patients. *Cells* **2020**, *9*, 522. [CrossRef]

66. Peng, H.; Lu, L.; Zhou, Z.; Liu, J.; Zhang, D.; Nan, K.; Zhao, X.; Li, F.; Tian, L.; Dong, H.; et al. CNV Detection from Circulating Tumor DNA in Late Stage Non-Small Cell Lung Cancer Patients. *Genes* **2019**, *10*, 926. [CrossRef] [PubMed]

67. FDA Summary of Safety and Effectiveness Data. FoundationOne CDx. 2020. Available online: https://www.accessdata.fda.gov/cdrh_docs/pdf17/P170019S011B.pdf (accessed on 8 March 2023).

68. NCCN Clinical Practice Guidelines in Oncology (NCCN Guidelines®) for Non-Small Cell Lung Cancer V.3.2023. Available online: https://www.nccn.org/professionals/physician_gls/pdf/nscl.pdf (accessed on 15 June 2023).

69. Souquet, P.; Kim, S.; Solomon, B.; Vansteenkiste, J.; Carbini, M.; Kenny, S.; Glaser, S.; Chassot Agostinho, A.; Wolf, J. P47.17 Capmatinib vs Docetaxel in Pretreated Patients With MET Exon 14 Skipping–mutated Stage IIIB/IIIC or IV NSCLC (GeoMETry-III). *J. Thorac. Oncol.* **2021**, *16*, S1104. [CrossRef]

70. Dagogo-Jack, I.; Moonsamy, P.; Gainor, J.F.; Lennerz, J.K.; Piotrowska, Z.; Lin, J.J.; Lenne, I.T.; Sequist, L.V.; Shaw, A.T.; Goodwin, K.; et al. A Phase 2 Study of Capmatinib in Patients With MET-Altered Lung Cancer Previously Treated With a MET Inhibitor. *J. Thorac. Oncol.* **2021**, *16*, 850–859. [CrossRef]

71. Lee, J.M.; Awad, M.M.; Saliba, T.R.; Caro, N.; Banerjee, H.; Kelly, K. Neoadjuvant and adjuvant capmatinib in resectable non–small cell lung cancer with *MET* exon 14 skipping mutation or high *MET* amplification: GEOMETRY-N trial. *J. Clin. Oncol.* **2022**, *40*, TPS8590. [CrossRef]

72. Wu, Y.-L.; Tsuboi, M.; He, J.; John, T.; Grohe, C.; Majem, M.; Goldman, J.W.; Laktionov, K.; Kim, S.-W.; Kato, T.; et al. Osimertinib in Resected *EGFR* -Mutated Non–Small-Cell Lung Cancer. *N. Engl. J. Med.* **2020**, *383*, 1711–1723. [CrossRef]

73. Sabari, J.K.; Leonardi, G.C.; Shu, C.A.; Umeton, R.; Montecalvo, J.; Ni, A.; Chen, R.; Dienstag, J.; Mrad, C.; Bergagnini, I.; et al. PD-L1 expression, tumor mutational burden, and response to immunotherapy in patients with MET exon 14 altered lung cancers. *Ann. Oncol.* **2018**, *29*, 2085–2091. [CrossRef]

74. Wolf, J.; Heist, R.; Kim, T.M.; Nishio, M.; Dooms, C.; Kanthala, R.R.; Leo, E.; Giorgetti, E.; Mardjuadi, F.I.; Corto, A. 994P Efficacy and safety of capmatinib plus spartalizumab in treatment-naïve patients with advanced NSCLC harboring MET exon 14 skipping mutation. *Ann. Oncol.* **2022**, *33*, S1007–S1008. [CrossRef]

75. Mok, T.S.K.; Cortinovis, D.L.; Majem, M.; Johnson, M.L.; Mardjuadi, F.I.; Zhao, X.; Siripurapu, S.V.; Jiang, Z.; Wolf, J. Efficacy and safety of capmatinib plus pembrolizumab in treatment (tx)-naïve patients with advanced non–small cell lung cancer (NSCLC) with high tumor PD-L1 expression: Results of a randomized, open-label, multicenter, phase 2 study. *J. Clin. Oncol.* **2022**, *40*, 9118. [CrossRef]

76. Dempke, W.C.M.; Fenchel, K. Has programmed cell death ligand-1 MET an accomplice in non-small cell lung cancer?—A narrative review. *Transl. Lung Cancer Res.* **2021**, *10*, 2667–2682. [CrossRef]

77. Sequist, L.V.; Han, J.-Y.; Ahn, M.-J.; Cho, B.C.; Yu, H.; Kim, S.-W.; Yang, J.C.-H.; Lee, J.S.; Su, W.-C.; Kowalski, D.; et al. Osimertinib plus savolitinib in patients with EGFR mutation-positive, MET-amplified, non-small-cell lung cancer after progression on EGFR tyrosine kinase inhibitors: Interim results from a multicentre, open-label, phase 1b study. *Lancet Oncol.* **2020**, *21*, 373–386. [CrossRef]

78. Ou, S.-H.I.; Young, L.; Schrock, A.B.; Johnson, A.; Klempner, S.J.; Zhu, V.W.; Miller, V.A.; Ali, S.M. Emergence of Preexisting MET Y1230C Mutation as a Resistance Mechanism to Crizotinib in NSCLC with MET Exon 14 Skipping. *J. Thorac. Oncol.* **2017**, *12*, 137–140. [CrossRef] [PubMed]

79. Fujino, T.; Suda, K.; Koga, T.; Hamada, A.; Ohara, S.; Chiba, M.; Shimoji, M.; Takemoto, T.; Soh, J.; Mitsudomi, T. Foretinib can overcome common on-target resistance mutations after capmatinib/tepotinib treatment in NSCLCs with MET exon 14 skipping mutation. *J. Hematol. Oncol.* **2022**, *15*, 79. [CrossRef]

80. Bahcall, M.; Sim, T.; Paweletz, C.P.; Patel, J.D.; Alden, R.S.; Kuang, Y.; Sacher, A.G.; Kim, N.D.; Lydon, C.A.; Awad, M.M.; et al. Acquired METD1228V Mutation and Resistance to MET Inhibition in Lung Cancer. *Cancer Discov.* **2016**, *6*, 1334–1341. [CrossRef] [PubMed]

81. Krebs, M.; Spira, A.I.; Cho, B.C.; Besse, B.; Goldman, J.W.; Janne, P.A.; Ma, Z.; Mansfield, A.S.; Minchom, A.R.; Ou, S.-H.I.; et al. Amivantamab in patients with NSCLC with MET exon 14 skipping mutation: Updated results from the CHRYSALIS study. *J. Clin. Oncol.* **2022**, *40*, 9008. [CrossRef]

82. Shu, C.A.; Goto, K.; Ohe, Y.; Besse, B.; Lee, S.-H.; Wang, Y.; Griesinger, F.; Yang, J.C.-H.; Felip, E.; Sanborn, R.E.; et al. Amivantamab and lazertinib in patients with EGFR-mutant non–small cell lung (NSCLC) after progression on osimertinib and platinum-based chemotherapy: Updated results from CHRYSALIS-2. *J. Clin. Oncol.* **2022**, *40*, 9006. [CrossRef]

83. Fu, Z.; Li, S.; Han, S.; Shi, C.; Zhang, Y. Antibody drug conjugate: The "biological missile" for targeted cancer therapy. *Signal Transduct. Target. Ther.* **2022**, *7*, 93. [CrossRef]

84. Camidge, D.R.; Bar, J.; Horinouchi, H.; Goldman, J.W.; Moiseenko, F.V.; Filippova, E.; Cicin, I.; Bradbury, P.A.; Daaboul, N.; Tomasini, P.; et al. Telisotuzumab vedotin (Teliso-V) monotherapy in patients (pts) with previously treated c-Met–overexpressing (OE) advanced non-small cell lung cancer (NSCLC). *J. Clin. Oncol.* **2022**, *40*, 9016. [CrossRef]

85. Oh, S.Y.; Lee, Y.W.; Lee, E.J.; Kim, J.H.; Park, Y.; Heo, S.G.; Yu, M.R.; Hong, M.H.; DaSilva, J.; Daly, C.; et al. Preclinical Study of a Biparatopic METxMET Antibody–Drug Conjugate, REGN5093-M114, Overcomes MET-driven Acquired Resistance to EGFR TKIs in EGFR-mutant NSCLC. *Clin. Cancer Res.* **2023**, *29*, 221–232. [CrossRef]

86. Goodwin, K.; Ledezma, B.; Heist, R.; Garon, E. MO01.04 Management of Selected Adverse Events With Capmatinib: Institutional Experiences From the GEOMETRY Mono-1 Trial. *J. Thorac. Oncol.* **2021**, *16*, S16–S17. [CrossRef]

Landscape of Savolitinib Development for the Treatment of Non-Small Cell Lung Cancer with MET Alteration—A Narrative Review

Xiaokuan Zhu [1,†], Yao Lu [2,†] and Shun Lu [1,*]

1   Department of Shanghai Lung Cancer Center, Shanghai Chest Hospital, Shanghai Jiao Tong University, Shanghai 200030, China
2   AstraZeneca China, Shanghai 201200, China
*   Correspondence: shunlu@sjtu.edu.cn
†   These authors contributed equally to this work.

Simple Summary: In this article, we outline updates on the clinical development of savolitinib, a novel, reversible c-MET kinase inhibitor conditionally approved in China for treatment of advanced non-small cell lung cancer (NSCLC) patients harboring *MET* exon 14 skipping mutation (*MET*ex14). Savolitinib was developed as a monotherapy for NSCLC with *MET* alterations, and in combination with epidermal growth factor receptor (*EGFR*) inhibitors for patients who developed resistance to EGFR–TKIs because of *MET* alterations. Savolitinib showed anti-tumor activity in preclinical models. The early phase I trial established the recommended phase II dose to be 600 mg once-daily. Savolitinib plus osimertinib showed beneficial efficacy and safety in *EGFR* mutant patients with acquired resistance due to *MET* amplification and/or c-MET overexpression. Benefits were noted with savolitinib in Chinese patients with pulmonary sarcomatoid carcinoma and other NSCLC subtypes positive for *MET*ex14 mutation. Results from phase III trials are awaited to further confirm the beneficial effects from early phase trials.

Citation: Zhu, X.; Lu, Y.; Lu, S. Landscape of Savolitinib Development for the Treatment of Non-Small Cell Lung Cancer with MET Alteration—A Narrative Review. *Cancers* **2022**, *14*, 6122. https://doi.org/10.3390/cancers14246122

Academic Editors: Jan Trøst Jørgensen and Jens Mollerup

Received: 8 November 2022
Accepted: 7 December 2022
Published: 12 December 2022

Publisher's Note: MDPI stays neutral with regard to jurisdictional claims in published maps and institutional affiliations.

Abstract: Non-small cell lung cancer (NSCLC) is increasingly being treated with targeted therapies. Savolitinib (Orpathys®) is highly selective mesenchymal epithelial transition (MET)–tyrosine kinase inhibitor (TKI), which is conditionally approved in China for advanced NSCLC with *MET* exon 14 skipping mutations (*MET*ex14). This article summarizes the clinical development of savolitinib, as a monotherapy in NSCLC with *MET*ex14 mutation and in combination with epidermal growth factor receptor (*EGFR*) inhibitor in post EGFR–TKI resistance NSCLC due to *MET*-based acquired resistance. Preclinical models demonstrated anti-tumor activities in *MET*-driven cancer cell line and xenograft tumor models. The Phase Ia/Ib study established an optimized, recommended phase II dose in Chinese NSCLC patients, while TATTON study of savolitinib plus osimertinib in patients with *EGFR* mutant, *MET*-amplified and TKI-progressed NSCLC showed beneficial efficacy with acceptable safety profile. In a pivotal phase II study, Chinese patients with pulmonary sarcomatoid carcinoma, brain metastasis and other NSCLC subtype positive for *MET*ex14 mutation showed notable responses and acceptable safety profile with savolitinib. Currently, results from ongoing clinical trials are eagerly anticipated to confirm the efficacious and safety benefits of savolitinib as monotherapy and in combination with EGFR–TKI in acquired resistance setting in advanced NSCLC and its subtypes with *MET* alterations.

Keywords: savolitinib; non-small cell lung cancer; *MET* aberrations; *EGFR*; tyrosine kinase inhibitor

1. Introduction

Non-small cell lung cancer (NSCLC) accounts for approximately 80% of all lung cancers with a low 5-year survival rate of about 22% [1,2]. Most NSCLC are usually diagnosed at an advanced stage with traditional chemotherapy and radiotherapy showing limited efficacy. However, recent advances in immune therapy and targeted therapy have radically improved the treatment paradigm of NSCLC over the past decade [2]. Molecular profiling of lung cancer samples for activated oncogenes, including epidermal growth factor receptor (*EGFR*), anaplastic lymphoma kinase (*ALK*) and c–ros oncogene 1 (*ROS1*), is considered as standard-of care to select the most appropriate up-front treatment [3]. However, the identification of new therapeutic targets remains a high priority. Recently, mesenchymal-epithelial transition (*MET*) exon 14 skipping mutations (*MET*ex14) and high-level *MET* amplification have emerged as one of the novel, actionable oncogenic alterations in NSCLC, sensitive to MET inhibitors [4,5].

MET is a receptor tyrosine kinase activated by binding ligand hepatocyte growth factor (HGF) which plays a key physiological role in the interaction between mesenchyme and epithelia during embryonic wound closure and embryogenesis [6–9]. At cellular levels, MET-TK activity transduces mitogenesis by activating Ras–Raf–MAPK signaling pathway and motogenic signals by activating phosphoinositide 3–kinase (PI–3K) pathway upon HGF binding [10]. Aberrant MET/HGF signaling promotes mitogenesis, invasion and angiogenesis, thus contributing towards tumorigenesis and progression of cancer [11]. Importantly, significant implications for tumorigenesis are observed due to crosstalk between downstream signal pathways of *MET* and *EGFR* [12]. The oncogenic role of *MET* was first discovered in 1984 as a part of an oncogenic fusion with the translocated promoter region gene in a mutagenized osteosarcoma cell line [13]. *MET* alterations, including amplification, mutations, gene fusion, MET/HGF protein over expression and the crosstalk between dysregulated *MET* and other signaling pathways, are associated with poor prognosis in cancers, and thus, molecularly targeted [4]. *MET*ex14 mutations are the most commonly reported oncogenic mutations. Exon 14 encodes the 47-amino acid juxtamembrane domain of the *MET* receptor, a key regulatory region that prevents *MET* over signaling. *MET*ex14 mutations include a heterogeneous group of mutations with base substitutions or indels that disrupt the branch point of intron 13, the 3′ splice site of intron 13 or the 5′ splice site of intron 14, producing a *MET* variant that lacks the exon 14 leading to disruption of cellular signaling [14]. The identification of *MET* oncogene and the journey leading to development of MET–TKIs is represented in Figure 1.

Clinical studies conducted earlier suggest that activation of *MET* can act as primary oncogenic driver, or secondary driver of acquired resistance to targeted therapy in subsets of lung cancer [9–11]. *MET*ex14 mutations occur in approximately 0.9 to 4% of NSCLC cases across all histologic subtypes [6] and are enriched in pulmonary sarcomatoid carcinoma (PSC) (20 to 31%), a rare subtype of poorly differentiated NSCLC [15,16]. Furthermore, 1 to 5% of NSCLC harbors de novo *MET* gene amplification, while 15% of cases in *MET*ex14-mutated NSCLC report *MET* amplification [17,18]. *MET* fusion is known to occur in 0.5% [18] and MET protein overexpression in 13.7 to 63.7% of NSCLC patients [17]. Significant cross talk between aberrant *MET* pathway and other signaling pathways, especially *EGFR* results in acquired resistance to EGFR tyrosine kinase inhibitors (TKIs) in patients with NSCLC [19]. Mechanistically, *MET* amplification causes EGFR–TKI resistance by activating *EGFR*-independent phosphorylation of *ErbB*3 and downstream activation of the PI3K/AKT pathway, providing a bypass pathway in the presence of an *EGFR* inhibitor [20]. Thus, concomitant inhibition of both *EGFR* and *MET* would be required to overcome resistance to *EGFR* inhibitors by *MET* amplification [19]. Approximately, 5–22% of NSCLC patients with first- or second-generation EGFR–TKI resistance [18,21] and 5–50% patients with third generation EGFR–TKI resistance harbor *MET* amplification [22], while *MET* amplification as a co-driver occurs in 2–11% *EGFR*-positive treatment-naïve NSCLC patients [23,24]. The incidence of high MET expression after EGFR–TKI resistance is as high as 30.4 to 37% [25]. The proportion of different *MET* alterations in NSCLC patients is summarized in Table 1.

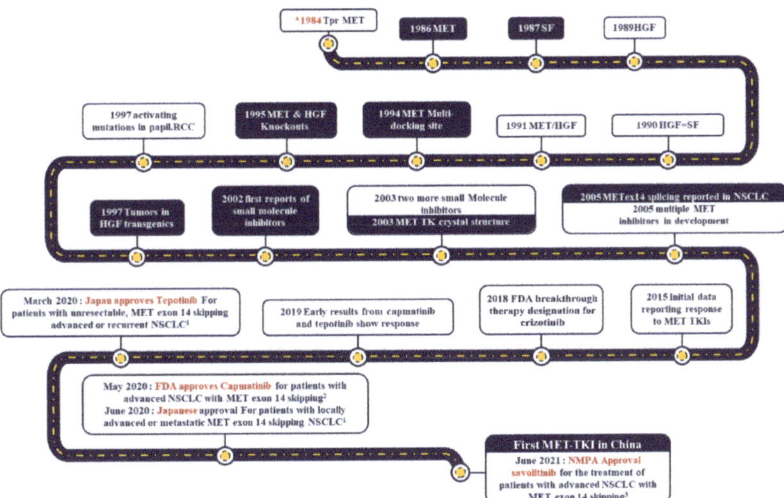

Figure 1. Exploration of *MET* as oncogene and the journey leading to the development of MET–TKI. [1] Based on overall NSCLC population. [2] Based on treatment-naïve NSCLC population. [3] Based on *EGFR*-positive treatment-naïve NSCLC population.

Table 1. Proportion of different *MET* alterations in NSCLC patients.

MET Alterations		Proportion, %	Publication [Reference]
MET ex14	NSCLC [1]	0.9–4	Davies KD et al. [6]
	PSC subtype	20–31.8	Mo HN et al. [15] Tong JH et al. [16]
MET Fusion [1]		0.5	Recondo G et al. [18]
MET Overexpression [1]		13.7–63.7	Guo R et al. [17]
MET Amplification [2]		1–5	Guo R et al. [17]
Secondary *MET* Amplification	1/2G EGFR–TKI resistance	5–22	Recondo G et al. [18] Bean J et al. [21]
	3G EGFR–TKI resistance	5–50	Wang Y et al. [22]
MET Amplification Co-occurrence with *EGFR* Mutation [3]		2–11	Li XM et al. [23] Lai GGY et al. [24]

[1] Based on overall NSCLC population. [2] Based on treatment-naïve NSCLC population. [3] Based on *EGFR*-positive treatment-naïve NSCLC population. EGFR, Epidermal Growth Factor Receptor; TKI, Tyrosine Kinase Inhibitor; MET, Mesenchymal Epithelial Transition; NSCLC, Non-Small Cell Lung Cancer; PSC, Pulmonary Sarcomatoid Carcinoma.

Currently the FDA approved MET–TKIs are capmatinib and tepotinib with crizotinib granted as breakthrough therapy designation, while savolitinib is conditionally approved in China [26–28]. Further, for *EGFR*-mutated NSCLC with *MET* amplification treatment, efficacy of combination of MET–TKIs with EGFR–TKIs has been preliminarily approved by several clinical trials [29,30]. The clinical development strategy for savolitinib is centered both as monotherapy for advanced *MET*ex14-altered NSCLC and in combination with EGFR–TKI for correction of *MET*-driven acquired resistance to EGFR–TKIs [31]. In this review, we briefly describe the major milestones achieved in the clinical development of savolitinib as standard of care for NSCLC with *MET*ex14 mutation and potential treatment for NSCLC with other *MET* alterations.

2. Savolitinib, in Brief

Savolitinib (Orpathys®) is an orally bioavailable and highly selective small molecule MET–TKI that has demonstrated profound efficacy in preclinical and clinical studies of various cancers, including NSCLC, papillary renal cell carcinoma (PRCC) and gastric carcinoma [32–34]. Figure 2 demonstrates the chemical structure of savolitinib.

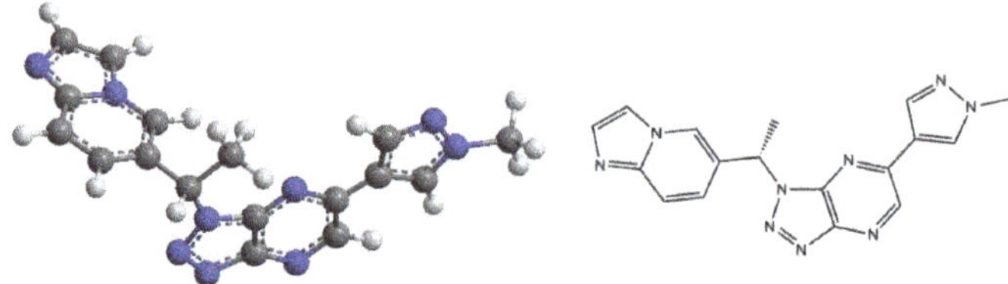

Figure 2. Chemical structure of Savolitinib.

Early on, in vitro studies have established inhibitory effect of savolitinib on growth of gastric cells lines, while in vivo studies observed anti-tumor activity in human xenograft tumor models of *MET*-amplified gastric cancer and PRCC [33–35]. Another study by Jones and colleagues related to pharmacokinetic-pharmacodynamic (PK-PD) model observed inhibition of phosphorylated-MET by savolitinib at an effective concentration (EC)50 of 0.35 ng/mL and EC90 of 3.2 ng/mL in a cell line-derived xenograft (CDX) mice model using human lung cancer (EBC-1) and gastric cancer (MKN-45) cells [36]. Furthermore, PK studies in healthy male Chinese volunteers administered with single oral savolitinib doses of 200, 400 and 600 mg following an overnight fast or a high-fat and high-calorie breakfast prior to dosing showed no clinically relevant impact on PK and bioavailability of savolitinib [37].

In NSCLC with *MET* aberrations, several clinical trials have shown the potential benefit of savolitinib as a monotherapy and in combination with EGFR–TKI [30]. Savolitinib received its first approval by The National Medical Products Administration (NMPA), China for patients with *MET*ex14-altered locally advanced or metastatic NSCLC with disease-progression following systemic treatment or unable to receive chemotherapy [28]. The approval was based on a phase II trial conducted in China in patients with *MET*ex14-altered NSCLC, including patients with the more aggressive PSC subtype [38]. The key milestones in the development of savolitinib for NSCLC treatment are demonstrated in Figure 3.

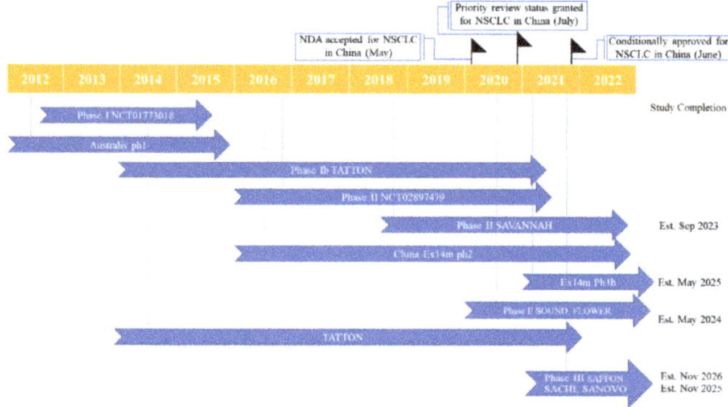

Figure 3. Key milestones and clinical trials in the development of savolitinib for non-small cell lung cancer.

3. First Steps towards the Development of Savolitinib as Mono and Combination Therapies

The availability of substantial evidence of anti-tumor activity and acceptable safety profile led to the development of savolitinib as a treatment for advanced NSCLC with *MET* aberrations. The high selectivity of savolitinib for *MET* was confirmed using a screening platform of more than 900 cell lines of which 111 represented NSCLC [39]. In vitro study by Henry and colleagues demonstrated the ability of savolitinib as a single agent to inhibit *MET* activity and reduce NSCLC cell viability in a dose dependent manner [39]. Further, anti-tumor efficacy was observed with savolitinib in vivo, in lung cancer PDX model with *MET*ex14 mutation. Savolitinib showed tumor regression (tumor volume reduction: 62%) with a dose of 25 mg/kg in all mice on treatment ($n = 9$) as well as 98% inhibition in tumor growth (TGI) with 5 mg/kg dose in 4 out of 9 mice of PDX model (Data on file). In addition, in vivo study using H1993 and EBC-1 tumor xenografts showed considerable decrease in tumor growth, with savolitinib achieving an optimal response at doses as low as 0.3 mg/kg and 2.5 mg/kg in H1993 and EBC-1 tumors, respectively [39]. Interestingly, the same group (Henry and colleagues) concurred that savolitinib resistance in NSCLC is partially driven by MYC overexpression in H1993 cells, suggesting potential mechanism and treatment strategies for future acquired resistance to MEK–TKI.

Savolitinib, in combination with erlotinib, a first-generation EGFR–TKI inhibitor, showed substantial tumor inhibition in H441, an *EGFR* wild type model with *MET* amplification [40,41]. In addition, savolitinib treatment exhibited substantial anti-tumor activity in vivo (tumor regression: 35%) in the NSCLC cancer cell line NCI-H820 harboring an activating *EGFR* mutation (Ex19del), a gefitinib/erlotinib resistant mutation (T790M) as well as hyperactivated *MET* (data on file). Osimertinib, a third-generation, irreversible EGFR–TKI, at either 25 mg/kg daily or 12.5 mg/kg daily exhibited minimal anti-tumor activity, with TGI of 24% and 4%, respectively. However, when treated in combination with savolitinib, 25 mg/kg of savolitinib plus osimertinib at either 25 mg/kg or 12.5 mg/kg daily resulted in 94% and 90% TGI, respectively. These preclinical results highlight the beneficial anti-tumor effect of osimertinib plus savolitinib combination at optimal doses of 0.3~1.5 mg/kg savolitinib combined with 10 mg/kg osimertinib. Further, another study analyzed different doses of savolitinib, ranging from 0.02 mg/kg to 15 mg/kg (15 mg/kg equivalent to 600 mg clinical dose), in combination with a fixed dose of 10 mg/kg osimertinib (equivalent to 80 mg clinical dose). Pan–CYP inhibitor 1–aminobenzotriazole was dosed along with savolitinib and osimertinib to prolong PK half-life by reducing elimination rate so that plasma concentration time profile matches clinical exposure of the drugs (data on file). The combination of osimertinib and savolitinib demonstrated strong anti-tumor activity leading to tumor regressions. The benefit of combination treatment was observed with as low as 0.3 mg/kg dose of savolitinib. Thus, these encouraging preclinical results led to the evaluation of savolitinib's efficacy and safety in clinical trials for NSCLC with *MET* aberrations.

4. Clinical Development of Savolitinib: Phase I Trials

A first in-human phase I clinical study (NCT01773018) was conducted in patients ($n = 48$) with locally advanced solid tumors from Australia [42]. The doses administered were 100–1000 mg once daily (OD) and 300–500 mg twice-daily (BID), and the maximum tolerated dose was 800 mg. Savolitinib showed preliminary efficacy in patients with papillary renal cell carcinoma with *MET* gene copy number changes. The most frequent adverse events (AE) were nausea (62.5%), vomiting (41.7%), fatigue (35.4%) and peripheral edema (27.1%). The tolerability profile of savolitinib was acceptable, and the recommended phase II dose (RP2D) was established as 600 mg OD [42]. In another open-label, multi-center, phase Ia/Ib study (NCT0198555) conducted in China in patients ($n = 85$) with advanced tumors bearing *MET* aberrations, savolitinib demonstrated a manageable safety profile and promising anti-tumor activity in NSCLC with *MET*ex14 mutation, apparent tumor shrinkage (55% and 27%) in target lesions was observed, although partial response (PR) was not achieved. The most common treatment-related AEs were nausea (29.4%), vomiting (27.1%) and peripheral edema (21.2%). The RP2D of savolitinib was established at 600 mg OD or 500 mg BID and was consistent with phase I first-in human study conducted in Australia [43]. There was certain comparability between the patients with NSCLC enrolled in the phase I study conducted in Australia and the phase I study conducted in China, and thus the results could be analyzed accordingly [42,43].

Savolitinib demonstrated the ability to overcome *MET*-mediated resistance in patients with *EGFR*-mutant, *MET*-amplified or c-MET overexpressed NSCLC when combined with osimertinib, and these benefits extended to those with disease that had previously progressed on a prior EGFR–TKI [29,44]. Part A of the multi-arm phase Ib TATTON study (NCT02143466) demonstrated the safety and tolerability of osimertinib plus savolitinib (n = 18) in patients with advanced NSCLC disease progression on a prior EGFR–TKI [44]. Doses of savolitinib applied were escalated from 600 to 800 mg OD with a fixed dose of osimertinib 80 mg. The most common AEs reported were nausea (67%), rash (56%) and vomiting (50%). The objective response rate (ORR) was 44% [44]. Furthermore, in the expansion cohorts of TATTON trial, investigators evaluated the safety and efficacy of osimertinib plus savolitinib in locally advanced or metastatic, *MET*-amplified, *EGFR* mutation-positive NSCLC patients who had progressed on EGFR–TKIs [29]. Part B (n = 138) was substratified into three cohorts: B1 included those who had previously received a third-generation EGFR-TKI; patients without prior third-generation EGFR–TKI treatment were separated into B2 with Thr790Met negative and B3 with Thr790Met positive, at the time of enrolment. These patients received 600 mg QD, although the protocol was later amended, causing patients who weighed lesser than 55 kg to receive a 300-mg dose of savolitinib. The Part D expansion cohort was comprised of patients (n = 42) who had not previously received a third-generation EGFR–TKI and were T790M negative, and these patients received osimertinib 80 mg plus savolitinib 300 mg OD. Objective partial responses (PR) were observed (by 4 March 2020) in 68 (49%) patients in total of part B, with 23 (33%) patients, 33 (65%) patients and 12 (67%) patients in B1, B2 and B3, respectively, while in 26 (62%) patients in part D [45]. Regarding safety, the 4 expanded cohorts had similar safety profiles with 28% in part B and 19% in part D experiencing AEs possibly related to savolitinib. Serious AEs of grade 3 or 4 were associated with 49% patients in part B and 38% patients in part D. The most common AEs of grade 1–2 in expanded cohorts included nausea (48%), peripheral edema (34%), decreased appetite (32%), vomiting (30%) and fatigue (28%). In part B cohorts, the most common grade 3 or higher AEs related to savolitinib were decreased neutrophil count (6%) and aminotransferase elevations (4%), while in part D, hypersensitivity (5%), diarrhea (5%) and myalgia (5%) are more frequent [45]. Generally, in the dose expansion cohorts of TATTON trial, savolitinib plus osimertinib showed promising anti-tumor activity in *MET*-amplified *EGFR* positive advanced NSCLC patients who received a prior third-generation EGFR–TKI. These results have now been further investigated in the phase II SAVANNAH trial.

In another phase Ib study (NCT02374645), the clinical evaluation of savolitinib plus gefitinib (a first-generation EGFR-TKI) demonstrated promising anti-tumor activity with acceptable safety profile in *EGFRm*, *MET*-amplified advanced NSCLC patients from China who had disease progression on EGFR-TKIs. Patients received savolitinib 600 or 800 mg plus gefitinib 250 mg orally OD for which no dose-limiting toxicities were reported in safety run-in. The most commonly reported AEs were vomiting (46%), nausea (40%) and increased aspartate aminotransferase (39%) [30]. ORR in *EGFR* T790M-negative and -positive patients were 52% and 9%, respectively, suggesting beneficial anti-tumor activity [30].

5. Clinical Development of Savolitinib: Phase II Trials

A pivotal open-label phase II clinical study (NCT02897479) conducted in China demonstrated encouraging efficacy and tolerable safety profile of savolitinib in overall and patient subsets stratified according to tumor type (PSC and other NSCLC), brain metastasis status and prior anti-tumor treatment (pretreated and treatment naïve) [38]. Unresectable or metastatic NSCLC patients (n = 70) harboring *MET*ex14 mutation were administered savolitinib monotherapy at recommended starting dose of 600 mg orally once daily (OD) for patients weighing $\geq$50 kg, or 400 mg OD for patients weighing <50 kg, until disease progression or unacceptable toxicity. The majority of patients were elderly with advanced

NSCLC on prior systemic therapy. In both the full analysis set (FAS) and the tumor response evaluable set (TRES), independent review committee (IRC) assessments were the main analyses, while investigators' (INV) assessments were supportive analyses. The IRC-assessed tumor response evaluable set (TRES) was comprised of 62 patients. The ECOG performance status of full analysis set (FAS) for majority of patients (81%) was 1 and in pre-specified subsets (PSC vs. other NSCLC subtypes, treatment naïve vs. previously treated), 78% to 88% patients had ECOG status of 1. Of the total PSC population ($n = 25$), pre-treated and treatment-naïve subsets were comprised of 29% and 46%, while brain metastasis and non-brain metastasis groups were comprised of 13% and 42% of PSC patients, respectively [38,46,47]. The primary efficacy end point was ORR (as assessed by IRC in TRES) defined as the proportion of patients with a confirmed complete response or partial response according to RECIST version 1.1. Secondary outcomes included duration of response (DoR), time to response (TTR), progression free survival (PFS), overall survival (OS) and safety. The latest results of the trial were presented at the 2022 ELCC conference and published in JTO Clinical and Research Reports [46,48]. The baseline characteristics are provided in Table 2.

Table 2. Baseline characteristics of phase II trial conducted in China [48].

		Age			Sex		Smoking History		ECOG Performance Status		
		Median Age, Years	<75 Years	≥75 Years	Female	Male	Non-Smokers	Smokers	0	1	3
Full Analysis Set ($n = 70$)		68.7 (51.7–85.0)	54 (77%)	16 (23%)	29 (41%)	41 (59%)	42 (60%)	28 (40%)	12 (17%)	57 (81%)	1 (1%)
Type of Primary Tumor	PSC ($n = 25$)	69.3 (54.1–84.8)	19 (76%)	6 (24%)	8 (32%)	17 (68%)	13 (52%)	12 (48%)	3 (12%)	22 (88%)	0
	Other NSCLC ($n = 45$)	68.1 (51.7–85.0)	35 (78%)	10 (22%)	21 (47%)	24 (53%)	29 (64%)	16 (36%)	9 (20%)	35 (78%)	1 (2%)
Prior Anti-tumor Treatment	Pre-treated ($n = 42$)	67.7 (51.7–84.8)	38 (90%)	4 (10%)	17 (40%)	25 (60%)	28 (67%)	14 (33%)	8 (19%)	34 (81%)	0
	Treatment-naïve ($n = 28$)	74.5 (56.0–85.0)	16 (57%)	12 (43%)	17 (40%)	16 (57%)	14 (50%)	14 (50%)	4 (14%)	23 (82%)	1 (4%)
Brain Metastases Status	Brain metastases ($n = 15$)	68.6 (51.7–84.8)	11 (73%)	4 (27%)	7 (47%)	8 (53%)	11 (73%)	4 (27%)	3 (20%)	12 (80%)	0
	Non-brain metastases ($n = 55$)	68.7 (51.9–85.0)	43 (78%)	12 (22%)	22 (40%)	33 (60%)	31 (56%)	24 (44%)	9 (16%)	45 (82%)	1 (2%)

		Histology					Prior Anti-tumor Treatment		Brain Involvement at Baseline
		Pulmonary sarcomatoid carcinoma	Other NSCLC subtypes				Yes	No	
			Adenocarcinoma	Squamous cell carcinoma	Adenosquamous carcinoma	NSCLC, not otherwise specified			
Full Analysis Set ($n = 70$)		25 (36%)	40 (57%)	3 (3%)	1 (1%)	1 (1%)	42 (60%)	28 (40%)	15 (21%)
Type of Primary Tumor	PSC ($n = 25$)	25 (100%)		-			12 (48%)	13 (52%)	2 (8%)
	Other NSCLC ($n = 45$)	-	40 (89%)	3 (7%)	1 (2%)	1 (2%)	30 (67%)	15 (33%)	13 (29%)
Prior Anti-tumor Treatment	Pre-treated ($n = 42$)	12 (29%)	27 (64%)	2 (5%)	1 (2%)	0	42 (100%)	-	11 (26%)
	Treatment-naïve ($n = 28$)	13 (46%)	13 (46%)	1 (4%)	0	1 (4%)	-	28 (100%)	4 (14%)
Brain Metastases Status	Brain metastases ($n = 15$)	2 (13%)	13 (87%)	0	0	0	11 (73%)	4 (27%)	15 (100%)
	Non-brain metastases ($n = 55$)	23 (42%)	27 (49%)	3 (5%)	1 (2%)	1 (2%)	33 (60%)	22 (40%)	-

Data in median (IQR) or n (%). NSCLC, Non-Small Cell Lung Cancer; PSC, Pulmonary Sarcomatoid Carcinoma; ECOG, Eastern Cooperative Oncology Group.

5.1. Efficacy Evidence

At a median follow-up of 17.6 months, the IRC and INV assessed ORR was 49.2% and 53.2 %, respectively in TRES subset, while ORR assessed in FAS set by IRC and INV was 42.9% and 47.1%, respectively. Further, the IRC and INV assessed disease control rate

(DCR) was 93.4% and 91.9%, respectively in TRES subset, while DCR assessed in FAS set by IRC and INV was 82.9% and 81.4%, respectively. The median time to response was 1.4 months across TRES and FAS sets as judged by IRC and INV. Median DoR for TRES and FAS as assessed by IRC and INV was 8.3 and 6.9 months. Savolitinib was associated with mOS of 12.5 months and a mPFS of 6.9 months in FAS at a median follow-up time of 28.4 months. The 18-month OS rate is 42.1%, dropping to 31.5% at 24 months [38,48].

In subgroup analyses (assessed in TRES set by INV, median follow-up of 28.4 months), for PSC (n = 20), 10 patients had partial response (ORR 50%) with a median duration of response of 12.4 months. In other NSCLC subtypes (n = 42), 23 patients had partial response (ORR 54.8%) with a median duration of response of 5.6 months and DCR of 92.9%. In pre-treated (n = 38) patients, partial response was observed in 20 patients (52.6%), while in treatment-naïve (n = 24) subgroup, partial response was observed in 13 patients (54.2%). Extracranial ORR for brain metastasis group was 64.3%. For survival outcomes, the PSC group showed a mPFS of 5.5 months, while with brain metastasis (n = 15), it was 7.0 months and without brain metastasis was 6.2 months. Similar values of mPFS were observed with pre-treated (6.9 months) and treatment-naïve (6.9 months) subgroups, respectively. The mOS for PSC and other NSCLC patients was 10.6 months and 17.3 months, respectively, with corresponding 24-month OS rates of 26% and 35%. Among brain metastases patients, the mOS was 17.7 months with the 24-month OS rate being 36%. The mOS for pre-treated and treatment-naïve patients was 19.4 months and 10.9 months, respectively, with corresponding 24-month OS rates of 38% and 22% [48]. However, this large difference in OS can be attributed to the higher proportion of patients with PSC in treatment-naïve population (46% vs. 29% in pre-treated patients) and a higher median age (74.5 vs. 67.7 in pre-treated patients). Patients with PSC had a short mOS vs other NSCLC patient (10.6 months vs. 17.3 months), likely due to the poor prognosis associated with PSC. These results confirmed savolitinib having beneficial efficacy towards NSCLC with *MET*ex14 mutation and its PSC subtype [32,38,48]. The PFS and OS results have been illustrated graphically in Table 3.

Table 3. Investigator-Assessed Responses in the Tumor-Response-Evaluable Set and the Full Analysis Set of Phase II Trial Conducted in China [48].

		ORR, n (%)	DCR, n (%)	Median DOR, Months [1]	Median TTR, Months [1]
	Total (n = 62)	33 (53.2%)	57 (91.9%)	6.9	1.4
	PSC (n = 20)	10 (50.0%)	18 (90.0%)	12.4	1.4
	Other NSCLC subtypes (n = 42)	23 (54.8%)	39 (92.9%)	5.6	1.4
Tumor-Response-Evaluable Set (n = 62)	Pretreated (n = 38)	20 (52.6%)	34 (89.5%)	10.9	1.4
	Treatment-naive (n = 24)	13 (54.2%)	23 (95.8%)	5.6	1.4
	Brain metastases (n = 14)	9 (64.3%)	14 (100.0%)	4.9	1.5
	Non-brain metastases (n = 48)	24 (50.0%)	43 (89.6%)	7.0	1.4
	Total (n = 70)	33 (47.1%)	57 (81.4%)	n/A	n/A
	PSC (n = 25)	10 (40.0%)	18 (72.0%)	n/A	n/A
	Other NSCLC subtypes (n = 45)	23 (51.1%)	39 (86.7%)	n/A	n/A
Full Analysis Set (n = 70)	Pretreated (n = 42)	20 (47.6%)	34 (81.0%)	n/A	n/A
	Treatment-naive (n = 28)	13 (46.4%)	23 (82.1%)	n/A	n/A
	Brain metastases (n = 15)	9 (60.0%)	14 (93.3%)	n/A	n/A
	Non-brain metastases (n = 55)	24 (43.6%)	43 (78.2%)	n/A	n/A

[1] DOR and TTR were analyzed in the tumor-response-evaluable set. DCR, disease control rate; DOR, duration of response; n/A, not applicable; NSCLC, non-small cell lung cancer; ORR, objective response rate; PSC, pulmonary sarcomatoid carcinoma; TTR, time to response.

An earlier study reported mOS of 6.7 months in patients with *MET*ex14 mutation NSCLC on chemotherapy treatment who did not receive prior targeted therapy [49]. In addition, mOS of PSC subset in NSCLC patients treated with chemotherapy has been reported to be 4 to 8 months [32,49–53]. With savolitinib, mOS of NSCLC patients reaches 12.5 months with 70% maturity. In PSC subset, higher OS is seen with savolitinib treatment compared to chemotherapy, with OS reaching 10.6 months. So far, literature related to MET inhibitor treatment with PSC population is available only for savolitinib [51–54].

P–glycoprotein (gp) and breast cancer resistance protein (BRCP) are efflux proteins located in the luminal membrane of brain capillary endothelium, preventing drugs from entering the central nervous system. Most MET inhibitors, such as crizotinib and tepotinib, are known substrates of the P–gp and BRCP efflux transport system [55–57]. Steady concentrations of savolitnib are readily maintained in an intracerebral area which may be attributed to it not being a substrate of P–gp and BRCP efflux transport system. Promising efficacy of savolitinib was observed in brain metastasis subgroup, with ORR at 64.3%, DCR at 100% and significant survival benefit (PFS, 7.0 months; OS, 17.7 months). These encouraging results provide a treatment option for this subgroup of patients with poor prognosis and few treatment options [38,47,48].

5.2. Safety Evidence

Savolitinib demonstrated tolerable safety profile consistent with previous trials; most AEs were grades 1–2 and resolved with dose adjustment and discontinuation. Adverse events that presented at rates of ≥30% are listed below (Table 4, median follow-up of 28.4 months). The incidence of grade3 or more AEs was 65.7%, while 50% of patients reported treatment-related serious adverse events (SAE). The top ≥ grade 3 AE was elevated AST (12.9%). The most common treatment-related AEs (TRAEs) (≥30%) are peripheral edema (55.7%), nausea (45.7%) and elevated aminotransferase (38.6% and 37.1%). The top ≥ grade 3 treatment related AE was elevated AST (12.9%) [48,58]. The common SAEs reported were abnormal liver function (4.3%, 3 patients), drug hypersensitivity reaction (2.9%, 2 patients) and fever (2.9%, 2 patients). Treatment related fatal SAE, tumor lysis syndrome was reported in one patient. Ten patients discontinued treatment due to AEs, of which drug-induced liver damage and drug hypersensitivity reactions were seen in 2.9% of patients (2 patients), respectively [38]. No occurrence of pulmonary interstitial pneumonia and interstitial lung disease (ILD) was observed with savolitinib, while ILD is seen with tepotinib (n = 2) and capmatinib (n = 1) [4,59].

Table 4. Adverse events (>30%) in the full analysis set of phase II trials conducted in China (n = 70) [48].

		Any Grade	≥Grade 3
	Any event	70 (100.0%)	46 (65.7%)
	Peripheral edema	40 (57.1%)	6 (8.6%)
	Nausea	37 (52.9%)	0
	Hypoalbuminemia	29 (41.4%)	1 (1.4%)
All-cause adverse events	Elevated alanine aminotransferase	27 (38.6%)	7 (10.0%)
	Elevated aspartate aminotransferase	27 (38.6%)	9 (12.9%)
	Decreased appetite	24 (34.3%)	0
	Vomiting	23 (32.9%)	0
	Pyrexia	21 (30.0%)	1 (1.4%)
	Any event	70 (100.0%)	32 (45.7%)
	Peripheral edema	39 (55.7)	6 (8.6)
	Nausea	32 (45.7)	0
Treatment-related adverse events	Hypoalbuminemia	16 (22.9)	0
	Elevated alanine aminotransferase	27 (38.6)	7 (10.0%)
	Elevated aspartate aminotransferase	26 (37.1)	9 (12.9%)
	Decreased appetite	14 (20.0%)	0
	Vomiting	18 (25.7%)	0
	Pyrexia	11 (15.7%)	1 (1.4%)

Data in n (%). Derived from latest safety analysis of phase II trial (NCT02897479) [48].

The updated results further confirm that savolitinib can benefit *MET*ex14-mutated NSCLC patients and each subgroup with acceptable safety profile [38,46–48]. Savolitinib thus displays promising efficacy and tolerability in PSC associated with *MET*ex14 mutation and holds potential to become the first approved treatment in this setting. In addition, the study showed that savolitinib can penetrate the blood–brain barrier and is effective in patients with brain metastases.

5.3. Brief Introduction of Other Phase II Trials

Other ongoing phase II trials include SAVANNAH, SOUND and FLOWERS trials. SAVANNAH trial (NCT03778229) continues to explore the sequence of savolitinib plus osimertinib with previous osimertinib monotherapy resistance. It is a phase II, single-arm study evaluating the efficacy of osimertinib in combination with savolitinib in 259 patients with *EGFRm* and *MET* amplified and/or c-MET overexpressed locally advanced or metastatic NSCLC who have progressed on osimertinib. Patients were treated with osimertinib (80 mg OD) and savolitinib (300 mg QD, 300 mg BID or 600 mg OD) until objective disease progression. Efficacy endpoints—such as ORR (primary endpoint), PFS, OS, DoR, HRQoL, pharmacokinetics, safety points such as AEs and patient related outcomes (PROs)—were studied. This is the first phase II clinical study of the third-generation EGFR–TKI osimertinib resistance in patients with advanced NSCLC with *MET* amplification and/or c-MET overexpression. *MET* detection was performed using fluorescence in situ hybridization (FISH) and immunohistochemistry (IHC) methods. The detection criteria were set to FISH, *MET* GCN $\geq$ 5 and/or *MET*/CEP7 $\geq$ 2; IHC, $\geq$50% tumor cells 3+. Sixty-two percent of osimertinib resistant patients was at low threshold [IHC50+ and/or FISH5+] as well as 34%—at the high threshold [IHC90+ and/or FISH10+] subgroups. Figure S1 provides the proportion of patient population with *MET* amplified and/or c-MET overexpressed in this study suggesting amplification and/or overexpression is the most common osimertinib resistance mechanism. The baseline characteristics are provided in Table S1. The overall median age of patients is 63 years, 62% were female, 54% were Asian and 34% were with brain metastases at baseline. On savolitinib 300 mg OD plus osimertinib 80 mg OD treatment, advanced NSCLC patients (n = 193) with high *MET* amplification and/or high threshold c-MET overexpression level show a trend toward better efficacy benefit, emphasizing the necessity of patients' selection according to appropriate *MET* detection criteria in this population. Among the overall population, ORR was 32%; median DoR was 8.3 months; and median PFS was 5.3 months, while among 108 patients who met the threshold for high *MET* amplification and/or high threshold c-MET overexpression level (IHC90+ and/or FISH10+), ORR was 49%; median DoR was 9.3 months; and median PFS was 7.1 months (Table S2). The safety results showed that the incidence of treatment-related AEs was 84%; treatment-related $\geq$grade 3 AEs at 20%; and treatment-related SAEs at 7% (Table S3). The incidence of hypersensitivity, ILD and pneumonia were 2% (4/196), and QT interval prolongation at 5% (10/196) [60].

In addition, the FLOWERS trial (NCT05163249) explores the efficacy and safety of osimertinib with or without savolitinib in patients with de novo *MET* amplified and/or c-MET overexpressed, *EGFR*-mutant advanced NSCLC. In SOUND trial (NCT05374603), an open-label, interventional, multi-center, exploratory trial, savolitinib combined with durvalumab will be evaluated in Chinese *EGFR* wild-type locally advanced or metastatic NSCLC patients with *MET* alterations. NSCLC patients from China with *MET* amplification (n = 30) and *MET*ex14 mutation (n = 30) will be treated with 1500 mg durvalumab and 300 to 600 mg savolitinib (OD) for 28-day/cycle till disease progression, death or toxicity. Efficacy endpoints will be PFS, ORR, DoR, DCR, 12 m OS rate and safety endpoints will be AEs and AEs of special interest (AESI) [61]. Further, phase III SAFFRON trial (NCT05261399) is investigating savolitinib plus osimertinib versus platinum-based doublet chemotherapy in participants with NSCLC (*EGFR* mutated, c-MET overexpressed and/or *MET* gene amplified) who have progressed on osimertinib treatment.

6. Ongoing Phase III Trials

Currently, four phase III trials evaluating savolitinib as a monotherapy and in combination with EGFR-TKIs are underway. The confirmatory phase IIIb clinical study (CTR20211151) is evaluating efficacy and safety of savolitinib in two cohorts from patients with locally advanced or metastatic NSCLC with *MET*ex14 mutation in China; patients of one cohort are with disease progression or toxicity intolerance after previous platinum-based chemotherapy regimens, and patients of another cohort are with no prior systemic antineoplastic therapy for advanced disease. The patients were treated until disease progression or intolerable toxicity. Phase III SACHI trial (CTR20211441) is a randomized, two-arm, open-label, multi-center study evaluating the efficacy and safety of savolitinib plus osimertinib versus chemotherapy in NSCLC patients from China with *MET* amplification who has progressed after first- to third-generation EGFR–TKI therapy and has already begun its recruitment in multiple centers. Another similar phase III trial SAFFRON is designed to evaluate the efficacy and safety of the same combined therapy as SACHI versus chemotherapy, but focus on global advanced NSCLC patients with *MET* amplification/c-MET overexpression that progressed after osimertinib treatment. SANOVO Phase III study is evaluating the efficacy and safety of savolitinib in combination with osimertinib in treatment-naïve patients with *EGFR* mutant positive and c-MET overexpression advanced NSCLC (NCT05009836).

7. Discussion

Savolitinib, an investigational MET highly selective agent, has shown pronounced efficacy in preclinical and clinical studies. Savolitinib demonstrated preclinical anti-tumor activity against *MET*-dependent cancer cell line growth and *MET*-driven tumor growth in xenograft models. Following which, data from a phase I clinical trial established recommended phase II dose in patients with *MET*ex14-mutated NSCLC. Further, the TATTON study established utility of savolitinib with osimertinib in advanced NSCLC with *MET*-mediated acquired resistance to EGFR-TKIs. Final results of the phase II study (NCT02897479) further confirmed the benefit of savolitinib in patients with *MET*ex14-mutated NSCLC across all predefined subgroups. In addition, phase IIIb clinical study CTR20211151 is confirming the result of phase II study on *MET*ex14-mutated NSCLC, while three ongoing phase II trials, SAVANNAH, SOUND and FLOWERS, as well as three phase IIIB trials, SAFFRON, SACHI and SANOVO, are actively exploring solutions for different types of savolitinib combination regimens against EGFR resistance mechanisms. Preliminary results of the SAVANNAH trial have demonstrated the beneficial efficacy of osimertinib plus savolitinib in *EGFRm* NSCLC patients with *MET* amplified and/or c-MET overexpressed, supporting the results of TATTON study and paving the way for phase III SACHI and SAFFRON study.

In the hallmark phase II registry trial, savolitinib displayed promising efficacy and tolerability in patients with *MET*ex14-altered advanced NSCLC, with mOS reaching 12.5 months. The effect of savolitinib was rapid, substantiated by time to response (TTR) of 1.4 months. Promising results with PFS of 5.5 months and OS of 10.6 months were also seen in the PSC subtype, which does not respond well to chemotherapy and has limited effective treatments. By now, savolitinib is the only MET inhibitor with data related to PSC associated with *MET*ex14 mutation and is becoming the first approved agent in this setting. For patients of treatment naïve population, the PFS and OS of savolitinib were 6.9 months and 10.9 months, respectively, while PFS and OS of prior treatment patients reached 6.9 months and 19.4 months, respectively. In the current scenario, the reported ORR of savolitinib is the highest in the prior treatment population compared to other treatments (52.6% vs. 44.0% of capmatinib, 49.5% of tepotinib and 21% of amivantamab) [48,62–64]. Savolitinib is also currently the only MET inhibitor that has recorded beneficial OS data in brain metastases, with PFS of 7.0 months and OS of 17.7 months. In addition, savolitinib has the best tumor response in brain metastasis population with ORR at 64.3% and DCR at 100% [38,48]. Based on these promising results, savolitinib received its first conditional approval by NMPA, China in June 2021, for patients with *MET*ex14-altered NSCLC after systemic treatment resistance or unable to receive chemotherapy. Post-marketing phase IIIb trial is now undergoing (HutchMed) in larger population of NSCLC patients and is expected to provide more clinical evidences for savolitinib in first-line therapy. Furthermore, latest *post hoc* analysis based on ctDNA detection suggests undetectable baseline *MET*ex14 or post-treatment clearance in ctDNA being relevant to favourable clinical outcomes, including better PFS and OS results, while secondary *MET* mutations and other acquired gene alterations after treatment (e.g., RTK–RASP–I3K pathway) may explain resistance mechanism to savolitinib [65].

Table 5 summarizes the data for MET-TKIs developed for *MET*ex14-altered advanced NSCLC population as well as subtypes [4,38,47,48,58,59,62–64,66–73]. Patient population of Chinese Phase II registry trial were from China. In other global trials, east Asian population varied from 15.9 to 50.9%. Proportion of NSCLC patients with brain metastases was higher (28.9%) in Chinese Phase II registry trial compared to other trials [48]. Tumor response of different types of MET–TKIs shows ORR (54.8%) and DCR (92.9%) to be highest with savolitinib. Among AEs, most commonly, elevated transaminases were seen with savolitinib, tepotinib and crizotinib; peripheral edema with savolitnib, capmatinib and tepotinib; ILD with capmitnib, tepotinib and crizotinib; difficulty in breathing in tepotinib, crizotinib and amivantamab [38,48,58,59,62–64,67–73].

Table 5. Data summary of MET inhibitors in *MET*ex14 mutation.

	Savolitinib [1]	Capmatinib [2]	Tepotinib [3]	Crizotinib [4]	Amivantamab [5]
Approval	China approved in June 2021	Approved in the US in 2020	Approved in Japan in 2020	FDA breakthrough therapy designation	Approved in the US in May 2021
Mechanism	METi Ib	METi Ib	METi Ib	ALK/ROS1/METi Ia	Anti-MET and EGFR antibody
n	45	160	313	25/69	46
Population	100% Chinese patients	20.2% Asian patients	33.9% Asian patients	Unknown/15.9% Asian patients	50.9% Asian patients
Proportion of brain metastases	28.9%	16.9%	18.2%	Unknown	18.2%
Dose	600 mg (BW $\geq$ 50 kg), or 400 mg (BW < 50 kg) OD	400 mg BID	500 mg OD	250 mg BID	1050 mg (<80 kg), or 1400 mg ($\geq$80 kg)
ORR	54.8%	52.5%	50.8%	12.0%/32.3%	32.6%
DCR	92.9%	88.1%	75.4%	44.0%/unknown	76.1%
Median PFS, Months	6.9	12.4/12.5/5.4/6.9	11.2	3.6/7.3	6.7
Common Grade 3/4 AEs	Elevated AST Elevated ALT Peripheral edema (No interstitial lung disease occurred in registry studies)	Peripheral edema Difficulty breathing Fatigue Elevated ALT Weak Pneumonia	Peripheral edema Generalized edema Vomit Nausea Interstitial lung disease	Elevated transaminases Difficulty breathing Hypophosphatemia Lymphopenia Pulmonary embolism Interstitial lung disease	Rash Hypoalbuminemia Difficulty breathing

[1] The number of patients and the proportion of patients with brain metastases are based on other types of NSCLC in general, and the ORR, DCR and median PFS data are derived from data from other types of NSCLC in the efficacy-evaluable set [48]; safety data is analyzed based on the overall patient (*n* = 70) [38,48]. [2] Data derived from the latest analysis of four different cohorts from GEOMETRY mono-1 study: cohort 4, expansion cohort 6, cohort 5b and expansion cohort 7. Number of patients, proportion of brain metastases, ORR and DCR represent four cohorts in total; proportion of population based on cohort 4, 5b and 7; mPFS reflect results of four cohorts, respectively [62,71]. [3] Data based on VISION study cohort A + cohort C latest overall analysis [63]. [4] Patient population, number of patients, ORR, DCR and median PFS data are derived from two different studies of AcSé [72] and PROFILE-1001 [73]; safety data is based on combination of these two trials. [5] Data from latest analysis of CHRYSALIS study [64]. EGFR, Epidermal Growth Factor Receptor; TKI, Tyrosine Kinase Inhibitor; MET, Mesenchymal Epithelial Transition; ALK, Anaplastic Lymphoma Kinase; ROS1, ROS proto-oncogene 1; OD, Once Daily; BID, twice daily; ORR, Objective Response Rate; DCR, Disease Control Rate; PFS, Progression Free Survival; AE, Adverse Event; AST, Aspartate aminotransferase; ALT, Alanine aminotransferase.

Bypass activation mediated by the *MET* signaling pathway is one of the important mechanisms leading to EGFR–TKI resistance. *MET*-driven resistance can be manifested as gene-level amplification or protein-level overexpression with previous treatment regimens such as chemotherapy, immunotherapy and targeted therapies including *EGFR, BRAF* and *MEK* [74,75]. The efficacy of tepotinib on NSCLC with T790M-negative *MET* amplification and/or c-MET overexpression after first/second-generation EGFR–TKI resistance is limited, with a mPFS of only 4.9 months [76,77]. The current immunotherapy efficacy for advanced NSCLC after EGFR–TKI resistance needs further improvement, and there is a lack of *MET* amplification and/or c-MET overexpression subgroup data. Nivolumab monotherapy has limited efficacy after EGFR–TKI resistance, with a mPFS of only 1.5–1.7 months [78,79]. In IMpower 150 and ORIENT-31 studies, EGFR–TKI resistance, followed by immunotherapy combined with bevacizumab and chemotherapy, showed a mPFS of 6.9–9.7 months but no subgroup data on *MET* amplification and/or c-MET overexpression was reported; meanwhile, safety of the combination therapy regimen needs attention [80,81]. *MET*-amplified and/or c-MET overexpressed advanced NSCLC patients with EGFR–TKI resistance have limited therapeutic effect with MET inhibitor monotherapy. Only 1 of 12 evaluable patients on inhibitor monotherapy reported an objective response [82]. Dual-target inhibition of *EGFR* and *MET* pathways may bring synergistic therapeutic benefit in *MET*-driven EGFR–TKI-resistant advanced NSCLC patients [83]. Meanwhile, efficacy of savolitinib combined with durvalumab in *EGFR* wild-type NSCLC with *MET* alterations is also under exploration in SOUND trial, as previously described.

The combination of EGFR inhibitor and MET-highly selective TKI possesses the potential to prevent or overcome *MET*-driven resistance to EGFR–TKIs. Acquired resistance to first- and second-generation EGFR–TKIs is often caused by the acquisition of the T790M mutation, which accounts for approximately 60% of resistant cases and has been overcome by third-generation EGFR–TKIs such as osimertinib. For first- and second-generation EGFR–TKIs, acquired resistance for *MET*-amplification is at least 5% (for example, gefitinib), while up to 25% of acquired resistance is observed with third generation EGFR–TKI (for example, osimertinib) [84]. TATTON study, set up in the back drop of acquired *MET* amplification associated with EGFR–TKI resistance offered explicit benefit with savolitinib in NSCLC patients without prior third-generation EGFR–TKI, while those who were administered with a prior third-generation EGFR–TKI had a relatively lower rate of response regardless of T790 status, possibly related to larger proportion of patients with ≥ 3 lines of treatment comprising the prior third-generation EGFR–TKI group (56.5% vs. 22.6% in partB2 + partD). Nonetheless, TATTON program demonstrated beneficial efficacy of savolitinib plus osimertinib combination in the *MET*-amplified, *EGFR* mutation–positive setting with acceptable safety profile which is a first in this setting [44]. Further, SAVANNAH phase II trial validates TATTON results with advanced NSCLC patients with *MET* amplification or c-MET overexpression due to osimertinib-acquired resistance. Initial results from the SAVANNAH trial show a trend toward improved response rates, with increasing level of *MET* amplified and/or c-MET overexpressed. Across all patients in this analysis, ORR was 32%; mDoR was 8.3 months; and mPFS was 5.3 months, while in high level *MET* amplification and/or c-MET overexpression subgroup, ORR was 49%; mDoR was 9.3 months; and mPFS was 7.1 months [60]. A summary of key data after EGFR–TKI resistance with secondary *MET* alterations treated with combination therapies available so far is provided in Table 6.

Table 6. Summary of key data after EGFR–TKI resistance with secondary *MET* alterations treated with combination therapies.

Combination	Publication [Reference]	*n*	Patient Population	*MET* Status	ORR	Median PFS, Months
-	Sequist LV et al. [29] Hartmaier RJ et al. [45] [1]	93	1/2G EGFR–TKI resistance, T790M-	FISH: *MET* GCN ≥ 5 or *MET*/CEP7 ≥ 2; IHC: 3+ in ≥50% tumor cell; NGS: ≥ 20% tumor cell, ≥200X seq, GCN ≥ 5	Part B2: 64.7%	Part B2: 9.1
					Part D: 61.9%	Part D: 9.0
	Hartmaier RJ et al. [45] [2]	69	3G EGFR–TKI (osimertinib) resistance	FISH: *MET* GCN ≥ 5 or *MET*/CEP7 ≥ 2; IHC: 3+ in ≥50% tumor cell; NGS: ≥ 20% tumor cell, ≥200X seq, GCN ≥ 5	33.3%	5.5
	Yu HA et al. [85]	17	3G EGFR–TKI (osimertinib) resistance	NGS: GCN range from 7 to 68	41.2%	Unknown
	Ahn MJ et al. [60]	193	3G EGFR–TKI (osimertinib) resistance	FISH: *MET* GCN ≥ 5 or *MET*/CEP7 ≥ 2; IHC: 3+ in ≥50% tumor cell	Overall: 32%	Overall: 5.3
					FISH10+ or IHC90+: 49.1% [3]	FISH10+ or IHC90+: 7.1
Capmatinib + Gefitinib	Wu YL et al. [86]	100	1/2G EGFR–TKI resistance, T790M-	FISH: GCN ≥ 4	4 ≤ GCN < 6: 22.2%	4 ≤ GCN < 6: 5.4
					GCN ≥ 6: 47.2%	GCN ≥ 6: 5.5
				IHC: 3+ in ≥50% tumor cell	IHC3+: 32.1%	IHC3+: 5.5
Tepotinib + Gefitinib	Wu YL et al. [87] Liam CK et al. [88]	31	1/2G EGFR–TKI resistance, T790M-	FISH: GCN ≥ 5 or *MET*/CEP7 ≥ 2	Overall: 45.2%	Overall: 4.9
					MET amp: 66.7%	*MET* amp: 16.6
				IHC: 2+ or 3+	IHC3+: 68.4%	IHC3+: 8.3
Amivantamab + Lazertinib	Bauml J et al. [76]	45	3G EGFR–TKI (osimertinib) resistance, without previous chemotherapy	No *MET* selection	Overall: 35.6%	Overall: 4.9
					EGFR/MET dependent: 47.1% [4]	*EGFR/MET* dependent: 6.7
					Unknown/non-*EGFR/MET*: 28.6%	Unknown/non-*EGFR/MET*: 4.1
					EGFR/MET IHC+: 90.0%	EGFR/MET IHC+: 12.5
Telisotuzumab vedotin + Osimertinib	Goldman JW et al. [89]	19	3G EGFR–TKI (osimertinib) resistance	IHC: 3+ in ≥25% tumor cell	57.9%	Unknown

[1] Data based on part B2 (*n* = 51) and part D (*n* = 42) of TATTON study. [2] Data based on part B1 of TATTON study. [3] Represents high *MET* amplification and/or high c-MET overexpression subgroup (*n* = 108); FISH10+: *MET* GCN ≥ 10; IHC90+: 3+ in ≥90% tumor cell. [4] *EGFR/MET* dependent, *EGFR/MET* dependent mechanism of resistance (*n* = 17); unknown/non-*EGFR/MET*, unknown mechanism or non-*EGFR/MET* mechanism of resistance to osimertinib (*n* = 28); EGFR/MET IHC+, high IHC results (combined EGFR + MET H score > 400) (*n* = 10). EGFR, Epidermal Growth Factor Receptor; TKI, Tyrosine Kinase Inhibitor; MET, Mesenchymal Epithelial Transition; FISH, Fluorescence In Situ Hybridization; IHC, Immunohistochemistry; GCN, Gene Copy Number; CEP7, Centromere 7; ORR, Objective Response Rate; PFS, Progression Free Survival.

MET amplification can be detected by using fluorescence in situ hybridization (FISH) and immunohistochemistry (IHC). With *MET* amplification, *MET/CEP7* ratio is as follows: low: ≥1.8 to ≤2.2; intermediate: >2.2 to <5; or high: ≥5 will be applied in clinical settings when treating patients with MET inhibitors [90]. The frequency of *MET* amplification in NSCLC ranges from 3% to 10% depending on the cut-off of *MET* copies per cell [91]. c-MET overexpression score of 2+ or 3+ as determined by IHC is considered as *MET* positive [60]. The TATTON study conducted an exploratory analysis of the relationship between the *MET* detection method and the dual-target efficacy after third-generation EGFR–TKI resistance: Based on FISH detection, the ORR value of *MET* local amplification was higher than that of *MET* polysomy patients although polysomy patients benefited from the treatment. In the *MET*-amplified population, patients with higher gene copy numbers detected by FISH had a better treatment benefit [44]. Further in SAVANNAH trial, promising clinical efficacy in a population with high *MET* amplification and/or high threshold c-MET overexpression level (IHC 90+ and/or FISH 10+) with an ORR 49%, mDoR of 9.3 months, and mPFS of 7.1 months was observed. The safety profile was acceptable, similar to that of TATTON study [60]. Further results of the SAVANNAH trial are awaited. However, the sample sizes of these studies are limited. Hence the need to interpret the results with caution is warranted, and further verification is required with larger clinical trials. Further phase III confirmatory trials, SAFFRON and SACHI have been initiated in patients whose disease progressed following treatment with any EGFR-TKI.

8. Conclusions

The conditional approval of savolitinib for the treatment of metastatic *MET*ex14-mutated NSCLC is based on encouraging results from phase 2 trial conducted in China including patients with the more aggressive PSC subtype and brain metastasis. Savolitinib is a potent, highly selective MET inhibitor with robust response in advanced NSCLC. Preclinical and clinical data have shown savolitinib as effective and tolerable treatment in advanced NSCLC patients with *MET*ex14 skipping mutations. When used in combination with EGFR-TKIs, savolitinib has the potential to overcome resistance to these treatments driven through *MET* amplifications and/or c-MET overexpression, with future clinical trials verification needed. In conclusion, savolitinib offer another promising targeted treatment in the paradigm of metastatic NSCLC.

Supplementary Materials: The following supporting information can be downloaded at: https://www.mdpi.com/article/10.3390/cancers14246122/s1, Figure S1: Proportion of *MET*-altered NSCLC patients using different detection methods in SAVANNAH trail; Table S1: Baseline characteristics and clinical demographics of SAVANNAH study; Table S2: Efficacy parameters of SAVANNAH study; Table S3: Safety results of SAVANNAH trial.

Author Contributions: Conceptualization, X.Z., Y.L. and S.L.; methodology, X.Z. and Y.L.; investigation, X.Z. and Y.L.; data curation, X.Z. and Y.L.; writing—original draft preparation, X.Z. and Y.L.; writing—review and editing, X.Z., Y.L. and S.L.; supervision, S.L. All authors have read and agreed to the published version of the manuscript.

Funding: This research received no external funding.

Conflicts of Interest: Yao Lu is an employee of AstraZeneca and works as a medical advisor in AstraZeneca, China.

References

1. Cancer.Net. Lung Cancer—Non-Small Cell: Statistics. Available online: https://www.cancer.net/cancer-types/lung-cancer-non-small-cell/statistics (accessed on 12 June 2022).
2. Yuan, M.; Huang, L.L.; Chen, J.H.; Wu, J.; Xu, Q. The emerging treatment landscape of targeted therapy in non-small-cell lung cancer. *Signal. Transduct. Target Ther.* **2019**, *4*, 61. [CrossRef] [PubMed]
3. Liam, C.K.; Stone, E.; Andarini, S.; Liam, Y.S.; Lam, D.C.L.; Lee, P. Molecular testing of metastatic non-small cell lung cancer in the Asia-Pacific region. *Respirology* **2020**, *25*, 685–687. [CrossRef] [PubMed]
4. Drilon, A.; Cappuzzo, F.; Ou, S.H.I.; Camidge, D.R. Targeting MET in Lung Cancer: Will Expectations Finally Be MET? *J. Thorac. Oncol.* **2017**, *12*, 15–26. [CrossRef] [PubMed]
5. Santarpia, M.; Massafra, M.; Gebbia, V.; D'Aquino, A.; Garipoli, C.; Altavilla, G.; Rosell, R. A narrative review of MET inhibitors in non-small cell lung cancer with MET exon 14 skipping mutations. *Transl. Lung Cancer Res.* **2021**, *10*, 1536–1556. [CrossRef] [PubMed]
6. Davies, K.D.; Ritterhouse, L.L.; Snow, A.N.; Sidiropoulos, N. MET Exon 14 Skipping Mutations: Essential Considerations for Current Management of Non-Small Cell Lung Cancer. *J. Mol. Diagn.* **2022**, *24*, 841–843. [CrossRef]
7. Stoker, M.; Gherardi, E.; Perryman, M.; Gray, J. Scatter factor is a fibroblast-derived modulator of epithelial cell mobility. *Nature* **1987**, *327*, 239–242. [CrossRef] [PubMed]
8. Weidner, K.M.; Behrens, J.; Vandekerckhove, J.; Birchmeier, W. Scatter factor: Molecular characteristics and effect on the invasiveness of epithelial cells. *J. Cell Biol.* **1990**, *111*, 2097–2108. [CrossRef]
9. Kaposi-Novak, P.; Lee, J.S.; Gòmez-Quiroz, L.; Coulouarn, C.; Factor, V.M.; Thorgeirsson, S.S. Met-regulated expression signature defines a subset of human hepatocellular carcinomas with poor prognosis and aggressive phenotype. *J. Clin. Investig.* **2006**, *116*, 1582–1595. [CrossRef]
10. Baldanzi, G.; Graziani, A. Physiological Signaling and Structure of the HGF Receptor MET. *Biomedicines* **2014**, *3*, 1–31. [CrossRef]
11. Raghav, K.; Bailey, A.M.; Loree, J.M.; Kopetz, S.; Holla, V.; Yap, T.A.; Wang, F.; Chen, K.; Salgia, R.; Hong, D. Untying the gordion knot of targeting MET in cancer. *Cancer Treat. Rev.* **2018**, *66*, 95–103. [CrossRef]
12. Jo, M.; Stolz, D.B.; Esplen, J.E.; Dorko, K.; Michalopoulos, G.K.; Strom, S.C. Cross-talk between epidermal growth factor receptor and c-Met signal pathways in transformed cells. *J. Biol. Chem.* **2000**, *275*, 8806–8811. [CrossRef] [PubMed]
13. Soman, N.R.; Correa, P.; Ruiz, B.A.; Wogan, G.N. The TPR-MET oncogenic rearrangement is present and expressed in human gastric carcinoma and precursor lesions. *Proc. Natl. Acad. Sci. USA* **1991**, *88*, 4892–4896. [CrossRef] [PubMed]
14. Socinski, M.A.; Pennell, N.A.; Davies, K.D. MET Exon 14 Skipping Mutations in Non-Small-Cell Lung Cancer: An Overview of Biology, Clinical Outcomes, and Testing Considerations. *JCO Precis. Oncol.* **2021**, *5*, 653–663. [CrossRef] [PubMed]
15. Mo, H.N.; Liu, P. Targeting MET in cancer therapy. *Chronic Dis. Transl. Med.* **2017**, *3*, 148–153. [CrossRef]

16. Tong, J.H.; Yeung, S.F.; Chan, A.W.H.; Chung, L.Y.; Chau, S.L.; Lung, R.W.M.; Tong, C.Y.; Chow, C.; Tin, E.K.Y.; Yu, Y.H.; et al. MET Amplification and Exon 14 Splice Site Mutation Define Unique Molecular Subgroups of Non-Small Cell Lung Carcinoma with Poor Prognosis. *Clin. Cancer Res.* **2016**, *22*, 3048–3056. [CrossRef]
17. Guo, R.; Luo, J.; Chang, J.; Rekhtman, N.; Arcila, M.; Drilon, A. MET-dependent solid tumours—Molecular diagnosis and targeted therapy. *Nat. Rev. Clin. Oncol.* **2020**, *17*, 569–587. [CrossRef]
18. Recondo, G.; Che, J.; Jänne, P.A.; Awad, M.M. Targeting MET Dysregulation in Cancer. *Cancer Discov.* **2020**, *10*, 922–934. [CrossRef]
19. Engelman, J.A.; Zejnullahu, K.; Mitsudomi, T.; Song, Y.; Hyland, C.; Park, J.O.; Lindeman, N.; Gale, C.M.; Zhao, X.; Christensen, J.; et al. MET amplification leads to gefitinib resistance in lung cancer by activating ERBB3 signaling. *Science* **2007**, *316*, 1039–1043. [CrossRef]
20. Zhang, Z.; Yang, S.; Wang, Q. Impact of MET alterations on targeted therapy with EGFR-tyrosine kinase inhibitors for EGFR-mutant lung cancer. *Biomark. Res.* **2019**, *7*, 27. [CrossRef]
21. Bean, J.; Brennan, C.; Shih, J.Y.; Riely, G.; Viale, A.; Wang, L.; Chitale, D.; Motoi, N.; Szoke, J.; Broderick, S.; et al. MET amplification occurs with or without T790M mutations in EGFR mutant lung tumors with acquired resistance to gefitinib or erlotinib. *Proc. Natl. Acad. Sci. USA* **2007**, *104*, 20932–20937. [CrossRef]
22. Wang, Y.; Li, L.; Han, R.; Jiao, L.; Zheng, J.; He, Y. Clinical analysis by next-generation sequencing for NSCLC patients with MET amplification resistant to osimertinib. *Lung Cancer* **2018**, *118*, 105–110. [CrossRef] [PubMed]
23. Li, X.M.; Li, W.F.; Lin, J.T.; Yan, H.H.; Tu, H.Y.; Chen, H.J.; Wang, B.C.; Wang, Z.; Zhou, Q.; Zhang, X.C.; et al. Predictive and Prognostic Potential of TP53 in Patients with Advanced Non-Small-Cell Lung Cancer Treated With EGFR-TKI: Analysis of a Phase III Randomized Clinical Trial (CTONG 0901. *Clin. Lung Cancer* **2021**, *22*, 100–109. [CrossRef] [PubMed]
24. Lai, G.G.Y.; Lim, T.H.; Lim, J.; Liew, P.J.R.; Kwang, X.L.; Nahar, R.; Aung, Z.W.; Takano, A.; Lee, Y.Y.; Lau, D.P.X.; et al. Clonal MET Amplification as a Determinant of Tyrosine Kinase Inhibitor Resistance in Epidermal Growth Factor Receptor-Mutant Non-Small-Cell Lung Cancer. *J. Clin. Oncol.* **2019**, *37*, 876–884. [CrossRef] [PubMed]
25. Camidge, D.R.; Barlesi, F.; Goldman, J.W.; Morgensztern, D.; Heist, R.; Vokes, E.; Angevin, E.; Hong, D.S.; Rybkin, I.I.; Barve, M.; et al. A Phase Ib Study of 25 in Combination with Nivolumab in Patients With NSCLC. *JTO Clin. Res. Rep.* **2022**, *3*, 100262. [CrossRef] [PubMed]
26. Mathieu, L.N.; Larkins, E.; Akinboro, O.; Roy, P.; Amatya, A.K.; Fiero, M.H.; Mishra-Kalyani, P.S.; Helms, W.S.; Myers, C.E.; Skinner, A.M.; et al. FDA Approval Summary: Capmatinib and Tepotinib for the Treatment of Metastatic NSCLC Harboring MET Exon 14 Skipping Mutations or Alterations. *Clin. Cancer Res.* **2022**, *28*, 249–254. [CrossRef]
27. Pfizer. Pfizer's XALKORI® (Crizotinib) Receives FDA Breakthrough Therapy Designation in Two New Indications. Available online: https://www.pfizer.com/news/press-release/press-release-detail/pfizer_s_xalkori_crizotinib_receives_fda_breakthrough_therapy_designation_in_two_new_indications-0 (accessed on 12 June 2022).
28. AstraZeneca. Orpathys Approved in China for Patients with Lung Cancer and MET Gene Alterations. Available online: https://www.astrazeneca.com/media-centre/press-releases/2021/orpathys-approved-in-china-for-patients-with-lung-cancer-and-met-gene-alterations.html (accessed on 12 June 2022).
29. Sequist, L.V.; Han, J.Y.; Ahn, M.J.; Cho, B.C.; Yu, H.; Kim, S.W.; Yang, J.C.; Lee, J.S.; Su, W.C.; Kowalski, D.; et al. Osimertinib plus savolitinib in patients with EGFR mutation-positive, MET-amplified, non-small-cell lung cancer after progression on EGFR tyrosine kinase inhibitors: Interim results from a multicentre, open-label, phase Ib study. *Lancet Oncol.* **2020**, *21*, 373–386. [CrossRef]
30. Yang, J.J.; Fang, J.; Shu, Y.Q.; Chang, J.H.; Chen, G.Y.; He, J.X.; Li, W.; Liu, X.Q.; Yang, N.; Zhou, C.; et al. A phase Ib study of the highly selective MET-TKI savolitinib plus gefitinib in patients with EGFR-mutated, MET-amplified advanced non-small-cell lung cancer. *Investig. New Drugs.* **2021**, *39*, 477–487. [CrossRef]
31. Markham, A. Savolitinib: First Approval. *Drugs* **2021**, *81*, 1665–1670. [CrossRef]
32. Hong, L.; Zhang, J.; Heymach, J.V.; Le, X. Current and future treatment options for MET exon 14 skipping alterations in non-small cell lung cancer. *Ther. Adv. Med. Oncol.* **2021**, *13*, 1758835921992976. [CrossRef]
33. Schuller, A.G.; Barry, E.R.; Jones, R.D.O.; Henry, R.E.; Frigault, M.M.; Beran, G.; Linsenmayer, D.; Hattersley, M.; Smith, A.; Wilson, J.; et al. The MET Inhibitor AZD6094 (Savolitinib, HMPL-504) Induces Regression in Papillary Renal Cell Carcinoma Patient-Derived Xenograft Models. *Clin. Cancer Res.* **2015**, *21*, 2811–2819. [CrossRef]
34. Gavine, P.R.; Ren, Y.; Han, L.; Lv, J.; Fan, S.; Zhang, W.; Xu, W.; Liu, Y.J.; Zhang, T.; Fu, H.; et al. Volitinib, a potent and highly selective c-Met inhibitor, effectively blocks c-Met signaling and growth in c-MET amplified gastric cancer patient-derived tumor xenograft models. *Mol. Oncol.* **2015**, *9*, 323–333. [CrossRef] [PubMed]
35. Gu, Y.; Sai, Y.; Wang, J.; Yu, M.; Wang, G.; Zhang, L.; Ren, H.; Fan, S.; Ren, Y.; Qing, W.; et al. Preclinical pharmacokinetics, disposition, and translational pharmacokinetic/pharmacodynamic modeling of savolitinib, a novel selective cMet inhibitor. *Eur. J. Pharm. Sci.* **2019**, *136*, 104938. [CrossRef] [PubMed]
36. Jones, R.D.O.; Grondine, M.; Borodovsky, A.; San Martin, M.; DuPont, M.; D'Cruz, C.; Schuller, A.; Henry, R.; Barry, E.; Castriotta, L.; et al. A pharmacokinetic-pharmacodynamic model for the MET tyrosine kinase inhibitor, savolitinib, to explore target inhibition requirements for anti-tumour activity. *Br. J. Pharmacol.* **2021**, *178*, 600–613. [CrossRef] [PubMed]
37. Ding, Q.; Ou, M.; Zhu, H.; Wang, Y.; Jia, J.; Sai, Y.; Chen, Q.; Wang, J. Effect of food on the single-dose pharmacokinetics and tolerability of savolitinib in Chinese healthy volunteers. *Fundam. Clin. Pharmacol.* **2022**, *36*, 210–217. [CrossRef] [PubMed]

38. Lu, S.; Fang, J.; Li, X.; Cao, L.; Zhou, J.; Guo, Q.; Liang, Z.; Cheng, Y.; Jiang, L.; Yang, N.; et al. Once-daily savolitinib in Chinese patients with pulmonary sarcomatoid carcinomas and other non-small-cell lung cancers harbouring MET exon 14 skipping alterations: A multicentre, single-arm, open-label, phase 2 study. *Lancet Respir. Med.* **2021**, *9*, 1154–1164. [CrossRef]
39. Henry, R.E.; Barry, E.R.; Castriotta, L.; Ladd, B.; Markovets, A.; Beran, G.; Ren, Y.; Zhou, F.; Adam, A.; Zinda, M.; et al. Acquired savolitinib resistance in non-small cell lung cancer arises via multiple mechanisms that converge on MET-independent mTOR and MYC activation. *Oncotarget* **2016**, *7*, 57651–57670. [CrossRef]
40. Dua, R.; Zhang, J.; Parry, G.; Penuel, E. Detection of hepatocyte growth factor (HGF) ligand-c-MET receptor activation in formalin-fixed paraffin embedded specimens by a novel proximity assay. *PLoS ONE* **2011**, *6*, e15932. [CrossRef]
41. Li, Y.; Dong, S.; Tamaskar, A.; Wang, H.; Zhao, J.; Ma, H.; Zhao, Y. Proteasome Inhibitors Diminish c-Met Expression and Induce Cell Death in Non-Small Cell Lung Cancer Cells. *Oncol. Res.* **2020**, *28*, 497–507. [CrossRef]
42. Gan, H.K.; Millward, M.; Hua, Y.; Qi, C.; Sai, Y.; Su, W.; Wang, J.; Zhang, L.; Frigault, M.M.; Morgan, S.; et al. First-in-Human Phase I Study of the Selective MET Inhibitor, Savolitinib, in Patients with Advanced Solid Tumors: Safety, Pharmacokinetics, and Antitumor Activity. *Clin. Cancer Res.* **2019**, *25*, 4924–4932. [CrossRef]
43. Wang, Y.; Liu, T.; Chen, G.; Gong, J.; Bai, Y.; Zhang, T.; Xu, N.; Liu, L.; Xu, J.; He, J.; et al. Phase Ia/Ib Study of the Selective MET Inhibitor, Savolitinib, in Patients with Advanced Solid Tumors: Safety, Efficacy, and Biomarkers. *Oncologist* **2022**, *27*, e342–e383. [CrossRef]
44. Oxnard, G.R.; Yang, J.C.H.; Yu, H.; Kim, S.W.; Saka, H.; Horn, L.; Goto, K.; Ohe, Y.; Mann, H.; Thress, K.S.; et al. TATTON: A multi-arm, phase Ib trial of osimertinib combined with selumetinib, savolitinib, or durvalumab in EGFR-mutant lung cancer. *Ann. Oncol.* **2020**, *31*, 507–516. [CrossRef] [PubMed]
45. Hartmaier, R.J.; Markovets, A.A.; Ahn, M.J.; Sequist, L.V.; Han, J.Y.; Cho, B.C.; Yu, H.A.; Kim, S.W.; Yang, J.C.; Lee, J.S.; et al. Osimertinib+Savolitinib to Overcome Acquired MET-Mediated Resistance in Epidermal Growth Factor Receptor Mutated MET-Amplified Non-Small Cell Lung Cancer: TATTON. *Cancer Discov.* **2022**, *20*, CD-22-0586. [CrossRef] [PubMed]
46. Lu, S.; Fang, J.; Li, X.; Cao, L.; Zhou, J.; Guo, Q.; Liang, Z.; Cheng, Y.; Jiang, L.; Yang, N.; et al. Final OS results and subgroup analysis of savolitinib in patients with MET exon 14 skipping mutations (METex14+) NSCLC. In Proceedings of the ELCC 2022 Virtual Meeting, Virtual, 30 March–2 April 2022.
47. Lu, S.; Fang, J.; Cao, L.; Li, X.; Guo, Q.; Zhou, J.; Cheng, Y.; Jiang, L.; Chen, Y.; Zhang, H.; et al. Abstract CT031: Preliminary efficacy and safety results of savolitinib treating patients with pulmonary sarcomatoid carcinoma (PSC) and other types of non-small cell lung cancer (NSCLC) harboring MET exon 14 skipping mutations. In Proceedings of the AACR Annual Meeting 2019, Atlanta, GA, USA, 29 March–3 April 2019.
48. Lu, S.; Fang, J.; Li, X.; Cao, L.; Zhou, J.; Guo, Q.; Liang, Z.; Cheng, Y.; Jiang, L.; Yang, N.; et al. Long-Term Efficacy, Safety and Subgroup Analysis of Savolitinib in Chinese Patients with Non-Small Cell Lung Cancers Harboring MET Exon 14 Skipping Alterations. *JTO Clin. Res. Rep.* **2022**, *3*, 100407. [CrossRef] [PubMed]
49. Gow, C.H.; Hsieh, M.S.; Wu, S.G.; Shih, J.Y. A comprehensive analysis of clinical outcomes in lung cancer patients harboring a MET exon 14 skipping mutation compared to other driver mutations in an East Asian population. *Lung Cancer* **2017**, *103*, 82–89. [CrossRef] [PubMed]
50. Chen, R.; Manochakian, R.; James, L.; Azzouqa, A.G.; Shi, H.; Zhang, Y.; Zhao, Y.; Zhou, K.; Lou, Y. Emerging therapeutic agents for advanced non-small cell lung cancer. *J. Hematol. Oncol.* **2020**, *13*, 58. [CrossRef]
51. Bae, H.M.; Min, H.S.; Lee, S.H.; Kim, D.W.; Chung, D.H.; Lee, J.S.; Kim, Y.W.; Heo, D.S. Palliative chemotherapy for pulmonary pleomorphic carcinoma. *Lung Cancer* **2007**, *58*, 112–115. [CrossRef]
52. Ung, M.; Rouquette, I.; Filleron, T.; Taillandy, K.; Brouchet, L.; Bennouna, J.; Delord, J.P.; Milia, J.; Mazières, J. Characteristics and Clinical Outcomes of Sarcomatoid Carcinoma of the Lung. *Clin. Lung Cancer.* **2016**, *17*, 391–397. [CrossRef]
53. Vieira, T.; Girard, N.; Ung, M.; Monnet, I.; Cazes, A.; Bonnette, P.; Duruisseaux, M.; Mazieres, J.; Antoine, M.; Cadranel, J.; et al. Efficacy of first-line chemotherapy in patients with advanced lung sarcomatoid carcinoma. *J. Thorac. Oncol.* **2013**, *8*, 1574–1577. [CrossRef]
54. Hong, J.Y.; Choi, M.K.; Uhm, J.E.; Park, M.J.; Lee, J.; Park, Y.H.; Ahn, J.S.; Park, K.; Han, J.H.; Ahn, M.J. The role of palliative chemotherapy for advanced pulmonary pleomorphic carcinoma. *Med. Oncol.* **2009**, *26*, 287–291. [CrossRef]
55. Cheng, H.; Perez-Soler, R. Leptomeningeal metastases in non-small-cell lung cancer. *Lancet Oncol.* **2018**, *19*, e43–e55. [CrossRef]
56. Deeken, J.F.; Löscher, W. The blood-brain barrier and cancer: Transporters, treatment, and Trojan horses. *Clin. Cancer Res.* **2007**, *13*, 1663–1674. [CrossRef] [PubMed]
57. Choueiri, T.K.; Heng, D.Y.C.; Lee, J.L.; Cancel, M.; Verheijen, R.B.; Mellemgaard, A.; Ottesen, L.H.; Frigault, M.M.; L'Hernault, A.; Szijgyarto, Z.; et al. Efficacy of Savolitinib vs Sunitinib in Patients With MET-Driven Papillary Renal Cell Carcinoma: The SAVOIR Phase 3 Randomized Clinical Trial. *JAMA Oncol.* **2020**, *6*, 1247–1255. [CrossRef] [PubMed]
58. Cortot, A.; Le, X.; Smit, E.; Viteri, S.; Kato, T.; Sakai, H.; Park, K.; Camidge, D.R.; Berghoff, K.; Vlassak, S.; et al. Safety of MET Tyrosine Kinase Inhibitors in Patients with MET Exon 14 Skipping Non-small Cell Lung Cancer: A Clinical Review. *Clin. Lung Cancer* **2022**, *3*, 195–207. [CrossRef] [PubMed]
59. Paik, P.K.; Felip, E.; Veillon, R.; Sakai, H.; Cortot, A.B.; Garassino, M.C.; Mazieres, J.; Viteri, S.; Senellart, H.; Meerbeeck, J.V.; et al. Tepotinib in Non-Small-Cell Lung Cancer with MET Exon 14 Skipping Mutations. *N. Engl. J. Med.* **2020**, *383*, 931–943. [CrossRef]

60. Ahn, M.J.; De Marinis, F.; Bonanno, L.; Cho, B.C.; Kim, T.M.; Cheng, S.; Novello, S.; Proto, C.; Kim, S.W.; Lee, J.S.; et al. MET Biomarker-based Preliminary Efficacy Analysis in SAVANNAH: Savolitinib+osimertinib in EGFRm NSCLC Post-Osimertinib. In Proceedings of the WCLC 2022, Vienna, Austria, 6–9 August 2022.

61. Li, A.; Chen, H.J.; Yang, J.J. Design and Rationale for a Phase II, Randomized, Open-Label, Two-Cohort Multicenter Interventional Study of Osimertinib with or Without Savolitinib in De Novo MET Aberrant, EGFR-Mutant Patients with Advanced Non-Small-Cell Lung Cancer: The FLOWERS Trial. *Clin. Lung Cancer* **2022**, *22*, 00205-4. [CrossRef]

62. Wolf, J.; Garon, E.B.; Groen, H.J.M.; Tan, D.S.W.; Robeva, A.; Le Mouhaer, S.; Carbini, M.; Chassot-Agostinho, A.; Heist, R.S. Capmatinib in *MET* exon 14-mutated, advanced NSCLC: Updated results from the GEOMETRY mono-1 study. In Proceedings of the 2021 ASCO Annual Meeting I, Virtual, 4–8 June 2021.

63. Thomas, M.; Garassino, M.; Felip, E.; Sakai, H.; Le, X.; Veillon, R.; Smit, E.; Mazieres, J.; Cortot, A.; Raskin, J.; et al. Tepotinib in Patients with MET Exon 14 (METex14) Skipping NSCLC: Primary Analysis of the Confirmatory VISION Cohort C. In Proceedings of the WCLC 2022, Vienna, Austria, 6–9 August 2022.

64. Krebs, M.; Spira, A.I.; Cho, B.C.; Besee, B.; Goldman, J.W.; Janne, P.A.; Ma, Z.; Mansfield, A.S.; Minchom, A.R.; Ou, S.H.I.; et al. Amivantamab in patients with NSCLC with MET exon 14 skipping mutation: Updated results from the CHRYSALIS study. In Proceedings of the 2022 ASCO Annual Meeting, Chicago, IL, USA, 3–7 June 2022.

65. Yu, Y.; Ren, Y.; Fang, J.; Cao, L.; Liang, Z.; Guo, Q.; Han, S.; Ji, Z.; Wang, Y.; Sun, Y.; et al. Circulating tumour DNA biomarkers in savolitinib-treated patients with non-small cell lung cancer harbouring MET exon 14 skipping alterations: A *post hoc* analysis of a pivotal phase 2 study. *Ther. Adv. Med. Oncol.* **2022**, *14*, 17588359221133546. [CrossRef]

66. Le, X.; Sakai, H.; Felip, E.; Veillon, R.; Garassino, M.C.; Raskin, J.; Cortot, A.B.; Viteri, S.; Mazieres, J.; Smit, E.F.; et al. Tepotinib Efficacy and Safety in Patients with MET Exon 14 Skipping NSCLC: Outcomes in Patient Subgroups from the VISION Study with Relevance for Clinical Practice. *Clin. Cancer Res.* **2022**, *28*, 1117–1126. [CrossRef]

67. Wolf, J.; Seto, T.; Han, J.Y.; Reguart, N.; Garon, E.B.; Groen, H.J.M.; Tan, D.S.W.; Hida, T.; de Jonge, M.; Orlov, S.V.; et al. Capmatinib in MET Exon 14-Mutated or MET-Amplified Non-Small-Cell Lung Cancer. *N. Engl. J. Med.* **2020**, *383*, 944–957. [CrossRef]

68. Paik, P.K.; Sakai, H.; Felip, E.; Veillon, R.; Garassino, M.C.; Raskin, J.; Viteri, S.; Mazieres, J.; Cortot, A.; Smit, E.; et al. Tepotinib in patients with MET exon 14 (METex14) skipping advanced NSCLC: Updated efficacy from VISION Cohort A. In Proceedings of the WCLC 2020, Virtual, 28–31 January 2021.

69. Spira, A.; Krebs, M.; Cho, B.C.; Besse, B.; Goldman, J.; Janne, P.; Lee, C.K.; Ma, Z.; Mansfield, A.; Minchom, A.; et al. Amivantamab in non-small cell lung cancer with MET exon 14 skipping muta- tion: Initial results from CHRYSALIS. In Proceedings of the WCLC 2021, Virtual, 8–14 September 2021.

70. Fujino, T.; Kobayashi, Y.; Suda, K.; Koga, T.; Nishino, M.; Ohara, S.; Chiba, M.; Shimoji, M.; Tomizawa, K.; Takemoto, T.; et al. Sensitivity and Resistance of MET Exon 14 Mutations in Lung Cancer to Eight MET Tyrosine Kinase Inhibitors In Vitro. *J. Thorac. Oncol.* **2019**, *14*, 1753–1765. [CrossRef]

71. Wolf, J.; Garon, E.B.; Groen, H.J.M.; Tan, D.S.W.; Robeva, A.; Le Mouhaer, S.; Carbini, M.; Yovine, A.; Heist, R. Capmatinib in treatment (Tx)-naive MET exon 14-mutated (METex14) advanced non-small cell lung cancer (aNSCLC): Updated results from GEOMETRY mono-1. In Proceedings of the ELCC 2022 Virtual Meeting, Virtual, 30 March–2 April 2022.

72. Moro-Sibilot, D.; Cozic, N.; Pérol, M.; Mazières, J.; Otto, J.; Souquet, P.J.; Bahleda, R.; Wislez, M.; Zalcman, G.; Guibert, S.D.; et al. Crizotinib in c-MET- or ROS1-positive NSCLC: Results of the AcSé phase II trial. *Ann. Oncol.* **2019**, *30*, 1985–1991. [CrossRef]

73. Drilon, A.; Clark, J.W.; Weiss, J.; Ou, S.I.; Camidge, D.R.; Solomon, B.J.; Otterson, G.A.; Villaruz, L.C.; Riely, G.J.; Heist, R.S.; et al. Antitumor activity of crizotinib in lung cancers harboring a MET exon 14 alteration. *Nat. Med.* **2020**, *26*, 47–51. [CrossRef] [PubMed]

74. Reis, H.; Metzenmacher, M.; Goetz, M.; Savvidou, N.; Darwiche, K.; Aigner, C.; Herold, T.; Eberhardt, W.E.; Skiba, C.; Hense, J.; et al. MET Expression in Advanced Non-Small-Cell Lung Cancer: Effect on Clinical Outcomes of Chemotherapy, Targeted Therapy, and Immunotherapy. *Clin. Lung Cancer* **2018**, *19*, e441–e463. [CrossRef] [PubMed]

75. Wood, G.E.; Hockings, H.; Hilton, D.M.; Kermorgant, S. The role of MET in chemotherapy resistance. *Oncogene* **2021**, *40*, 1927–1941. [CrossRef] [PubMed]

76. Bauml, J.; Cho, B.C.; Park, K.; Lee, K.H.; Cho, E.K.; Kim, D.W.; Kim, S.W.; Haura, E.B.; Sabari, J.K.; Sanborn, R.E.; et al. Amivantamab in combination with lazertinib for the treatment of osimertinib-relapsed, chemotherapy-naïve EGFR mutant (EGFRm) non-small cell lung cancer (NSCLC) and potential biomarkers for response. In Proceedings of the 2021 ASCO Annual Meeting I, Virtual, 4–8 June 2021.

77. Park, K.; Zhou, J.; Kim, D.; Ahmad, A.R.; Soo, R.A.; Bruns, R.; Straub, J.; Johne, A.; Scheele, J.; Yang, J.C.; et al. Tepotinib plus gefitinib in patients with MET-amplified EGFR-mutant NSCLC: Long-term outcomes of the INSIGHT study. In Proceedings of the ESMO Asia Congress 2019, Singapore, 22–24 November 2019.

78. Haratani, K.; Hayashi, H.; Tanaka, T.; Kaneda, H.; Togashi, Y.; Sakai, K.; Hayashi, K.; Tomida, S.; Yonesaka, K.; Nonagase, Y.; et al. Tumor immune microenvironment and nivolumab efficacy in EGFR mutation-positive non-small-cell lung cancer based on T790M status after disease progression during EGFR-TKI treatment. *Ann. Oncol.* **2017**, *28*, 1532–1539. [CrossRef] [PubMed]

79. Hayashi, H.; Sugawara, S.; Fukuda, Y.; Fujimoto, D.; Miura, S.; Ota, K.; Ozawa, Y.; Hara, S.; Tanizaki, J.; Azuma, K.; et al. A Randomized Phase II Study Comparing Nivolumab with Carboplatin-Pemetrexed for EGFR-Mutated NSCLC with Resistance to EGFR Tyrosine Kinase Inhibitors (WJOG8515L. *Clin. Cancer Res.* **2022**, *28*, 893–902. [CrossRef] [PubMed]

80. Reck, M.; Mok, T.; Socinski, M.A.; Jotte, R.M.; Lim, D.W.; Cappuzzo, F.; Orlandi, F.J.; Stroyakovskiy, D.; Nogami, N.; Rodriguez-Abreu, D.; et al. IMpower150: Updated efficacy analysis in patients with EGFR mutations. In Proceedings of the ESMO Virtual Congress 2020, Virtual, 19–21 September 2020.

81. Lu, S.; Wu, L.; Jian, H.; Chen, Y.; Wang, Q.; Fang, J.; Wang, Z.; Hu, Y.; Sun, M.; Han, L.; et al. Sintilimab plus bevacizumab biosimilar IBI305 and chemotherapy for patients with EGFR-mutated non-squamous non-small-cell lung cancer who progressed on EGFR tyrosine-kinase inhibitor therapy (ORIENT-31): First interim results from a randomised, double-blind, multicentre, phase 3 trial. *Lancet Oncol.* **2022**, *23*, 1167–1179. [CrossRef] [PubMed]

82. Baldacci, S.; Mazieres, J.; Tomasini, P.; Girard, N.; Guisier, F.; Audigier-Valette, C.; Monnet, I.; Wislez, M.; Pérol, M.; Dô, P.; et al. Outcome of EGFR-mutated NSCLC patients with MET-driven resistance to EGFR tyrosine kinase inhibitors. *Oncotarget* **2017**, *8*, 105103–105114. [CrossRef]

83. Baldacci, S.; Kherrouche, Z.; Cockenpot, V.; Stoven, L.; Copin, M.C.; Werkmeister, E.; Marchand, N.; Kyheng, M.; Tulasne, D.; Cortot, A.B. MET amplification increases the metastatic spread of EGFR-mutated NSCLC. *Lung Cancer* **2018**, *125*, 57–67. [CrossRef]

84. Wang, Q.; Yang, S.; Wang, K.; Sun, S.Y. MET inhibitors for targeted therapy of EGFR TKI-resistant lung cancer. *J. Hematol. Oncol.* **2019**, *12*, 63. [CrossRef]

85. Yu, H.A.; Ambrose, H.; Baik, C.; Cho, B.C.; Cocco, E.; Goldberg, S.B.; Goldman, J.W.; Kraljevic, S.; de Langen, A.J.; Okamoto, I.; et al. ORCHARD osimertinib + savolitinib interim analysis: A biomarker-directed phase II platform study in patients (pts) with advanced non-small cell lung cancer (NSCLC) whose disease has progressed on first-line (1L) osimertinib. In Proceedings of the ESMO Congress 2021, Paris, France, 16–21 September 2021.

86. Wu, Y.L.; Zhang, L.; Kim, D.W.; Liu, X.; Lee, D.H.; Yang, J.C.H.; Ahn, M.J.; Vansteenkiste, J.F.; Su, W.C.; Felip, E.; et al. Phase Ib/II Study of Capmatinib (INC280) Plus Gefitinib After Failure of Epidermal Growth Factor Receptor (EGFR) Inhibitor Therapy in Patients With EGFR-Mutated, MET Factor-Dysregulated Non-Small-Cell Lung Cancer. *J. Clin. Oncol.* **2018**, *36*, 3101–3109. [CrossRef]

87. Wu, Y.L.; Cheng, Y.; Zhou, J.; Lu, S.; Zhang, Y.; Zhao, J.; Kim, D.W.; Soo, R.A.; Kim, S.W.; Pan, H.; et al. Tepotinib plus gefitinib in patients with EGFR-mutant non-small-cell lung cancer with MET overexpression or MET amplification and acquired resistance to previous EGFR inhibitor (INSIGHT study): An open-label, phase Ib/2, multicentre, randomised trial. *Lancet Respir Med.* **2020**, *8*, 1132–1143. [CrossRef] [PubMed]

88. Liam, C.K.; Ahmad, A.R.; Hsia, T.C.; Zhou, J.; Kim, D.W.; Soo, R.A.; Cheng, Y.; Lu, S.; Shin, S.W.; Yang, J.C.H.; et al. Tepotinib + gefitinib in patients with *EGFR*-mutant NSCLC with *MET* amplification: Final analysis of INSIGHT. In Proceedings of the 2022 AACR Annual Meeting, New Orleans, LA, USA, 8–13 April 2022.

89. Goldman, J.W.; Horinouchi, H.; Cho, B.C.; Tomasini, P.; Dunbar, M.; Hoffman, D.; Parikh, A.; Blot, V.; Camidge, D.R. Phase 1/1b study of telisotuzumab vedotin (Teliso-V) + osimertinib (Osi), after failure on prior Osi, in patients with advanced, c-Met overexpressing, *EGFR*-mutated non-small cell lung cancer (NSCLC. In Proceedings of the 2022 ASCO Annual Meeting, Chicago, IL, USA, 3–7 June 2022.

90. Yin, W.; Cheng, J.; Tang, Z.; Toruner, G.; Hu, S.; Guo, M.; Robinson, M.; Medeiros, L.J.; Tang, G. MET Amplification (MET/CEP7 Ratio $\geq$ 1.8) Is an Independent Poor Prognostic Marker in Patients with Treatment-naive Non-Small-cell Lung Cancer. *Clin. Lung Cancer* **2021**, *22*, e512–e518. [CrossRef] [PubMed]

91. Fang, L.; Chen, H.; Tang, Z.; Kalhor, N.; Liu, C.H.; Yao, H.; Hu, S.; Lin, P.; Zhao, J.; Luthra, R.; et al. MET amplification assessed using optimized FISH reporting criteria predicts early distant metastasis in patients with non-small cell lung cancer. *Oncotarget* **2018**, *9*, 12959–12970. [CrossRef] [PubMed]

MDPI

St. Alban-Anlage 66

4052 Basel

Switzerland

www.mdpi.com

Cancers Editorial Office

E-mail: cancers@mdpi.com

www.mdpi.com/journal/cancers

www.ingramcontent.com/pod-product-compliance
Lightning Source LLC
LaVergne TN
LVHW070101170726
843515LV00007B/1583